Dinosaur Tracks of Mesozoic Basins in Brazil

Ismar de Souza Carvalho · Giuseppe Leonardi
Editors

Dinosaur Tracks of Mesozoic Basins in Brazil

Impact of Paleoenvironmental
and Paleoclimatic Changes

 Springer

Editors
Ismar de Souza Carvalho 🄳
Department of Geology
Universidade Federal do Rio de Janeiro
Rio de Janeiro, Rio de Janeiro, Brazil

Giuseppe Leonardi 🄳
Istituto Cavanis
Venice, Italy

ISBN 978-3-031-56354-6   ISBN 978-3-031-56355-3 (eBook)
https://doi.org/10.1007/978-3-031-56355-3

This Springer imprint is published by the registered company Springer Nature Switzerland AG
The registered company address is: Gewerbestrasse 11, 6330 Cham, Switzerland

Paper in this product is recyclable.

This book is dedicated to all the invisibles of Science, people who believe and support scientific research.

Foreword

Dinosaur ichnology has long been regarded as "nice... but useless" by geologist and vertebrate dinosaur researchers. After the earliest scientific reports by the founder of dinosaur ichnology, Edward Hitchcock in the early 30ties of the nineteenth century there was a long haul with almost no interest in the study of vertebrate tracks. In South America, particularly in Brazil, it was and is the merit of Giuseppe Leonardi to explore and document dinosaur tracks in the most remote places of the country. I first met him in the early 90thies of the last century when Martin Lockley and I visited him in Venice, and then on a visit to the famous Lavini di Marco tracksite in Trentino.

People like Edward Hitchcock in the late nineteenth century and Giuseppe Leonardi in the last century planted the seed that led to a new generation of ichnologists. This started the "Dinosaur track renaissance" in the 80ties of the last century, Martin G. Lockley got the ball rolling. And we all try to follow the footsteps of this gentle giant who left us too soon.

In Brazil, Ismar de Souza Carvalho, today a professor of geology and palaeontology at the Federal University of Rio de Janeiro, started in the early 80ties of the last century, together with his students and other scholars to explore many regions looking for dinosaur tracks. The result was a school of Brazilian ichnology, the present book shows that in an impressive way. Twelve eminent ichnologists, vertebrate scholars, and geologists from all over the country, the two editors included from many Brazilian institutions present in ten chapters the state of the art of dinosaur ichnology in Brazil.

The book includes the development of vertebrate ichnology and its studies before and after that renaissance. Our field has seen a dramatic increase of reports almost weekly on a global scale and is nowadays regarded as a valuable source of information on the palaeoecology and palaeobiology of dinosaurs, even by "hard core" vertebrate scholars and geologists alike.

The contributors of this amazing collection of scientific articles have traveled from the flood plains in the south to the Amazon basin to areas where no skeletal remains of dinosaurs have been found but in the search for their tracks. Every major sedimentary basin of Brazil is covered in the book, always preceded by an introduction to the local geological history. But not only this, they include stratigraphic information as well as paleobiogeographic and palaeoecological interpretations that are too often neglected in books and papers on dinosaur footprints. On top of that, the importance of the substrate and the role of microbial mats for the preservation of tracks are discussed. Equally important is the teaming up of the scholars with prominent visual artist Guilherme Gehr. His reconstructions of the past landscapes are the artistic transformation of the data extracted from the fossil record by the specialists of this book.

This collection of scientific articles is a holistic approach and a "homage" for the record of dinosaur tracks and trackways and their palaeoecological significance. Finally, a comprehensive book that summarizes more than 50 years of research on dinosaur ichnology in Brazil, long awaited and an important contribution to vertebrate ichnology. But it is more than that; it is a database for scholars and future researchers alike.

January 2024   Prof. Christian A. Meyer
Department of Environmental Sciences
University of Basel
Basel, Switzerland

Preface

*There is no branch of detective science
which is so important and so much
neglected as the art of tracing footsteps.*

*—Sir Arthur Conan Doyle,
A Study in Scarlet.*

The unearthing of a new fossil is the discovery of past life diversity. Nonetheless, the finding of an ichnofossil opens the possibility to understand the relationships among the organisms and aspects of their paleoautoecology. It is the possibility of bringing life from Earth's biological past to the present time.

The vertebrate ichnology, especially the study of dinosaur footprints and trackways, allows insights in the many biological remarks unable to observe directly through the osteology studies. Number of specimens, diversity, behavioral aspects, such as walking gaits, gregariousness, nidification, and diseases, are some matters that can be analyzed through the study of the dinosaur tracks and other traces. It allows a great insight into paleobiology and paleoecological interpretations.

The proposal of the book *Dinosaur Tracks of Mesozoic Basins in Brazil: Impact of Paleoenvironmental and Paleoclimatic Changes* is to present the diversity of dinosaur tracks from the Mesozoic Brazilian basins and the paleoenvironmental contexts where the tracks are found. It is a book with 10 chapters written by Brazilian researchers from universities and the Geological Survey of Brazil. All chapters have regular research on geology and paleontology subjects. Each chapter will include information concerning the geological context, the dinosaur tracks in their diversity and paleobiological interpretation; the paleogeographical distribution of the footprints, paleoenvironmental and paleoclimatic contexts, and extensive references about the Brazilian basins and their dinosaur footprints. The chapters discuss in detail the dinosaur footprints throughout Mesozoic basins in Brazil, including the Triassic and Jurassic deposits of Paraná Basin, the desert Cretaceous dinosaurs from Botucatu, Caiuá, and Sanfranciscana paleodeserts, and the Cretaceous dinosaur tracks found in interior and marginal basins linked to the events of the South America—Africa break up. The chapters are illustrated with geological maps, stratigraphic charts, countless images of the dinosaur footprints and trackways, and life reconstructions of the environments in which the tracks were produced. The target audience is geoscientists and paleontologists, including researchers on evolution subjects.

It is a unique book regarding dinosaur tracks from the Brazilian basins, with useful information to paleoenvironmental and paleoclimatic interpretations, allowing an overview of the environmental changes and the biological evolution throughout Mesozoic. There are exclusive paleoenvironmental color reconstructions of Mesozoic environments which illustrates the diversity of the dinosaur fauna.

Dinosaur Tracks of Mesozoic Basins in Brazil is a comprehensive volume where it is possible to evaluate the impact of paleoenvironmental and paleoclimatic changes throughout the Mesozoic in the basins of the Brazilian Gondwana. A new insight in tracking dinosaurs, since their origin, until their last steps.

After 50 years of systematic work in this specialized area, this book represents the fruit of a broad collaboration between Brazilian tetrapod ichnologists and between academic and scientific institutions of Brazil, including the collaboration of a Brazilian artist, who worked in the reconstruction of old life environments, based on the careful information provided by the team of authors.

This book is somehow also the story of men and women who went on the adventure of hunting dinosaurs also where had been found no bones; to witness their presence in flawed environments for body fossils. Sometimes, in this way, finding for the first time fossil material in unexplored basins, sometimes managing to re-date stratigraphic units, previously considered Paleozoic, for lack of fossils or lack of study. They traveled, in the great and wide country that is Brazil, on the one hand huge paths in time, seeking the tracks of the oldest dinosaurs from their first steps, in the Carnian, until their fall, at the end of the Cretaceous. On the other hand, Brazilian tetrapod ichnologists traveled great distances on the ground, visiting, in search of fossil footprints, virtually all the sedimentary basins of Brazil, from the southern flooded plains, to the rain forest through the Amazonian rivers.

Our primary audience is vertebrate paleontologists, students, and professors, who are engaged in research on dinosaurs and, more generally, tetrapods. Moreover, we wish that this book can be able to provide a useful database also for the paleontologists of invertebrates and paleobotanists. Above all, we hope that geologists, sedimentologists, biostratigraphers, and paleoecologists will be interested in this topic, which is not always recognized. Paleontology teachers, science professors in schools, interested enthusiasts, and evolutionary researchers will also find materials of their interest. Finally, we hope that news reporters, who have the very important mission of correctly informing the public about dinosaurs, a theme that comes up very often, in the press, and which is always in fashion, will also find this book interesting.

In this moment we would like to be magicians and predict how many millions of fossil tracks are buried in the most diverse sediments, not yet discovered, throughout this immense country that is Brazil. In the great syneclises of the Amazon region and in the small intracratonic and marginal basins of northeast, maybe hidden in private collections or in the storage drawers of sleeping museums, they whisper: "Study me!" We would still like to have the joy, even if the age advances, of being able to say, during a field activity, as it has happened to us many times in life: "I am the first who has the privilege of seeing and touching this dinosaur footprint, which was buried and hidden from light so many million years ago"!

This book also wants to be an open window on the "gondwanan" subcontinent, Brazil, sometimes little known and occasionally forgotten by the insiders. To the readers of this book, a grateful and cordial invitation to join us in the exciting adventure in the footsteps of dinosaurs of Brazil.

Rio de Janeiro, Brazil   Ismar de Souza Carvalho
Venice, Italy   Giuseppe Leonardi

Acknowledgements

The author of Chap. 1 is in debt to Bruno de Tolvo Borsoni, Giuseppe Leonardi, Gustavo Prado, Jaime Joaquim Dias, Marcelo Adorna Fernandes, Michel Godoy, Rafael Costa da Silva, and Rafael Matos Lindoso for the support in the manuscript review and photographs. The financial support for this study was provided by Fundação Carlos Chagas Filho de Amparo à Pesquisa do Estado do Rio de Janeiro (Faperj E-26/200.828/2021) and Conselho Nacional de Desenvolvimento Científico e Tecnológico (CNPq 303596/2016-3). Chapter 2 author thanks to Jorge Ferigolo and Ana Maria Ribeiro, Fundação Zoobotânica do Rio Grande do Sul (FZB-RS), for the access to the paleontological collections. To Giuseppe Leonardi and Ismar de Souza Carvalho, for their invaluable review of the manuscript. This study was supported by Fundação Carlos Chagas Filho de Amparo à Pesquisa do Estado do Rio de Janeiro (Proc. E-26/210.294/2021, R.C.S., Brazil) and by Conselho Nacional de Desenvolvimento Científico e Tecnológico (CNPq 407158/2022-7, R.C.S., Brazil) to field, laboratory and office activities. Chapter 3 authors are in debt to the students and staff of the Laboratório de Geologia e Paleontologia of the Universidade Federal do Rio Grande (FURG) and the Laboratório de Paleontologia de Vertebrados of the Universidade Federal do Rio Grande do Sul (UFRGS) for all field assistance during years of research. Jorge Arigony and Ana Paula Forgiarini (both FURG) are thanking for field support and VANT operation. Claiton Scherer (UFRGS) and Adriano Reis (UnB) provided interesting insights on the Guará Formation depositional context. Letícia Freitas from the Rosário do Sul City Hall is warmly acknowledged by her great contribution in the popularization of the science made in this municipality. The comments made by the editors Giuseppe Leonardi and Ismar de Souza Carvalho greatly improved the early draft of this chapter. This work was funded by CNPq, with grants to HF (309463/2022-1), DDC (161161/2023-5) and PD-D (312018/2021-5), and by FAPERGS (19/2551-0002016-3). Chapter 4 authors show gratitude to Giuseppe Leonardi for the careful review of the text. To Maria Izabel Lima de Manes for giving in and authorizing the use of photos of dinosaur footprints from Nioaque, Mato Grosso do Sul. To Department of Ecology and Evolutionary Biology of Federal University of São Carlos—UFSCar, for providing support for the storage of sandstone slabs, enabling the safeguarding of the material and the development of research. To the Museum of Archeology and Paleontology of Araraquara, for providing a place for the exhibition of the ichnofossils. To the Science Museum of São Carlos "Prof. Mário Tolentino", and Paulo Roberto Milanez, coordinator, for providing the public exhibition of the ichnofossils of the Botucatu Formation. To Giuseppe Leonardi and Ismar de Souza Carvalho for the opportunity to contribute to this publication and promote the dissemination of an important record of the Botucatu Formation footprints. Chapter 5 authors thank to Patricia Colombo Mescolotti for the information concerning the Sanfranciscana Basin and images of the footprints from the Três Barras Formation. We acknowledge financial support provided by Fundação Carlos Chagas Filho de Amparo à Pesquisa do Estado do Rio de Janeiro (Faperj E-26/200.828/2021) and Conselho Nacional de Desenvolvimento Científico e Tecnológico (CNPq 303596/2016-3). Chapter 6 authors acknowledge Fernando Luiz Kilesse Salgado, Flávia Alessandra Figueiredo, Penélope Bosio, and Rone Pacheco Ribeiro for their collection management in the Macrofossil Collection (IGEO/UFRJ). Financial support provided by Fundação Carlos Chagas Filho de Amparo à Pesquisa do Estado do Rio de Janeiro (Faperj E-26/200.828/2021) and Conselho Nacional de Desenvolvimento Científico e Tecnológico (CNPq 303596/2016-3). Chapter 7 authors are grateful to Maria de Fátima Cavalcante Ferreira dos Santos and Luiz Carlos da Silva Gomes for their support in the research field in the Rio do Peixe basins. Franco Capone for providing and allowing to publish several photographs. Financial support was provided by Fundação Carlos Chagas Filho de Amparo à Pesquisa do Estado do Rio de Janeiro (Faperj E-26/200.828/2021), Conselho Nacional de Desenvolvimento Científico e Tecnológico (CNPq 303596/2016-3), Cavanis Institute of Venice (Italy) and Ponta Grossa (Brazil). Chapter 8 authors acknowledge Francisco Pinheiro, Leonardo Menezes, Mário A. T. Dantas, Marco Avanzini, and Jasmin Lanker Godenzi for cordially providing data and images. Manoel Pantaleão de Arruda for his support during our field activities in Potiguar Basin. Luana dos Santos Lima for producing the stratigraphic image. Financial support was provided by Fundação Carlos Chagas Filho de Amparo à Pesquisa do Estado do Rio de Janeiro (FAPERJ E-26/201.881/2020). Chapter 9 authors thank Flávia Alessandra Figueiredo, Penélope Saliveros Bosio, Rone Pacheco Ribeiro, and Fernando Luiz Kilesse Salgado for the support to access the Macrofossil Collection of Rio de Janeiro Federal University. Financial support provided by Fundação Carlos Chagas Filho de Amparo à Pesquisa do Estado do Rio de Janeiro (Faperj E-26/200.828/2021) and Conselho Nacional de Desenvolvimento Científico e Tecnológico (CNPq 303596/2016-3).

We also express our warmest thanks to all the other friends and collaborators who have partnered with this book in other ways, and who are too numerous to be remembered in name. They are the invisibles of science. We refer to manual workers who operated in excavations, in soil preparation, in the transport of material; to the numerous mining workers operating in different regions of the country and particularly on the vast Araripe Basin (Ceará State) and in the outcrops of the Botucatu Formation (São Paulo State); the drivers who led us in the hunt for dinosaurs along the endless roads of Brazil and on the stony, rocky, sandy and dusty or muddy tracks of the hinterland. Thanks to friends interested in paleontology and knowledgeable about the places, that accompanied and guided us in areas that are sometimes wild and even potentially dangerous, difficult to locate, and on hard roads, which have sometimes challenged the technical possibilities of their cars; as well as the boat owners who accompanied us on Tocantins and other Amazon rivers. We warmly thank the technicians of the laboratories, the preparers and curators, the designers, graphic designers, photographers, typists, once, and the operators of computers more recently, who collaborated in the preparation of the manuscripts and more. Our thanks to surveyors who have made fossil tracks locatable in inaccessible or wooded regions; professional photographers, journalists who have made known our research and our successes, making it easier for us to receive research funds. Often, groups of students of the courses of geology and paleontology have voluntarily collaborated in the research of the countryside, for excavation, collection of material, its accommodation and transport and have supported us with their youthful enthusiasm. Without them this enterprise would not have been possible.

We also thank the public administrators who, at times, helped us with vehicles, fuel, provided us with the collaboration of operators of municipalities, and even welcomed us with free hospitality in local hotels and restaurants, having understood and appreciated and therefore supported our discoveries. We have nominally thanked them in other publications.

Our thanks are extensive to many fellow geologists and paleontologists who supported us in this phase of preparation of the book with their sympathetic moral support, replacing us in some of our tasks, advising us, and informing us. And we thank our respective families, who in these months of intense work and sometimes stress have endured and encouraged us.

All authors thank to Guilherme Gher for the environmental sceneries that illustrate this book and to the Springer publishing staff, especially to Mano Priya Saravanan and Saranyaa Vasuki Balasubramanian, for their support during the editorial process.

Contents

About the Authors

Ismar de Souza Carvalho Bachelor of Sciences in Geology at the Coimbra University (Portugal), Master and Doctor of Sciences (Geology) at Federal University of Rio de Janeiro (Brazil). Full Professor at the Federal University of Rio de Janeiro (UFRJ) and director of the Cultural Center for Science and Technology at UFRJ (Casa da Ciência). Researcher in the Geosciences Center at University of Coimbra (Portugal), the Brazilian National Research Council (CNPq), and the Rio de Janeiro State Scientist Program (FAPERJ). e-mail: ismar@geologia.ufrj.br

Giuseppe Leonardi Doctor in Sciences, specialized in Paleontology, at the University "La Sapienza" of Rome (Italy). Visiting Professor of Historical Geology, Geology of Brazil, and Vertebrate Paleontology at the Federal University of Paraná. Researcher of the National Council for Research (CNPq) of Brazil. Senior Associate Researcher at the Geology Department of the Federal University of Rio de Janeiro (Brazil) and Associate Curator of the Museo della Scienza of Trento (Italy). Author of ichnology textbooks about South America and European ichnosites and its ichnofauna.

Chapter 1
Dinosaur Footprints Throughout Mesozoic Basins in Brazil

Ismar de Souza Carvalho

1.1 Introduction

The major paleogeographic configurations at the end of Paleozoic, related to global tectonics, induced environmental changes, including climate and distribution of land-masses and seas. The worldwide transformations resulted in mass extinctions and provided new possibilities to the organism's evolution. The influence of such changes on terrestrial ecosystems allowed the emergence of the dinosaurs during Triassic. This group is the result of ecological opportunities after the Permian-Triassic and a sustained long-term adaptive response to climatic shifts that lasted for ca. 57 Ma (Simões et al. 2022).

Dinosaurs were rare and geographically restricted during the first steps of their diversification in Late Triassic and the extinction of co-occurring groups such as aetosaurs, rauisuchians, and non-mammalian therapsids (Benton 1983; Brusatte et al. 2008; Langer and Godoy 2022; Langer et al. 2010; Dunne et al. 2023). Changes in the global climate played an important role in the dinosaur's distribution throughout the Mesozoic, especially during Triassic and the Triassic to Jurassic transition (Tucker and Benton 1982; Benton 1983; Whiteside et al. 2015; Brusatte et al. 2008; Olsen et al. 2022). After the end-Triassic mass extinction event they presented a wider distribution and greater abundance, an opportunistic expansion model (Dunne et al. 2023), in which a Triassic-Jurassic climatic crisis enabled their global abundance (Dunne et al. 2023). Part of this history is possible to observe from the fossils and ichnofossils found in the Brazilian sedimentary basins, that include large syneclises and rift basins throughout Mesozoic.

I. S. Carvalho (✉)
CCMN/IGEO, Departamento de Geologia, Universidade Federal do Rio de Janeiro, 21.910-200 Cidade Universitária, Ilha do Fundão, Rio de Janeiro, Estado do Rio de Janeiro, Brazil
e-mail: ismar@geologia.ufrj.br

Centro de Geociências, Universidade de Coimbra, Rua Sílvio Lima, 3030-790 Coimbra, Portugal

© The Author(s), under exclusive license to Springer Nature Switzerland AG 2024
I. S. Carvalho and G. Leonardi (eds.), *Dinosaur Tracks of Mesozoic Basins in Brazil*,
https://doi.org/10.1007/978-3-031-56355-3_1

During the end of Mesozoic, the tectonic events linked to the Gondwana break-up caused a new geographic position of the South American continent. Throughout an innovative cycle of great changes, occurred many extinction and diversification events of the flora and fauna as a response to the environmental transfigurations (Bittencourt and Langer 2011; Bronzati et al. 2015; Dunhill et al. 2016; Gorscak and O'Connor 2016). Concerning the Southern hemisphere, the South Atlantic origin has driven deep modifications in climate, geographic configuration, distribution of land and seas (Arai 2014a, b) with a direct influence in the biota, which were deeply modified by the South Atlantic tectonic scenarios. Nevertheless, in spite of the wide distribution in the Brazilian intracratonic and marginal sedimentary basins, the genesis of rocks originated in continental environments and the diversity of their fossils are poorly understood.

The dinosaur tracks are important elements to the reconstruction of terrestrial and coastal environments improving paleoenvironmental interpretation and the knowledge of the biota diversity. An overview of the spatial and temporal changes in the environments during the Mesozoic are important to understand the evolution and territorial distribution of the dinosaurs, improving the data obtained from fossils. Footprints are temporal markers of subaerial surfaces throughout the Mesozoic basins, recording cyclical changes in the environmental conditions during the deposition. They are produced in an exposed substrate or in a flooded area, resulting in distinct patterns of tracks. If there is a waterlogged substrate there will occur liquefaction of the sediments and local deformation, or in the case of more cohesive sediments, the morphological aspects of the footprint will be recorded, including features such as pads and claws, which enable the knowledge of the trackmaker. Besides behavioral insights into the trackmaker's biology, substrate properties, and environmental factors, footprints are also an important tool for the reconstruction of the terrestrial ecosystems during the Gondwanan Mesozoic throughout the Brazilian territory.

1.2 Geological Context

Preservation of animal footprints in the fossil record is strongly dependent on taphonomic processes, although it is the grain size and the sedimentation regime that determines if preservation will take place and if a footprint will be incorporated into the sedimentary record. The possibility of preservation is minimal during long-lasting periods of exposure without any sedimentation, and preservation is favored by rapid and significant sedimentation events. Thus, footprints are most commonly preserved in environments of cyclic sedimentation (Lockley 1991; Carvalho and Leonardi 2021; Carvalho et al. 2021a).

Dinosaur footprints and trackways are found in many sedimentary basins throughout the Brazilian Mesozoic. The basin's classification of sedimentary areas in Brazil is grouped in two wide groups of intracratonic (syneclises) and rift

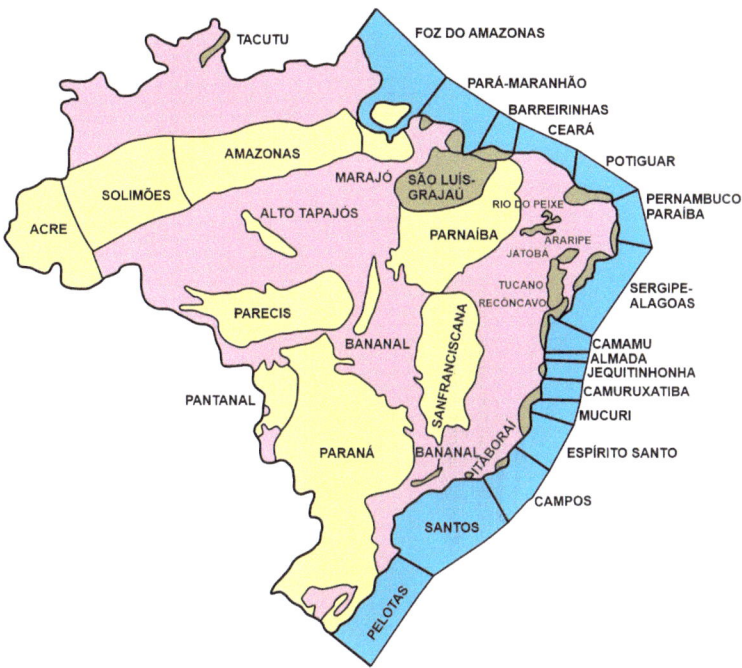

Fig. 1.1 Brazilian sedimentary basins with the large intracratonic (syneclises) and the rift basins (modified from Lucchesi 1998)

basins (Fig. 1.1). The main intracratonic basins are Paraná, Parecis, Sanfranciscana, Parnaíba and Amazonas basin whose geological history spans from Paleozoic to Cenozoic in a syneclise tectonic context, although initially developed on Cambro-Ordovician rift systems (Toczeck et al. 2019). During Mesozoic the South Atlantic opening created new basins in the present Atlantic margin and in the interior Proterozoic belts.

1.2.1 Paraná Basin

The Paraná Basin records a span time of sedimentation (Fig. 1.2) throughout most all of the Phanerozoic (Henrique-Pinto et al. 2021). It is a sag-type intracratonic depression developed on the South American platform covering an area of around 1.4 million km^2 (Milani 1992; Milani and Ramos 1998; Milani and Zalán 1999). Milani et al. (2007) divided the Paraná Basin into six supersequences: the Rio Ivaí (Sandbian-Aeronian, 455–438 Ma), Paraná (Pridoli-Famennian, 420–360 Ma), Gondwana I (Pennsylvanian–Lower Triassic, 323–247 Ma), Gondwana II (Anisian–Norian, 247–208 Ma), Gondwana III (Upper Jurassic–Berriasian, 149–139 Ma) and Bauru-Caiuá (Turonian–Maastrichtian, 93–66 Ma).

Fig. 1.2 Simplified stratigraphic chart of the Paraná Basin (modified from Milani et al. 2007; Teramoto et al. 2020) and occurrences of dinosaur footprints and trackways

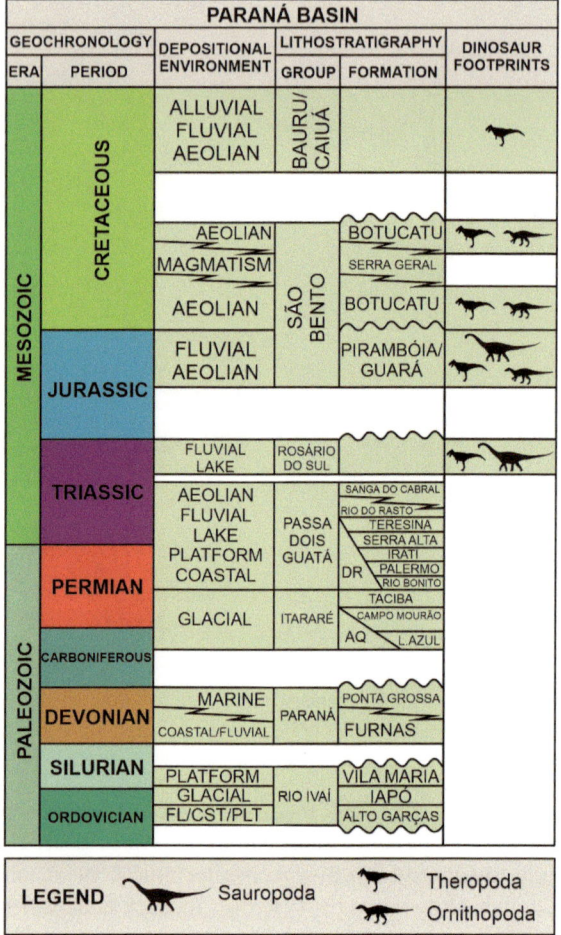

The last three upper supersequences of Milani et al. (2007) are characterized by continental sedimentary rocks. The exclusive Triassic sedimentation is recorded with the Rosário do Sul Group (Gondwana II) followed by the Gondwana III-cycle (Henrique-Pinto et al. 2021). In the Carnian deposits of Santa Maria Formation (Rosário do Sul Group) there are dinosaur footprints (Fig. 1.3) interpreted as done by theropod and prosauropod trackmakers (Silva et al. 2007, 2008). Also in the Rio Grande do Sul State, during the Late Jurassic, there are footprints in sand bars of aeolian and fluvial deposits interpreted as ornithopod, theropod and sauropod trackmakers (Guará Formation, Francischini et al. 2015), accumulated in a semi-arid climate (Scherer and Lavina 2005, 2006). A correlated deposit is the Pirambóia Formation (São Paulo State), a Late Jurassic fluvial-aeolian unit, in which dinosaur footprints in cross section were recognized in wet interdune deposits (Christofo-letti et al. 2021). A more arid climate during the last stages of Upper Jurassic and

Lower Cretaceous enable the establishment of extensive field dunes covering all the basin (Botucatu Formation, São Bento Group), where dinosaur tracks occur in aeolian dunes of the Botucatu paleodesert (Fernandes and Carvalho 2007; Leonardi et al. 2007). These deposits are coeval with the Early Cretaceous magmatic activity (Scherer 2000, 2002; Scherer et al. 2002; Brückmann et al. 2014) grouped as Serra Geral Formation (Mizusaki and Thomaz Filho 2004) or Serra Geral Group (Rossetti et al. 2018).

The upper continental sedimentation of the Bauru Supersequence (Bauru and Caiuá groups) is a post-volcanic section accumulated in the flexural depression loaded by the Serra Geral Group (Milani and De Wit 2008; Henrique-Pinto et al. 2021). There is a very distinct basin area and change of the depocenters when compared with the Paraná Basin. Then, this supersequence is considered to encompass other basin, the

Fig. 1.3 **a** Triassic footprints from Rosário do Sul Group (Caturrita Formation, Norian, Paraná Basin). **a** Dinoturbation in a sandstone bedding plane at Faxinal do Soturno, Rio Grande do Sul State; **b** a theropod footprint. Photograph **b** by Michel Godoy

Bauru Basin, covering an area of approximately 379,362 km^2 (Menegazzo et al. 2016). This basin includes a large area of Paraná, São Paulo, Mato Grosso, Mato Grosso do Sul, Goiás, and Minas Gerais states, as well as a part of Paraguay with sedimentation from the Turonian until the Maastrichtian, in semi-arid conditions (Batezelli 2010, 2017; Dias-Brito et al. 2001; Arai and Fernandes 2023). The Bauru Basin developed in the back-bulge province of a retroarc foreland system in response to Andean orogenic events with supracrustal load (Menegazzo et al. 2016). There is one reference of an isolated tridactyl footprint in the Maastrichtian deposits (Bauru Group, Marília Formation) of this basin (Riff et al. 2018) and few footprints and trackways in the Caiuá Group (Leonardi 1977; Fernandes et al. 2008).

1.2.2 Parecis Basin

It is located in the central-west region of Brazil, in the southwest sector of the Amazon Craton. During the Early Paleozoic, the Amazon region was affected by an extensional event, when a system of intracontinental rifts was established (Siqueira 1989). A syneclise was then developed over this Lower Paleozoic rift system (Pedreira and Bahia 2000). Part of the Silurian and Devonian history of the Parecis Basin is related to the Paraná Basin. The Mesozoic deposits are linked to an extensional event (Lower Jurassic), connected to the separation between South America and Africa. There are sandstones (Rio Ávila Formation), interpreted as deposition in aeolian environments, followed by basaltic flows with approximately 198 Ma (Marzolli et al. 1999). The Cretaceous Supersequence (Parecis Formation) is composed of conglomerates and sandstones, deposited in fluvial and aeolian environments (Pedreira da Silva et al. 2003). So far, no record of fossil footprints has been found in the Mesozoic deposits of this basin.

1.2.3 Sanfranciscana Basin

The Sanfranciscana Basin is a 220,000 km^2 syneclise basin (Fig. 1.4) established in the São Francisco Craton divided in two sub basins: Abaeté (south) and Urucuia (north). It is located in central-eastern Brazil (Minas Gerais, Goiás, Bahia, Tocantins, Piauí and Maranhão states), oriented in the N–S direction with approximately 1,100 km in length and 200 km in width (Cabral et al. 2021).

The sedimentary successions of Sanfranciscana Basin include the Santa Fé (Permian-Carboniferous), Areado (Lower Cretaceous), Mata da Corda (Upper Cretaceous) and Urucuia groups (Campos and Dardenne 1997a, b; Sgarbi 2000; Sgarbi et al. 2001, 2004).

The oldest sedimentary succession is the Santa Fé Group deposited during Late Carboniferous and Permian. The deposits are divided into Floresta (conglomerates and coarse sandstones) and Tabuleiro (mudstones, shales, siltstones and sandstones)

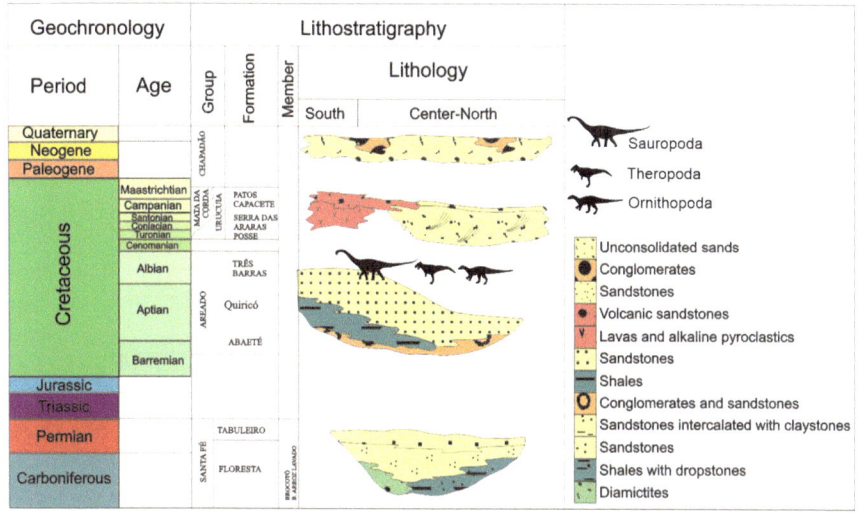

Fig. 1.4 Stratigraphic chart of Sanfranciscana Basin (modified from Carmo et al. 2004; Leite and Carmo 2021) and occurrence of dinosaur footprints and trackways

formations. It is a glaciogenic sequence that represents a gondwanan glaciation record (Campos and Dardenne 1997a).

The Lower Cretaceous Areado Group (Berriasian-Aptian) comprises the Abaeté, Quiricó, and Três Barras formations. The Abaeté Formation (up to 30 m thickness) is composed of matrix-supported and clast-supported conglomerates interpreted as deposited by braided rivers and alluvial fans (Campos and Dardenne 1997a). The Quiricó Formation (up to 100 m thickness) are constituted of fine-grained sediments, including shales, siltstones and fine grained siltstones with some levels of evaporates, that record a lacustrine sedimentation (Campos and Dardenne 1997a, b; Mescolotti et al. 2019). The ostracod data and palynologic content of the Quiricó Formation indicate a Barremian to Aptian age (Arai et al. 1995; Carmo et al. 2004). The Três Barras Formation (maximum thickness of 150 m) is mainly sandstones, with conglomerates and fine-grained siltstones and shales (Campos and Dardenne 1997a). Mescolotti et al. (2019) recognized within Três Barras Formation an unconformity (at least Cenomanian to Coniacian) separating the sedimentary succession into a lower (wet aeolian system) and an upper stratigraphic unit (dry eolian system). This is a record of the desertification events in the interior of southeast Brazil during the Cretaceous revealing prevailing winds from northeast, validating models of global paleo-circulation during the Cretaceous in Gondwana (Mescolotti et al. 2019). There was also the deposition in fluvial and deltaic environments.

The paleoclimate in the Sanfranciscana Basin during the Lower Cretaceous (Berriasian-Aptian) is interpreted as in the context of a tropical-equatorial hot arid belts (Skelton et al. 2003). Despite this basin is in the interior of Gondwana Supercontinent there are deposits of shallow lakes (Quiricó Formation) with the signs of a wide biota represented by palynomorphs, gymnosperms and angiosperms, annelids, insects, ostracods, spinicaudatans, elasmobranchs, actinopterygians, coelacanthiforms and dinosaurs (Barbosa 1965; Duarte 1968; Santos 1971; Lima 1979; Arai et al. 1995; Duarte 1997; Carvalho and Kattah 1998; Delício et al. 1998; Carmo et al. 2004; Gallego and Martins-Neto 2006; Carvalho and Maisey 2008; Zaher et al. 2011, 2020; Leite et al. 2018; Fragoso et al. 2019; Brito et al. 2020; Bittencourt et al. 2015, 2018; Ribeiro et al. 2018; Coimbra 2020; Carvalho and Santucci 2021). The existence of shallow lake environments points to more humid conditions on the southern edge of the arid belt, between 20° and 30° latitude at south (Mescolotti et al. 2019; Nascimento et al. 2022). The dinosaur footprints and trackways of Sanfranciscana Basin occur into the lower portion of the Três Barras Formation. This lower succession is interpreted as moist aeolian systems. Then the dinosaur footprints are settled in an environment with more humid conditions probably related to the Quiricó lakes.

The Mata da Corda Group comprises volcanic alkaline rocks (80 Ma U–Pb average isotopic ages) that overlain the deposits of the Areado Group (Sgarbi et al. 2004). It includes the Patos (alkaline volcanic rocks) and Capacete (epiclastic sediments) formations. The Urucuia Group (Upper Cretaceous) is a 200-m-thick unit and covers an area of approximately 76,000 km^2 (Campos and Dardenne 1997a), composed of sandstones, divided into Posse and Serra das Araras formations. This unit is interpreted as dune field deposits of dry aeolian systems (Mescolotti et al. 2019) followed by an upper succession of fluvial sediments deposited by sheet flows (Spigolon and Alvarenga 2002). The last lithostratigraphic unit is the Chapadão Group, a Quaternary unit that represents the recent sandy, unconsolidated, covers of talus, residual or alluvium origin (Campos and Dardenne 1997a).

1.2.4 Parnaíba Basin

The Parnaíba Basin is a large Paleozoic syneclise in northeastern Brazil, located partially in Tocantins, Ceará, Piauí, Maranhão and Pará states (Fig. 1.5). It is a sag basin up to 3.5 km thick, 1,000 km long and 970 km wide, nearly circular-shaped area (Cordani et al. 1984). The Precambrian crystalline basement comprises a complex lithostructural and tectonic framework formed during the Neoproterozoic–Eopaleozoic Brasilian–Pan African orogenic collage (Almeida et al. 2000; Brito Neves and Fuck 2013; Castro et al. 2013; Porto et al. 2022). This area presents 600,000 km^2 and due the polycyclic tectonic evolution and distinctive sedimentation, Góes (1995) proposed the term Parnaíba Province, an area with four depositional centers (Góes and Feijó 1994): Parnaíba, Alpercatas, Grajaú and Espigão-Mestre basins. The Parnaíba Basin is filled with Ordovician to Early Triassic sediments, mostly of marine, but also fluvial, deltaic and desert environments. The Alpercatas Basin (Jurassic to Cretaceous age) encompasses fluvial-lacustrine and aeolian sedimentary rocks alternated between basaltic flows. The Grajaú and Espigão Mestre basins, both of Cretaceous age are filled with rocks deposited in closed marine environments (Grajaú Basin) and aeolian sandstones (Espigão Mestre Basin) of the northern

Fig. 1.5 Geotectonic units of the Parnaíba Province, which includes distinct depositional areas (Pedreira da Silva et al. 2003)

extension of the Urucuia domain of the Sanfranciscana Basin (Pedreira da Silva et al. 2003).

The chronostratigraphy and lithostratigraphy of the Parnaíba Basin (Fig. 1.6) present as the oldest Paleozoic deposits a pre-Silurian (Cambro-Ordovician) sequence filling up graben-like structures, attributed to the Jaibaras Group (Oliveira and Mohriak 2003; Cerri et al. 2020). After this first deposition there are three depositional supersequences (Pedreira da Silva et al. 2003): Silurian (Serra Grande Group), Devonian-Carboniferous (Canindé Group) and Carboniferous-Triassic (Balsas Group). The Serra Grande Group (Ipu, Tianguá and Jaicós formations) comprises conglomerates, sandstones and shales. The deposits of this unit are interpreted as fluvial and glacial, fluvial and marine deposits of Silurian age. The Canindé Group (Itaim, Pimenteiras, Cabeças, Longá and Poti formations) is composed of sandstones and shales of marine, glacial and fluvial environments. The Carboniferous-Triassic is the Balsas Group (Piauí, Pedra-de-Fogo, Motuca and Sambaíba formations), composed of sandstones, shales, carbonates and stromatolites interpreted as aeolian dunes and tidal flats (Santos and Carvalho 2009).

Fig. 1.6 Stratigraphic chart of Parnaíba Basin (adapted from Vaz et al. 2007; Araújo 2017; Pereira et al. 2021) and occurrences of dinosaur footprints and trackways

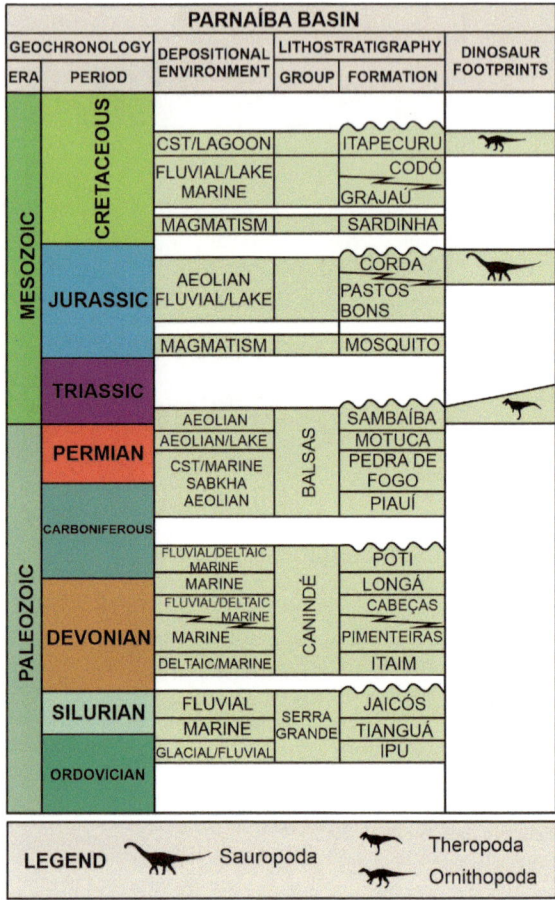

During the Mesozoic, the main regional tectonic elements were the Xambioá (E–W) Arch, located in the center of the basin, and the Ferrer–Urbano Santos Arch delimiting the small marginal basins associated with the opening of the Equatorial South Atlantic (Araújo 2017). Throughout the Jurassic and Cretaceous, magma flows and diabase dykes indicate the effects of the break-up of Pangea (Sardinha Formation) and Gondwana (Mosquito Formation).

The Alpercatas Basin is a rift system with the Jurassic supersequence of the Mearim Group (Pastos Bons and Corda formations) limited by the basalts (Góes and Feijó 1994) of Mosquito (Jurassic) and Sardinha (Lower Cretaceous) formations. The Pastos Bons Formation are a succession of shales and sandstones interpreted as fluvial and aeolian environments. The Corda Formation is bimodal sandstones, with some mudstone levels, is interpreted as a desert environment.

The Grajaú Basin (Góes and Rossetti 2001) is isolated from the São Luís Basin by the Ferrer-Urbano Santos structural arch. This basin is filled up by the Cretaceous supersequence (Aptian-Albian), including the Codó, Grajaú and Itapecuru formations (Rossetti and Truckenbrodt 1997; Rossetti 2001). The Grajaú Formation deposits are sandstones intercalated with the shales, carbonates and evaporites of the Codó Formation, interpreted as fluvial and lagoon environments. The Itapecuru Formation is a succession of sandstones, shales and mudstones interpreted as fluvial and deltaic deposits. The Espigão-Mestre Basin is located in the southern portion of the Parnaíba Province and it is the north area of the Urucuia sub-basin of the Sanfranciscana Basin (Pedreira da Silva et al. 2003). There are sandstones interpreted as aeolian deposits.

The footprints and dinosaur tracks found in the Parnaíba Province (or in a more common sense, Parnaíba Basin) are located in the Parnaíba, Espigão-Mestre and Grajaú basins. In the Parnaíba Basin - Sambaíba Formation, a dubious Triassic unit, there are many isolated theropod footprints (Fortaleza dos Nogueira locality, Maranhão State) identified by Assis et al. (2010). In the Espigão-Mestre Basin there are seven sauropod trackways (Leonardi 1980, 1994; De Valais et al. 2015; Lopes et al. 2021) in the Corda Formation (Barremian), São Domingos county, Itaguatins locality, Tocantins State (Fig. 1.7). In the Grajaú Basin is found an isolated footprint in the Itapecuru Formation (Aptian), identified as a *Caririchnium* footprint (Menezes et al. 2019), Itapecuru River, Maranhão State.

1.2.5 *Amazonas Basin*

The Amazon region comprises many sedimentary basins grouped as the Amazonas Province, in which the larger basins are Acre, Solimões, Alto Tapajós, Amazonas and Marajó. These basins are isolated by structural arches and throughout their geological history they present some episodes of environmental connections, especially during the Devonian marine transgressions and later through the Cenozoic fluvial systems.

The most complete stratigraphic succession occurs in the Amazonas Basin. It presents an area of 515,000 km^2 with 5,000 m thickness (Cunha et al. 1994) overlapping Precambrian magmatic and metamorphic rocks (Fig. 1.8). The oldest Paleozoic deposits are conglomerates and sandstones, probably Cambrian-Ordovician—the Prosperança Formation. Subsequently, the Ordovician-Devonian Supersequence (Trombetas Group) presents clastic rocks deposited in marine environments (Cunha et al. 1994). Follow the deposition of the Devonian-Carboniferous Supersequence, comprising the Urupadi and Curuá groups, which represent fluvial and deltaic environments and also include a glacial interval. The Carboniferous-Permian Supersequence (Tapajós Group) includes both continental and restricted marine environments. At the beginning of the Jurassic occurred extensive basaltic magmatism as part of the CAMP Province—Central Atlantic Magmatic Province (Marzolli et al. 1999). The final sedimentation cycle in the Amazonas Basin are two continental sequences: Upper Cretaceous (Alter do Chão Formation) and Cenozoic (Solimões

Fig. 1.7 **a** Theropod footprint from the Sambaíba Formation (Parnaíba Basin), a probable Late Triassic-Jurassic lithostratigraphic unit. Fortaleza dos Nogueiras locality, Maranhão State. Photograph by Rafael Matos Lindoso; scale bar: 3 cm. **b** Three sauropod trackways from the Corda Formation (Parnaíba Basin), Lower Cretaceous. Itaguatins locality, Tocantins State. Photograph by Giuseppe Leonardi

and Içá formations) sequences, deposited in the context of fluvial and lacustrine environments (Mendes et al. 2012).

The Acre Basin is in the same geological context of Amazonas Basin and they share a common Paleozoic history. It is a retroarch basin of the Andes mountain range (Milani and Thomaz Filho 2000). The Jurassic Supersequence (Juruá-Mirim Formation) is constituted by sandstones, evaporites and basalt flows in the context of terrestrial environments. The Cretaceous Supersequence (Moa, Rio Azul, Divisor and Ramón formations) presents sandstones and shales deposited in fluvial and lakes environments.

It was not found until now any dinosaur footprint in the Mesozoic deposits of the Amazon Province (Acre, Solimões, Alto Tapajós, Amazonas and Marajó basins). Despite there are Jurassic and Cretaceous outcrops in the Amazonas and Acre basins, there are few geological studies in this region and no data concerning dinosaur footprints.

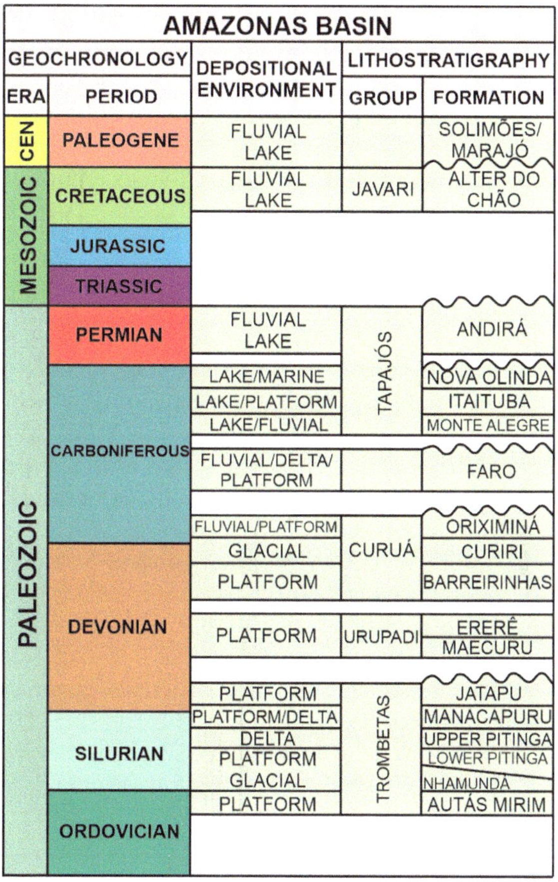

Fig. 1.8 Stratigraphic chart of the Amazonas Basin (adapted from Cunha et al. 1994, 2007). Although the wide area of this basin there is no occurrences of dinosaur footprints and trackways

AMAZONAS BASIN				
GEOCHRONOLOGY		DEPOSITIONAL ENVIRONMENT	LITHOSTRATIGRAPHY	
ERA	PERIOD		GROUP	FORMATION
MESOZOIC — CEN	PALEOGENE	FLUVIAL LAKE		SOLIMÕES/ MARAJÓ
	CRETACEOUS	FLUVIAL LAKE	JAVARI	ALTER DO CHÃO
	JURASSIC			
	TRIASSIC			
PALEOZOIC	PERMIAN	FLUVIAL LAKE	TAPAJÓS	ANDIRÁ
	CARBONIFEROUS	LAKE/MARINE		NOVA OLINDA
		LAKE/PLATFORM		ITAITUBA
		LAKE/FLUVIAL		MONTE ALEGRE
		FLUVIAL/DELTA/ PLATFORM		FARO
		FLUVIAL/PLATFORM	CURUÁ	ORIXIMINÁ
		GLACIAL		CURIRI
		PLATFORM		BARREIRINHAS
	DEVONIAN	PLATFORM	URUPADI	ERERÊ
				MAECURU
		PLATFORM	TROMBETAS	JATAPU
		PLATFORM/DELTA		MANACAPURU
	SILURIAN	DELTA		UPPER PITINGA
		PLATFORM		LOWER PITINGA
		GLACIAL		NHAMUNDÁ
		PLATFORM		AUTÁS MIRIM
	ORDOVICIAN			

1.2.6 Rift Basins

The Brazilian rift basins are related to the tectonic events during the breakup of Gondwana. During the Late Jurassic, intense tectonic activity fragmented the crust and created small half grabens with a great accumulation of sediments. Small lakes that captured the drainage network (Machado et al. 1990) were the main continental environments, with an eventual physical linkage, where the dinosaur footprints are found. These occurrences are not synchronous and did not coincide temporally throughout the Cretaceous, as the beginning of the South Atlantic occurred during three diachronous tectono-sedimentary domains (Popoff 1988): Austral (southern), tropical (midlatitude), and equatorial (northern). Then, there are many small interior basins in Northeastern Brazil (tropical tectonic domain), bordering the Atlantic margin (tropical and equatorial tectonic domain), and also in the Amazon region (equatorial tectonic domain). The outcrops indicate deposition in a wide variety of geological settings, including fluvial, lacustrine and seashore environments.

The majority of ichnosites are found in the Early Cretaceous intracontinental basins of Sousa, Triunfo (both included in the Rio do Peixe basins), Lima Campos, Malhada Vermelha and Araripe. In the marginal Atlantic basins there are few occurrences, generally cross-section and isolated footprints. The exception is the São Luís Basin in the equatorial margin, where a large number of theropod, sauropod and ornithopod footprints are distributed in six ichnosites located in the São Luís and Alcântara counties, Maranhão State.

1.2.6.1 Intracontinental Basins of Northeast Brazil

These basins are intracratonic areas (grabens and half-grabens) in which the sedimentation was controlled by the reactivation of Precambrian tectonic structures during the first steps of the South America and Africa drifting (Ponte 1992; Mabesoone 1994; Valença et al. 2003). The region of the Precambrian basement (Meso- and Neo-Proterozoic), where they are found, is known as Borborema Province (Santos and Brito Neves 1984).

The Borborema Province shows diverse tectonic, metamorphic, and magmatic events that are interpreted to belong to a larger Precambrian paleocontinent expanding into Africa (Trompette et al. 1993). This area was periodically affected by the formation of intracontinental rifts (Matos 1992; Córdoba et al. 2008) and the reactivated fault movements within the ancient Precambrian fault lines (E-W and SW–NE oriented) created several sedimentary basins (Fig. 1.9).

These basins lie in the western of Paraíba, Rio Grande do Norte, Piauí, Pernambuco, and in the southern Ceará states, Northeast Brazil. There is a diversity of

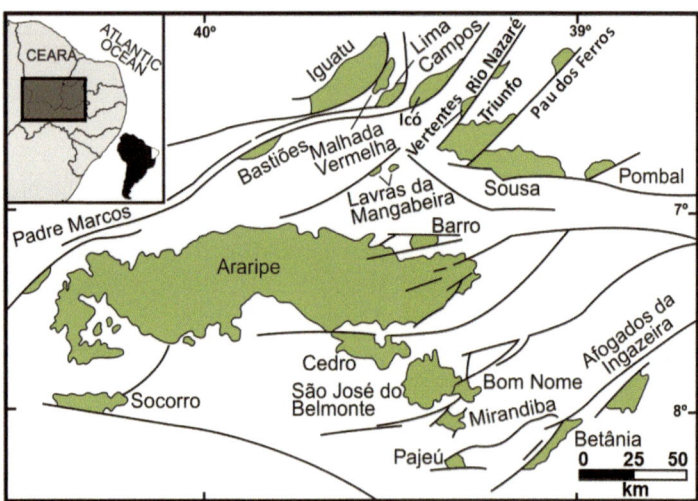

Fig. 1.9 Location map of the interior rift basins in the Northeast Brazil (modified from Carvalho et al. 2013a, b)

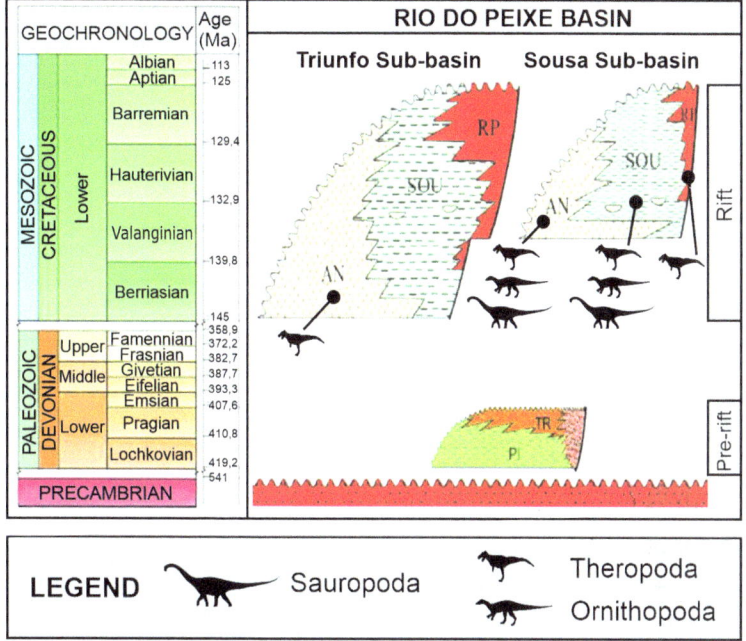

Fig. 1.10 Stratigraphic chart of Rio do Peixe basins. Santa Helena Group (PI—Pilões Formation; TR—Triunfo Formation) and Rio do Peixe Group (AN—Antenor Navarro Formation; SO—Sousa Formation; RP—Rio Piranhas Formation). 2021 Modified from Rapozo et al. (2021)

vertebrate and invertebrate ichnofossils, especially Early Cretaceous dinosaur footprints in the Rio do Peixe (Leonardi 1979a, b, 1989; Carvalho 1996a; Carvalho et al. 1993, 1995, Carvalho and Leonardi 2021, 2022; Leonardi and Carvalho 2021) and Araripe basin's (Carvalho 2004; Carvalho et al. 1994a, b, 2021a, b). During the Barremian-Berriasian in the Rio do Peixe basins (Fig. 1.10), the dinosaur footprints (Figs. 1.11 and 1.12) occur in the Rio do Peixe Group, that includes the Antenor Navarro (alluvial fans/braided channels), Sousa (shallow lacustrine/floodplain), and Rio Piranhas (alluvial fans/braided channels) formations (Srivastava and Carvalho 2004). In the ?Barremian, Aptian and Cenomanian of Araripe Basin they are found in the Mauriti (fluvial), Rio da Batateira (deltaic/floodplain), Crato (alkaline lake) and Exu formations (Viana et al. 1993; Carvalho 2000a, b, 2004; Carvalho et al. 2021a, b).

1.2.6.2 Marginal Atlantic Basins

The south to north breakup of Western Gondwana started in the southern region of South America during the Late Jurassic reaching the equatorial margin by late Aptian-early Albian. In general there are four megasequences of sedimentation (Fig. 1.13)

Fig. 1.11 *Sousaichnium pricei*, an ornithopod trackway from the Sousa Basin (Sousa Formation, Sousa Basin) preserved in a floodplain succession.

controlled by the evolution of the rifting events (Cainelli and Mohriak 1999). The southern basins, such as Pelotas, Santos, Campos and Espírito Santo are offshore regions and there are no outcrops of the Mesozoic successions. The Recôncavo-Tucano-Jatobá is an aborted rift system filled by the pre-rift and continental megasequences, in which there are some dinosaur footprints in the continental deposits of the pre-rift (Late Jurassic) and continental megasequences (Aptian). They are found as cross section footprints in the Aliança and Sergi formations (Recôncavo Basin, Carvalho and Borghi 2008) and as isolated tridactyl footprints in the Tucano Basin (São Sebastião Formation, Dantas et al. 2019).

The South Atlantic marginal basins of Sergipe-Alagoas and Potiguar present outcrops of the rifting evolution events. This allows the observation of bedding planes with dinosaur footprints in the Aptian succession of Sergipe-Alagoas Basin (Maceió Formation, Carvalho and Souza-Lima 2008) and the Aptian?–Cenomanian of the Potiguar Basin (Açu Formation, Leonardi et al. 2021).

In the Equatorial Atlantic margin, the dinosaur footprints of São Luís Basin occur in Cenomanian fine-grained quartz's sandstones of the Alcântara Formation (Fig. 1.14). The tracksites are located in the early equatorial seashore of the Atlantic Ocean, in the environmental setting of an estuary that occupied a low gradient coastal

Fig. 1.12 A theropod footprint from Sousa Basin (Antenor Navarro Formation) preserved in reddish sandstones of alluvial fan deposits from Serrote do Letreiro, Sousa County (Paraíba State). Scale bar: 4 cm

plain, under a dry and hot climate. Distinctive dinosaur communities are found in these environments (Carvalho 1995, 2001; Carvalho and Pedrão 1998).

1.2.6.3 Tacutu Basin

The Tacutu Basin is a NE-SW asymmetric graben system (4,500 km^2), with an extension of 300 km and up to 50 km wide, located in the borders of Brazil and Guyana Republic in the Amazon region. The sedimentary succession (Pedreira da Silva et al. 2003; Castro et al. 2021) comprises an initial pre-rift Jurassic Supersequence (Apoteri and Manari formations) with volcanic rocks and reddish siltstones of lacustrine environments (Eiras et al. 1994; Marzolli et al. 1999). Later, during a rift phase, there are fluvial, playa lake and deltaic deposition of evaporites, shales, sandstones and conglomerates (Rupununi and Pirara formations) followed by Cretaceous siltstones and sandstones (Tacutu and Serra do Tucano formations). Barros et al. (2023) reported dinosaur footprints in the Serra do Tucano Formation (Barremian-Albian), probably in floodplain deposits. The brief description indicates that there are a large number of footprints interpreted as sauropod, ornithopod, theropod and thyreophoran trackmakers.

Fig. 1.13 Stratigraphic columns of Mesozoic Brazilian marginal basins (Santos, Espírito Santo and Sergipe-Alagoas basins). The chart presents the local stages Dom João (Tithonian), Rio da Serra (Berriasian-lower Hauterivian), Aratu (Hauterivian-lower Barremian), Buracica (upper Barremian), Jiquiá (upper Barremian-lower Aptian) and Alagoas (upper Aptian). Modified from Cainelli and Mohriak (1999) and Menezes et al. (2016)

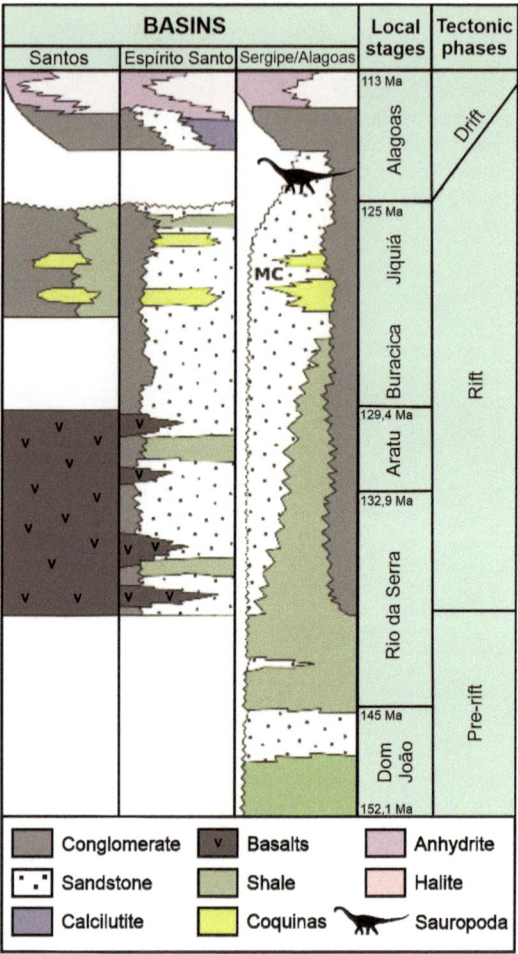

1.3 Paleogeographic and Paleoenvironmental Distribution of the Footprints

In the Paraná Basin (South Brazil), the Triassic records (Rosário do Sul Group, Anisian–Norian, ~247–208 Ma) are reddish sandstones and shales from fluvial and lake deposits. There are many fossils of permineralized logs and a wide diversity of bone remains (Langer 2003; Ferigolo and Langer 2006; Bittencourt and Langer 2011; Langer and Ferigolo 2013; Langer et al. 1999; Mestriner et al. 2023; Pacheco et al. 2019; Müller and Garcia 2023) and footprints (Silva et al. 2007, 2008) of the oldest dinosaurs. The prevailing climatic condition was hot and dry, with intermittent wet periods. Petrified logs indicate a subtropical to tropical climate, with irregular seasonality and short-term droughts, suggesting semi-arid conditions (Scaramuzza

Fig. 1.14 Footprints from de Cenomanian (Alcântara Formation, São Luís Basin) preserved in a tidal flat environment at Ponta da Guia, São Luís County (Maranhão State)

dos Santos et al. 2023). A similar climate is also found during the Late Jurassic deposition of the Guará and Pirambóia formations (Francischini et al. 2015; Christofoletti et al. 2021). Sand bars, fluvial and interdune deposits accumulated in a semi-arid climate (Scherer and Lavina 2005, 2006). From then on, a broad process of aridization occurred in the interior of the Gondwana. At the beginning of the Cretaceous, there was hyperarid conditions, resulting in a 1,300,000 km^2 desert (Botucatu Paleodesert) covering the center-south of Brazil, as well as the west of Uruguay, east of Paraguay and northeast of Argentina (Almeida 1953; Bertolini et al. 2021). The origin of the Botucatu Paleodesert and even other large aeolian deposits, such as those in the Sanfranciscana Basin, resulted from the Pangea geographic configuration of the Late Permian. The distribution of continents, low sea level and atmospheric currents, resulted in vast aridity during almost all of the Mesozoic (Almeida and Carneiro 1998). This climate model was changed after the West Gondwana break up, during the Early Cretaceous. There were changes in the atmospheric circulation, humidity and temperature and, therefore, to the disappearance of this great desert.

The deposits of this paleodesert are the sandstones of the Botucatu Formation, which overlap and are sometimes interspersed with the basalt flows of the Serra Geral Formation and the Caiuá Group (Leonardi 1977). They are interpreted as extensive dune fields and humid interdune areas, where the footprints of mammals, dinosaurs (Figs. 1.15 and 1.16) and small invertebrates were preserved (Leonardi 1977; Leonardi and Carvalho 2002; Francischini et al. 2015). In the Rio Paraná Formation (Caiuá Group) there is an association of footprints of dinosaurs and small mammals preserved in aeolian sandstones, indicating animals adapted to aridity (Fernandes et al. 2008).

At the end of the Jurassic and Early Cretaceous, an important event is the extrusion of basalts (ages between 127 and 137 million years, Brückmann et al. 2014) indicative of intense fissure eruptions (Serra Geral volcanism) associated with the South America and Africa drifting (Assine et al. 2004). The basalt flows (Serra Geral Formation) are interspersed with sandstones from the Botucatu Paleodesert (Scherer

Fig. 1.15 The geographic configuration of Gondwana resulted in hyperarid conditions originating a 1,300,000 km^2 desert (Botucatu Paleodesert, Paraná Basin) where dinosaur footprints are found in dunes and interdune areas. **a** The arrow indicates a surface in a dune deposit of the Botucatu Formation with an ornithopod trackway; **b** an isolated theropod footprint in a flagstone of the Botucatu Formation. Scale bar: 3 cm. Both images from the São Bento Quarry, Araraquara County (São Paulo State). Photographs by Marcelo Adorna Fernandes

Fig. 1.16 **a** Outcrop in the Catingueiro creek of the Lower Cretaceous aeolian sandstones of Caiuá Group (Bauru Basin); **b** theropod footprints in the Catingueiro creek, Cianorte County (Paraná State). Scale bar 5 cm. Photographs by Giuseppe Leonardi

2000, 2002; Scherer et al. 2002). Due to changes in the depocenters, after the Serra Geral volcanism, a new depositional setting was established named as Bauru Basin.

In the Parnaíba Basin, the aridity was striking since the early Mesozoic. During the Triassic, deposition occurred mainly in lacustrine and aeolian environments. A wide area of reddish sandstones, with large cross stratifications (Sambaíba Formation), is interpreted as the deposition of aeolian dunes. At the end of the Triassic and Early Jurassic (approximately 202–200 Ma) there is a record of intense volcanism that reflects the breakup of Pangea and the birth of the Central Atlantic Ocean (Nogueira et al. 2021). Throughout the Cretaceous, continental environments dominated, although at the end of the Early Cretaceous, there were marine ingressions related to the Brazilian equatorial margin opening.

A major landmark of the Mesozoic is the opening of the Atlantic Ocean, with the definitive rupture of West Gondwana. This is the South Atlantic Event, which comprises a set of tectonic, sedimentary and magmatic events, from the Permian-Triassic onwards. The spreading of the ocean floor occurred in three different moments (approximately 130 Ma, 113 Ma, and 110 Ma) starting from south to north, and reaching the Brazilian equatorial margin at the end of the Early Cretaceous (Chang et al. 1992; Matos 1992). Only after 110 million years ago there was a continuous and stable oceanic crust (Matos et al. 2021a, b).

The structural lineaments of the Brazilian and Pan-African provinces were intensely reactivated, with expressive vertical movements. In the interior of the Borborema Province and along the current Brazilian continental margin there were crustal ruptures and the origin of new sedimentary basins (Pedreira da Silva et al. 2003), such as the Rio do Peixe basins, in Northeast Brazil. On the continental margin, the tectonics controlled the phases of subsidence, sea level changes and sedimentation, with diverse environments in the course of the opening of the Atlantic Ocean (Chang and Kowsmann 1987; Petri 1987).

Throughout the pre-rift events (syn-rift I, Upper Jurassic), a great depression known as the Afro-Brazilian Depression, allowed the accumulation of continental sediments from rivers and temporary lakes (Ponte 1972) and consequently a higher humidity in the continental interior. Along the borders of this depression, topographic barriers played an important role in reducing wind velocity from the southeast, establishing climatic zones that probably controlled the flora and fauna distribution (Golonka et al. 1994). Despite the fact that it dominated a semi-arid climate, the locally higher precipitation led to the growth of abundant vegetation, mainly along the northern margin of the Afro-Brazilian Depression (García and Wilbert 1994; Da Rosa and Garcia 2000).

Later, there was a divergent movement between the South American and African plates, inducing a 3,500 km rift system on the present Brazilian continental margin. Thus, during the Early Cretaceous, alluvial fans, rivers and lakes were the main environments in the rift basins (Guardado et al. 1989; Lima Filho et al. 1999; Mabesoone et al. 1979, 2000; Córdoba et al. 2008). Transitional environments occurred between the rift and the drift phases, ranging from continental to marine environments (upper Aptian-lower Albian).

The current paleogeographical models demonstrate that despite the south to north tectonic opening, the first marine ingressions originated from the northern region, linked to the Tethys Sea (Azevedo 2004; Dias-Brito 1987, 2000; Arai 2009, 2014a, b, 2016; Tucker and Dias-Brito 2017; Araripe et al. 2022; Fauth et al. 2023; Lemos et al. 2023). Some sauropod dinosaurs from South America and Africa revealed that terrestrial connections persisted until about 100 million years ago (Calvo and Salgado 1996), but at the end of the Turonian (89.8 Ma, Late Cretaceous) dominated settled the open sea, approaching the conditions that still exist today. Hot weather were widespread, although there was probably a wide range of humidity (Petri 1983, 1998; Lima 1983; Lima and Coelho 1987; Carvalho 1996b; Carvalho and Carvalho 1990; Skelton 2003; Souza-Lima and Silva 2018; Degani-Schmidt et al. 2023) and marked climatic cycles in some regions (Gomes et al. 2021; Guerra-Sommer et al. 2021a, b). A humidification process, changing from arid to a tropical climate, with the onset of the equatorial humid belt, took place due the origin of the Atlantic Ocean during the Gondwana breakup (Carvalho et al. 2022; Salgado-Campos et al. 2021, 2022; Luft-Souza et al. 2022; Scaramuzza dos Santos et al. 2020, 2021, 2022, 2023; Dummann et al. 2023).

The tectonic evolution of the Mesozoic rift basins appears to be a strong constraint on the distribution of sedimentary environments and, consequently, on the possibility of preservation of fossil footprints. In addition, the distribution of outcrops in the marginal basins represents a limitation for the number of dinosaur footprints in a time interval favorable to their frequency.

1.4 Conclusions

The distribution of dinosaur footprints through time and space in the Brazilian territory covers all the Mesozoic era in the intracratonic and marginal rift basins with distinct geological history.

Since the Triassic (Carnian) of the Paraná Basin the first dinosaurs are recognized by their footprints, including theropods and prosauropods. The continental environments, including fluvial floodplains and lakes in an arid climate, were populated by a great number of the first dinosaur lineages. During almost all the Mesozoic, the land masses distribution, low stand eustatic level and atmospheric currents were responsible for an arid climate with episodes of hyperaridity. In the Early Cretaceous, the paleogeographic setting allowed an extreme arid climate and the origin of the Botucatu Paleodesert. The trackmakers that lived in this environment showed specific adaptations to this environment.

During the Gondwana breakup, the atmospheric circulation, humidity and temperature changed following the ending of the Botucatu Paleodesert. Furthermore, new ecological spaces were available with the origin of the South Atlantic Ocean throughout the late Early Cretaceous. From this moment new groups of dinosaurs left their footprints in coastal environments during the early tectonic stages of the rift basins.

References

Almeida FFM (1953) Botucatu, a Triassic desert of South America. In: 19th International Geological Congress, Argel, 1953. Comptes Rendus, Argélia, 1953, XIX Sessão, fasc. VII, pp 9–24

Almeida FFM, Carneiro CDR (1998) Botucatu: o grande deserto brasileiro. Ciência Hoje 24(143):36–43

Almeida FFM, Brito Neves BB, Carneiro CDR (2000) The origin and evolution of the South American Platform. Earth-Sci Rev 50:77–111

Arai M (2009) Paleogeografia do Atlântico Sul no Aptiano: um novo modelo a partir de dados micropaleontológicos recentes. Bol Geociênc Petrobras 17(2):331–351

Arai M (2014a) Aptian/Albian (Early Cretaceous) paleogeography of the South Atlantic: a pale-ontological perspective. Braz J Geol 44(2):339–350. https://doi.org/10.5327/Z2317-488920140 0020012

Arai M (2014b) Reconstituições paleo-oceanográfica e paleoclimática do Oceano Atlântico no Cretáceo, baseadas em dinoflagelados. In: Carvalho IS, Garcia MJ, Lana CC, Strohschoen O Jr (eds) Paleontologia: Cenários de Vida - Paleoclimas, vol 5, 1st edn. Editora Interciência, Rio de Janeiro, pp 45–62

Arai M (2016) Reply to the comments of Assine et al. (Comments on paper by M. Arai "Aptian/Albian (Early Cretaceous) paleogeography of the South Atlantic: a paleontological perspective)" Braz J Geol 46(1):9–13

Arai M, Fernandes LA (2023) Lower Campanian palynoflora from the Araçatuba Formation (Bauru Group), Southeastern Brazil. Cretac Res 150:105586. https://doi.org/10.1016/j.cretres.2023. 105586

Arai M, Dino R, Milhomem PS, Sgarbi GNC (1995) Micropaleontologia da Formação Areado, Cretáceo da Bacia Sanfranciscana: estudo dos ostracodes e palinologia. In: 14° Congresso Brasileiro de Paleontologia, Atas, Uberaba, Sociedade Brasileira de Paleontologia, pp 1–2

Araripe RC, Lemos FAP, Prado LAC et al (2022) Upper Aptian–lower Albian of the southern-central Araripe Basin, Brazil: microbiostratigraphic and paleoecological inferences. J S Am Earth Sci 116:103814. https://doi.org/10.1016/j.jsames.2022.103814

Araújo DB (2017) Bacia do Parnaíba. Sumário Geológico e Setores em Oferta. Agência Nacional do Petróleo. Superintendência de Definição de Blocos SDB, 19 p

Assine ML, Piranha JM, Carneiro CDR (2004) Os paleodesertos Pirambóia e Botucatu. In: Mantesso-Neto V, Bartorelli A, Carneiro CDR, Brito-Neves BB (eds) Geologia do Continente Sul Americano: Evolução da Obra de Fernando Flávio Marques de Almeida, vol 5. Beca, São Paulo, pp 77–93

Assis FP, Macambira JB, Leonardi G (2010) Dinossauros terópodes do Ribeirão das Lajes, primeiro Registro fóssil da Formação Sambaíba (Neotriássico-Eojurássico), Bacia do Parnaíba: Fortaleza dos Nogueiras, Maranhão-Brasil. In: 45° Congresso Brasileiro de Geologia, Belém, PA, Brasil, Anais, p 1720. http://sbg.sitepessoal.com/anais_digitalizados/2010-BEL%C3%89M/ 2010-BEL%C3%89M.zip

Azevedo RLM (2004) Paleoceanografia e a evolução do Atlântico Sul no Albiano. Bol Geociênc Petrobras 12(2):231–249

Barbosa O (1965) Geologia econômica e aplicada a uma parte do Planalto Central brasileiro. In: 19° Congresso Brasileiro de Geologia, Anais, Rio de Janeiro, pp 1–11

Barros LS, Vieira CEL, Souza V, Pinheiro FL (2023) Resultados preliminares da primeira ocorrência de icnofósseis de Dinosauria na porção Nordeste da Amazônia, Cretáceo Inferior da Bacia do Tacutu, Roraima. In: 12° Simpósio Brasileiro de Paleontologia de Vertebrados, Sociedade Brasileira de Paleontologia. Boletim de Resumos, Santa Maria, pp 28–29

Batezelli A (2010) Arcabouço tectono-estratigráfico e evolução das Bacias Caiuá e Bauru no Sudeste brasileiro. Rev Bras Geociênc 40:265–285

Batezelli A (2017) Continental systems tracts of the Brazilian Cretaceous Bauru Basin and their relationship with the tectonic and climatic evolution of South America. Basin Res 29:1–25

Benton MJ (1983) Dinosaur success in the Triassic: a noncompetitive ecological model. Q Rev Biol 58:29–55

Bertolini G, Marques JC, Hartley AJ, Basei MAS, Frantz JC, Santos PR (2021) Determining sediment provenance history in a Gondwanan erg: Botucatu Formation, Northern Paraná Basin, Brazil. Sediment Geol. https://doi.org/10.1016/j.sedgeo.2021.105883

Bittencourt JS, Kuchenbecker M, Vasconcelos AG, Meyer KEB (2015) O registro fóssil das coberturas sedimentares do cráton do São Francisco em Minas Gerais. Geonomos 23(2):39–62

Bittencourt JS, Langer MC (2011) Mesozoic dinosaurs from Brazil and their biogeographic implications. An Acad Bras Ciênc 83(1):23–60. ISSN 0001-3765

Bittencourt JS, Rohn R, Gallego OF, Monferra MD, Uhlein A (2018) The morphology and systematics of the clam shrimp *Platyestheria* gen. nov. *abaetensis* (Cardoso) (Crustacea, Spinicaudata) from the Lower Cretaceous of the Sanfranciscana Basin, southeast Brazil. Cretac Res 91:274–286

Brito PM, Figueiredo FJ, Leal MEC (2020) A revision of *Laeliichthys ancestralis* Santos, 1985 (Teleostei: Osteoglossomorpha) from the Lower Cretaceous of Brazil: phylogenetic relationships and biogeographical implications. PLoS ONE 15(10):e0241009. https://doi.org/10.1371/journal.pone.0241009

Brito Neves BB, Fuck RA (2013) Neoproterozoic evolution of the basement of the South American Platform. J S Am Earth Sci 47:72–89

Bronzati M, Montefeltro FC, Langer MC (2015) Diversification events and the effects of mass extinctions on Crocodyliformes evolutionary history. R Soc Open Sci 2:140385. https://doi.org/10.1098/rsos.140385

Brückmann M, Hartmann LA, Tassinari CCG, Sato K, Baggio SB (2014) The duration of magmatism in the Serra Geral Group, Paraná volcanic province. In: Hartmann LA, Sergio B, Baggio B (eds) Metallogeny and crustal evolution of the Serra Geral Group, 1st edn. Gráfica da UFRGS, IGeo, Porto Alegre, pp 507–518

Brusatte SL, Benton MJ, Ruta M, Lloyd GT (2008) Superiority, competition, and opportunism in the evolutionary radiation of dinosaurs. Science 321:1485–1488. https://doi.org/10.1126/science.1161833

Cabral VC, Mescolotti PC, Varejão FG (2021) Sedimentary facies and depositional model of the Lower Cretaceous Quiricó Formation (Sanfranciscana Basin, Brazil) and their implication for the occurrence of vertebrate fauna at the Coração de Jesus region. J S Am Earth Sci 112:103632. https://doi.org/10.1016/j.jsames.2021.103632

Cainelli C, Mohriak WU (1999) Some remarks on the evolution of sedimentary basins along the eastern Brazilian continental margin. Episodes 22(3):206–216. https://doi.org/10.18814/epiiugs/1999/v22i3/008

Calvo JO, Salgado L (1996) A land bridge connection between South America and Africa during Albian-Cenomanian times based on sauropod dinosaur evidences. In: 39° Congresso Brasileiro de Geologia, Salvador, Sociedade Brasileira de Geologia, Anais, vol 7, pp 392–393

Campos JEG, Dardenne MA (1997a) Estratigrafia e sedimentação da Bacia Sanfranciscana: uma revisão. Rev Bras Geol 27(3):269–282

Campos JEG, Dardenne MA (1997b) Origem e evolução tectônica da Bacia Sanfranciscana. Rev Bras Geociênc 27:283–294

Carmo DA, Tomassi HZ, Oliveira SBSG (2004) Taxonomia e distribuição estratigráfica dos ostracodes da Formação Quiricó, Grupo Areado (Cretáceo Inferior), Bacia Sanfranciscana, Brasil. Rev Bras Paleontol 7(2):139–149

Carvalho IS (1995) As pistas de dinossauros da Ponta da Guia (Bacia de São Luís, Cretáceo Superior – Maranhão, Brasil). An Acad Bras Ciênc 67(4):413–431

Carvalho IS (1996a) As pegadas de dinossauros da bacia de Uiraúna-Brejo das Freiras (Cretáceo Inferior, estado da Paraíba). In: 4° Simpósio Sobre o Cretáceo do Brasil, Rio Claro, São Paulo, UNESP, Brasil, Boletim, pp 115–121

Carvalho IS (1996b) Paleogeographic distribution of esthereliidean conchostracans on the Cretaceous rift interior basins of Northeastern Brazil. In: 39° Congresso Brasileiro de Geologia, Salvador, Bahia, Brazil, Anais, vol 7, pp 387–389

Carvalho IS (2000a) Geological environments of dinosaur footprints in the intracratonic basins from Northeast Brazil during South Atlantic opening (Early Cretaceous). Cretac Res 21:255–267. https://doi.org/10.1006/cres.1999.0194

Carvalho IS (2000b) Huellas de saurópodos Eocretácicas de la cuenca de Sousa (Serrote do Letreiro, Estado de Paraíba, Brasil). Ameghiniana 37(3):353–362

Carvalho IS (2001) Pegadas de dinossauros em depósitos estuarinos (Cenomaniano) da Bacia de São Luís (MA), Brasil. In: Rossetti DF, Góes AM, Truckenbrodt W (eds) O Cretáceo na Bacia de

São Luís-Grajaú. Museu Paraense Emílio Goeldi. Coleção Friedrich Katzer. MPEG Editoração, pp 245–264

Carvalho IS (2004) Dinosaur footprints from Northeastern Brazil: taphonomy and environmental setting. Ichnos 11:311–321

Carvalho IS, Borghi L (2008) Estruturas de Dinoturbação na Bacia do Recôncavo: Implicações Paleoambientais. In: 44º Congresso Brasileiro de Geologia, Curitiba, Paraná, Sociedade Brasileira de Geologia, Anais, vol 5, p 815

Carvalho IS, Carvalho MGP (1990) O significado paleoambiental dos conchostráceos da Bacia de Sousa. In: 1º Simpósio Sobre a Bacia do Araripe e Bacias Interiores do Nordeste, Crato, Ceará, Sociedade Brasileira de Paleontologia, Brazil, pp 329–333

Carvalho IS, Kattah SS (1998) As pegadas fósseis do Paleodeserto da Bacia Sanfranciscana (Jurássico Superior-Cretáceo Inferior, Minas Gerais). An Acad Bras Ciênc 70:53–67

Carvalho IS, Pedrão E (1998) Brazilian Theropods from the Equatorial Atlantic margin: behavior and environmental setting. Gaia 15:369–378

Carvalho IS, Souza-Lima W (2008) Processos de Dinoformação na Formação Maceió (Cretáceo Inferior), Bacia de Sergipe-Alagoas. In: 7º Simpósio Brasileiro de Paleontologia de Vertebrados, Boletim de Resumos, Paleontologia em Destaque, Edição Especial, pp 61–62

Carvalho IS, Leonardi G (2021) Fossil footprints as biosedimentary structures for paleoenvironmental interpretation: examples from Gondwana. J S Am Earth Sci 106:102936. https://doi.org/10.1016/j.jsames.2020.102936

Carvalho IS, Leonardi G (2022) Dinosaur tracks from the Sítio Pereiros ichnosite, Triunfo Basin (Lower Cretaceous) and the dinosaur diversity in the Rio do Peixe basins, Northeastern Brazil. Cretac Res 144:105446. https://doi.org/10.1016/j.cretres.2022.105446

Carvalho IS, Viana MSS, Lima Filho MF (1993) Bacia de Cedro: a icnofauna cretácica de vertebrados. An Acad Bras Ciênc 65:459–460

Carvalho IS, Viana MSS, Lima Filho MF (1994) Dinossauros do Siluriano: um anacronismo crono-geológico nas bacias interiores do Nordeste? In: 38º Congresso Brasileiro de Geologia, Camboriú. Boletim de Resumos Expandidos, Camboriú, Santa Catarina, SBG, 1994, vol 3, pp 213–214

Carvalho IS, Viana MSS, Lima Filho MF (1995) Os icnofósseis de dinossauros da Bacia do Araripe (Cretáceo Inferior, Ceará – Brasil). An Acad Bras Ciênc 67(4):433–442

Carvalho IS, Borghi L, Leonardi G (2013a) Preservation of dinosaur tracks induced by microbial mats in the Sousa Basin (Lower Cretaceous), Brazil. Cretac Res 44:112–121

Carvalho IS, Mendes JC, Costa T (2013b) The role of fracturing and mineralogical alteration of basement gneiss in the oil exhsudation in the Sousa Basin (Lower Cretaceous), Northeastern Brasil. J S Am Earth Sci 47:47–54

Carvalho IS, Bertolino LC, Borghi LF, Duarte L, Carvalho MSS, Cassab RCT (1994b) Range charts of the fossils of the Cretaceous interior basins—the São Francisco Basin. In: Beurlen G, Campos DA, Viviers MC (eds) Stratigraphic range of Cretaceous mega- and microfossils of Brazil. UFRJ, Rio de Janeiro, pp 333–352

Carvalho IS, Leonardi G, Rios-Netto AM, Borghi L, Paula Freitas A, Andrade JA, Freitas FI (2021a) Dinosaur trampling from the Aptian of Araripe Basin, NE Brazil, as tools for paleoenvironmental interpretation. Cretac Res 117:104626. https://doi.org/10.1016/j.cretres.2020.104626

Carvalho IS, Leonardi G, Andrade JAFG, Freitas FI, Borghi L, Rios-Netto AM, Figueiredo SMD, Cunha PP (2021b) Dinoturbation in the Exu Formation (Cenomanian, Upper Cretaceous) from the Araripe Basin, Brazil. In: 3rd Virtual Paleontological Congress, 2021. Book of abstracts 3rd Paleontological Virtual Congress, 2021, vol 1, pp 193–193

Carvalho JC, Santucci RM (2021) New fish remains from the Quiricó Formation (Lower Cretaceous, Sanfranciscana Basin), Minas Gerais, Brazil. J S Am Earth Sci 111:103430. https://doi.org/10.1016/j.jsames.2021.103430

Carvalho MA, Lana CC, Sá NP, Santiago G, Giannerini MCS, Bengtson P (2022) Influence of the intertropical convergence zone on Early Cretaceous plant distribution in the South Atlantic. Sci Rep 12:12600. https://doi.org/10.1038/s41598-022-16580-x

Carvalho MSS, Maisey JG (2008) New occurrence of *Mawsonia* (Sarcopterygii: Actinistia) from the Early Cretaceous of the Sanfranciscana Basin, Minas Gerais, southeastern Brazil. Geol Soc Spec Pub 295:109–144

Castro DL, Fuck RA, Phillips JD, Vidotti RM, Bezerra FHR, Dantas EL (2013) Crustal structure beneath the Paleozoic Parnaíba Basin revealed by airborne gravity and magnetic data, Brazil. Tectonophysics 614(2014):128–145. https://doi.org/10.1016/j.tecto.2013.12.009

Castro R, Giorgioni M, Souza V, Ramos M, Feitoza LM, Dino R, Antonioli L (2021) Facies analysis, petrography, and palynology of the Pirara Formation (Upper Jurassic-Lower Cretaceous) - Tacutu Basin (Roraima, Brazil). J S Am Earth Sci 112:103574. https://doi.org/10.1016/j.jsames.2021.103574

Cerri RI, Warren LV, Varejão FG, Marconato A, Luzivotto GL, Assine ML (2020) Unraveling the origin of the Parnaíba Basin: testing the rift to sag hypothesis using a multi-proxy provenance analysis. J S Am Earth Sci 101:102625. https://doi.org/10.1016/j.jsames.2020.102625

Chang HK, Kowsmann RO (1987) Interpretação genética das sequências estratigráficas das bacias da margem continental brasileira. Rev Bras Geociênc 17(2):74–80

Chang HK, Kowsmann RO, Figueiredo AMF, Bender A (1992) Tectonics and stratigraphy of the East Brazil Rift system: an overview. Tectonophysics 213:97–138

Christofoletti B, Peixoto BCPM, Warren LV, Inglez L, Fernandes MA, Alessandretti L, Perinotto JAJ, Simões MG, Assine ML (2021) Dinos among the dunes: dinoturbation in the Pirambóia Formation (Paraná Basin), São Paulo State and comments on cross-section tracks. J S Am Earth Sci 109:103252. https://doi.org/10.1016/j.jsames.2021.103252

Coimbra JC (2020) The Genus Cypridea (Crustacea, Ostracoda) and the age of the Quiricó Formation, SE Brazil: a critical review. Rev Bras Paleontol 23(2):90–96. https://doi.org/10.4072/rbp.2020.2.02

Cordani UG, Brito Neves BB, Fuck RA, Porto R, Thomaz Filho A, Cunha FMB (1984) Estudo preliminar de integração do Pré-Cambriano com os eventos tectônicos das bacias sedimentares brasileiras. Ciência Técnica Petróleo, Seção Exploração de Petróleo, Rio de Janeiro, vol 15, 70 p

Córdoba VC, Antunes AF, Jardim de Sá EF, Silva AN, Sousa DC, Lins FAPL (2008) Análise estratigráfica e estrutural da Bacia do Rio do Peixe, Nordeste do Brasil: integração a partir do levantamento sísmico pioneiro 0295_RIO_ DO_ PEIXE_2D. Bol Geociênc Petrobras 16:53–68

Cunha PRC, Gonzaga FG, Coutinho LFC, Feijó FJ (1994) Bacia do Amazonas. Bol Geociênc Petrobras 8(1):47–55

Cunha PRC, Melo JHG, Silva OB (2007) Bacia do Amazonas. Bol Geociênc Petrobras 15(2):227–251

Da Rosa AAS, Garcia AJV (2000) Palaeobiogeographic aspects of northeast Brasilian basins during the Berriasian before the break up of Gondwana. Cretac Res 21:221–239

Dantas MAT, Teixeira FAP, Santos DB, Macedo LAL, Aureliano T, Ghilardi AM (2019) Dinosaur footprints from the Lower Cretaceous (Aptian, Tucano Basin) of Canindé de São Francisco, Sergipe, Brazil. In: 26° Congresso Brasileiro de Paleontologia, Uberlândia, Sociedade Brasileira de Paleontologia, Paleontologia em Destaque, pp 270–271. ISBN 1807-2550

Degani-Schmidt I, Guerra-Sommer, M, Carvalho IS (2023) Stomatal numbers of *Pseudofrenelopsis capillata* (Cheirolepidiaceae, Coniferales) in the peri-equatorial late Aptian Crato Formation (Santana group, Araripe Basin, Brazil) and their paleoclimatic and paleoenvironmental significance. J S Am Earth Sci 126:104331

Delicio MP, Barbosa EM, Coimbra JC, Vilella RA (1998) Ocorrência de conchostraceos e ostracodes em sedimentos Pós-Paleozóicos da Bacia do Alto Sanfranciscana, Olhos d´Água, noroeste de Minas Gerais, Brasil. Acta Geol Leopoldensia 46(47):13–20

De Valais S, Candeiro CR, Tavares LF, Alves YM, Cruvinel C (2015) Current situation of the ichnological locality of São Domingos from the Corda Formation (Lower Cretaceous), northern Tocantins State, Brazil. J S Am Earth Sci 61:142–146. https://doi.org/10.1016/j.jsames.2014.09.023

Dias-Brito D (1987) A Bacia de Campos no Mesocretáceo: uma contribuição à paleoceanografia do Atlântico Sul primitivo. Rev Bras Geociênc 17(2):162–167

Dias-Brito D (2000) Global stratigraphy, palaeobiogeography and palaeoecology of Albian-Maastrichtian pithonellid calcispheres: impact on Tethys configuration. Cretac Res 21:315–349

Dias-Brito D, Musacchio EA, Castro JC, Maranhão MSAS, Suárez JM, Rodrigues R (2001) Grupo Bauru: uma unidade continental do Cretáceo do Brasil e concepções baseadas em dados micropaleontológicos, isotópicos e estratigráficos. Rev Paléobiol 20(1):245–304

Duarte L (1968) Restos vegetais fósseis da Formação Areado. In: 22° Congresso Brasileiro de Geologia, Anais, Belo Horizonte, Sociedade Brasileira de Geologia, p 68

Duarte L (1997) Vegetais do Cretáceo Inferior (Aptiano) da Formação Areado, município de Presidente Olegário, Estado de Minas Gerais. An Acad Bras Ciênc 69:495–503

Dummann W, Hofmann P, Herrle JO, Frank M, Wagner T (2023) The early opening of the Equatorial Atlantic gateway and the evolution of Cretaceous peak warming. Geology 51(5):476–480. https://doi.org/10.1130/G50842.1

Dunne EM, Farnsworth A, Benson RBJ, Godoy PL, Greene SE, Valdes PJ, Lunt DJ, Butler RJ (2023) Climatic controls on the ecological ascendancy of dinosaurs. Curr Biol 33:1–9. https://doi.org/10.1016/j.cub.2022.11.064

Dunhill AM, Bestwick J, Narey H, Sciberras J (2016) Dinosaur biogeographical structure and Mesozoic continental fragmentation: a network-based approach. J Biogeogr 43(9):1691–1704. https://doi.org/10.1111/jbi.12766

Eiras JF, Kinoshita EM, Feijó FJ (1994) Bacia do Tacutu. Bol Geociênc Petrobras 8(1):83–89

Fauth G, Kern HP, Villegas-Martín J, Mota MAL et al (2023) Early Aptian marine incursions in the interior of northeastern Brazil following the Gondwana breakup. Sci Rep 13:6728. https://doi.org/10.1038/s41598-023-32967-w

Ferigolo J, Langer MC (2006) A Late Triassic dinosauriform from south Brazil and the origin of the ornithischian predentary bone. Hist Biol 19:1–11

Fernandes LA, Sedor FA, Silva RC, Silva LR, Azevedo AA, Siqueira AG (2008) Ichnofossils of the Porto Primavera Power Plant, State of São Paulo—dinosaur and mammal footprints in rocks from the Caiuá neocretaceous desert. In: Winge M, Schobbenhaus C, Souza CRG, Fernandes ACS, Berbert-Born M, Queiroz ET (eds) Sítios Geológicos e Paleontológicos do Brasil. Available on line since 26/09/2008 at the address: http://www.unb.br/ig/sigep/sitio013/sitio013english.pdf

Fernandes MA, Carvalho IS (2007) Pegadas fósseis da Formação Botucatu (Jurássico Superior–Cretáceo Inferior): o registro de um grande dinossauro Ornithopoda na Bacia do Paraná. In: Carvalho IS, Cassab RCT, Schwanke C, Carvalho MA, Fernandes ACS, Rodrigues MAC, Carvalho MSS, Arai M, Oliveira MEQ (eds) Paleontologia: Cenários de Vida, vol 1. Interciência, Rio de Janeiro, pp 425–432

Fragoso LGC, Bittencourt JS, Mateus ALD, Cozzuol MA, Richter M (2019) Shark (Chondrichthyes) microremains from the Lower Cretaceous Quiricó Formation, Sanfranciscana Basin, Southeast Brazil. Hist Biol. https://doi.org/10.1080/08912963.2019.1692830

Francischini H, Dentzien-Dias PC, Fernandes MA, Schultz CL (2015) Dinosaur ichnofauna of the Upper Jurassic/Lower Cretaceous of the Paraná Basin (Brazil and Uruguay). J S Am Earth Sci 63:180–190

Gallego OF, Martins-Neto RG (2006) The Brazilian Mesozoic conchostracan faunas: its geological history as an alternative tool for stratigraphic correlations. Geociências 25:231–239

García, AJ, Wilbert A (1994) Palaeogeography evolution of Mesozoic pre-rift sequences in coastal and interior basins of northeastern Brasil. In: Embry AF, Beauchamp B, Glass DJ (eds) Pangea: global environments and resources. Memoir of Canadian Society of Petroleum Geology, vol 17, pp 123–130

Góes AM (1995) A Formação Poti (Carbonífero Superior) da Bacia do Parnaíba. Tese de Doutorado, Universidade de São Paulo, São Paulo, 171 p

Góes AM, Rossetti DF (2001) Gênese da Bacia de São Luís–Grajaú, Meio-Norte do Brasil. In: Rossetti DF, Góes AM, Truckenbrodt W (eds) O Cretáceo na Bacia de São Luís–Grajaú. Museu Paraense Emílio Goeldi, Belém, pp 15–29

Góes AMO, Feijó FJ (1994) Bacia do Parnaiba. Bol Geociênc Petrobras 8:57–67

Golonka J, Ross, MI, Scotese CR (1994) Phanerozoic palaeogeographic and palaeoclimatic modeling maps. In: Embry AF, Beauchamp B, Glass DJ (eds) Pangea: global environments and resources. Canadian Society of Petroleum Geology, Memoir, vol 17, pp 1–47

Gomes JMP, Rios-Netto AM, Borghi L, Carvalho IS, Filho JGM, Sabaraense LD, Araújo BC (2021) Cyclostratigraphic analysis of the early Cretaceous laminated limestones of the Araripe Basin, NE Brazil: estimating sedimentary depositional rates. J S Am Earth Sci 112(1):103563

Guerra-Sommer M, Siegloch AM, Degani-Schmidt I, Santos ACS, Carvalho IS, Andrade JAFG, Freitas FI (2021a) Climate change during the deposition of the Aptian Santana Formation (Araripe Basin, Brazil): preliminary data based on wood signatures. J S Am Earth Sci 111:103462

Guerra-Sommer M, Degani-Schmidt I, Mendonça J, Mendonça Filho JG, Lopes FD, Salgado-Campos VMJ, Araújo B, Carvalho IS (2021b) Multidisciplinary approach as a key for pale-oenvironmental interpretation in a Weichselia-dominant interval from the late Aptian Codó Formation (Parnaíba Basin, Brazil). J S Am Earth Sci 111:103490. https://doi.org/10.1016/j.jsa mes.2021.103490

Guardado LR, Gamboa LAP, Luchesi CF (1989) Petroleum geology of the Campos Basin, a model for a producing Atlantic-type basin. In: Edwards JD, Santogrossi PA (eds) Divergent/passive margin basins: AAPG Memoir, vol 48, pp 3–79

Gorscak E, O'Connor PM (2016) Time-calibrated models support congruency between Cretaceous continental rifting and titanosaurian evolutionary history. Biol Lett 12:20151047. https://doi.org/10.1098/rsbl.2015.1047

Henrique-Pinto R, Basei MAS, Santos PR, Saad AR, Milani EJ, Cingolani CA, Frugis GL (2021) Paleozoic Paraná Basin transition from collisional retro-foreland to pericratonic syneclise: implications on the geodynamic model of Gondwana proto-Andean margin. J S Am Earth Sci 111:103511. https://doi.org/10.1016/j.jsames.2021.103511

Langer MC (2003) The pelvic and hind limb anatomy of the stem sauropodomorph *Saturnalia tupiniquim* (Late Triassic, Brazil). PaleoBios 23:1–30

Langer MC, Ferigolo J (2013) The Late Triassic dinosauromorph *Sacisaurus agudoensis* (Caturrita Formation; Rio Grande do Sul, Brazil): anatomy and affinities. Geol Soc, Lond, Spec Publ 379(1):353–392. https://doi.org/10.1144/SP379.16

Langer MC, Godoy PL (2022) So volcanoes created the dinosaurs? a quantitative characterization of the early evolution of terrestrial pan-aves. Front Earth Sci 10. https://doi.org/10.3389/feart. 2022.899562

Langer MC, Abdala F, Richter M, Benton M (1999) A sauropodomorph dinosaur from the Upper Triassic (Carnian) of southern Brazil. C R Acad Sci 329:511–517

Langer MC, Ezcurra MD, Bittencourt JS, Novas FE (2010) The origin and early evolution of dinosaurs. Biol Rev Camb Philos Soc 85:55–110. https://doi.org/10.1111/j.1469-185X.2009. 00094.x

Leite AM, Carmo DA (2021) Description of the stratotype section and proposal of hypostratotype section of the Lower Cretaceous Quiricó formation, São Francisco Basin, Brazil. An Acad Bras Ciências 93(Suppl. 2):e20201296. https://doi.org/10.1590/0001-3765202120201296

Leite AM, Carmo DA, Ress CB, Pessoa M, Caixeta GM, Denezine M, Adorno RR, Antonietto LS (2018) Taxonomy of limnic Ostracoda (Crustacea) from the Quiricó Formation, Lower Cretaceous, São Francisco basin, Minas Gerais State, Southeast Brazil. J Paleontol 1–20. https://doi.org/10.1017/jpa.2018.1

Lemos FAP, Asakura Y, Antunes RL, Araripe RVC, Prado LAC, Tome METR, Oliveira DH, Nasci-mento LRSL, Ng C, Barreto AMF (2023) Calcareous nannofossils, biostratigraphy, and paleo-biogeography of the Aptian/Albian Romualdo Formation in the Araripe Basin, North-Eastern Brazil. Braz J Geol 53(2):e20220054. https://doi.org/10.1590/2317-4889202320220054

Leonardi G (1977) Two new ichnofaunas (Vertebrates and Invertebrates) in the Eolian Cretaceous Sandstones of the Caiuá Formation in Northwest Paraná. In: 1° Simpósio Geologia Regional, Sociedade Brasileira de Geologia, Núcleo São Paulo, Atas, pp 112–128, 16 figs

Leonardi G (1979a) Nota Preliminar Sobre Seis Pistas de Dinossauros Ornithischia da Bacia do Rio do Peixe (Cretáceo Inferior) em Sousa, Paraíba, Brasil. An Acad Bras Ciênc 51(3):501–516

Leonardi G (1979b) New archosaurian trackways from the Rio do Peixe Basin, Paraíba, Brasil. Ann Univ Ferrara, N.S., S. IX 5(14):239–249

Leonardi G (1980) Ornithischian trackways of the Corda Formation (Jurassic) Goiás, Brazil. In: 1st Congreso Latinoamericano de Paleontología, Buenos Aires, Argentina, Abstract, vol 1, pp 215–222

Leonardi G (1989) Inventory and statistics of the South American dinosaurian ichnofauna and its paleobiological interpretation. In: Gillette DD, Lockley MG (eds) Dinosaur tracks and traces. Cambridge University Press, New York, pp 165–178

Leonardi G (1994) Annotated atlas of South America tetrapod footprints (Devonian to Holocene). CPRM, Brasília

Leonardi G, Carvalho IS (2002) Jazigo icnofossilífero do Ouro, Araraquara, SP: ricas pistas de tetrápodes do Jurássico. In: Schobbenhaus C, Campos DA, Queiroz ET, Winge M, Berbert-Born MLC (eds) vol 1, 1ª ed. DNPM/CPRM. Comissão Brasileira de Sítios Geológicos e Paleobiológicos (SIGEP), Brasília, pp 39–48

Leonardi G, Carvalho IS (2021) Dinosaur tracks from Brazil: a lost world of Gondwana. Indiana University Press, Bloomington, 445 p

Leonardi G, Carvalho IS, Fernandes MA (2007) The desert ichnofauna from Botucatu Formation (Upper Jurassic–Lower Cretaceous), Brazil. In: Carvalho IS, Cassab RCT, Schwanke C, Carvalho MA, Fernandes ACS, Rodrigues MAC, Carvalho MSS, Arai M, Oliveira MEQ (eds) Paleontologia: Cenários de Vida, vol 1. Interciência, Rio de Janeiro, pp 379–391

Leonardi G, Santos MFCF, Barbosa FHS (2021) First dinosaur tracks from the Açu Formation, Potiguar Basin (mid-Cretaceous of Brazil). An Acad Bras Ciênc 93(Suppl. 2):e20210635. https://doi.org/10.1590/0001-3765202120210635

Lima MR (1979) Palinologia dos calcários laminados da Formação Areado, Cretáceo de Minas Gerais. In: 2º Simpósio Regional de Geologia, Atas, Rio Claro, Sociedade Brasileira de Geologia, pp 203–216

Lima MR (1983) Paleoclimatic reconstruction of the Brazilian Cretaceous based on palynology data. Rev Bras Geociênc 13:223–228

Lima MR, Coelho MPCA (1987) Estudo palinológico da sondagem estratigráfica de Lagoa do Forno, Bacia do Rio do Peixe, Cretáceo do Nordeste do Brasil. Bol Inst Geociênc–USP, Sér Cient 18:67–83

Lima Filho MF, Mabesoone JM, Viana MSS (1999) Late Mesozoic history of sedimentary basins in NE Brasilian Borborema Province before the final separation of South America and Africa 1: tectonic-sedimentary evolution. In: 5º Simpósio Sobre o Cretáceo do Brasil, UNESP Rio Claro, Brazil, Boletim, pp 605–611

Lopes RF, Lima CV, Candeiro CRA (2021) The paleontological heritage of Northern Tocantins State and Southwest Maranhão State, Brazil: a preliminary synthesis. Terr@Plural, Ponta Grossa 15: e2117157

Lucchesi CF (1998) Petróleo. Estud Avançados 12(33):17–40

Lockley MG (1991) Tracking dinosaurs: a new look at an ancient world. Cambridge University Press, Cambridge, MA, 238 p

Luft-Souza F, Fauth G, Bruno MDR, Mota MAL, Vázquez-García B, Santos Filho MAB, Terra GJS (2022) Sergipe-Alagoas Basin, Northeast Brazil: a reference basin for studies on the early history of the South Atlantic Ocean. Earth-Sci Rev 229:104034. https://doi.org/10.1016/j.earscirev.2022.104034

Mabesoone JM (1994) Sedimentary basins of Northeast Brasil. Federal University Pernambuco, Geology Department, Special Publication 2

Mabesoone JM, Lima PJ, Ferreira EMD (1979) Depósitos de cones aluviais antigos, ilustrados pelas formações Quixoá e Antenor Navarro (Nordeste do Brasil). In: 9º Simpósio de Geologia do Nordeste, Recife, Sociedade Brasileira de Geologia/Núcleo Nordeste, Brazil, Anais, vol 7, pp 225–235

Mabesoone JM, Viana MSS, Neumann VH (2000) Late Jurassic to Mid-Cretaceous lacustrine sequences in the Araripe-Potiguar Depression of Northeastern Brasil. In: Gierlowski-Kordesch EH, Kelts KR (eds) Lake basins through space and time. AAPG Studies in Geology, vol 46, pp 197–208

Machado DL, Dehira LK, Carneiro CDR, Almeida FFM (1990) Reconstruções paleoambientais do Juro-Cretáceo no Nordeste Oriental Brasileiro. Rev Bras Geociênc 19:470–485

Marzolli A, Renne PR, Picirillo EM, Ernesto M, Bellieni G, Min A (1999) Extensive 200-million-year-old continental flood basalts of the Central Atlantic Magmatic Province. Science 284:616–618

Matos RMD (1992) The Northeast Brazilian Rift System. Tectonics 11:766–791

Matos RMD, Krueger A, Norton I, Casey K (2021a) The fundamental role of the Borborema and Benin-Nigeria provinces of NE Brazil and NW Africa during the development of the South Atlantic Cretaceous Rift System. Mar Pet Geol 127:104872

Matos RMD, Medeiros WE, Almeida CB, Jardim de Sá EJ, Córdoba VC (2021b) A solution to the Albian fit challenge between the South American and African plates based on key magmatic and sedimentary events late in the rifting phase in the Pernambuco and Paraíba basins. Mar Pet Geol 128:105038

Mendes AC, Truckenbrod W, Nogueira ACR (2012) Análise faciológica da Formação Alter do Chão (Cretáceo, Bacia do Amazonas), próximo à cidade de Óbidos, Pará, Brasil. Rev Bras Geociênc 42(1):39–57

Menegazzo MC, Catuneanu O, Chang HK (2016) The South American retroarc foreland system: the development of the Bauru Basin in the back-bulge province. Mar Pet Geol 73:131–156. https://doi.org/10.1016/j.marpetgeo.2016.02.027

Menezes MN, Araújo-Júnior HI, Dal'Bó PF, Medeiros MA (2019) Integrating ichnology and paleopedology in the analysis of Albian alluvial plains of the Parnaíba Basin, Brazil. Cretac Res 96:210–226. https://doi.org/10.1016/j.cretres.2018.12.013

Menezes PTL, Travassos JM, Medeiros MAM, Takayama P (2016) High-resolution facies modeling of presalt lacustrine carbonates reservoir analog: Morro do Chaves Formation example, Sergipe-Alagoas Basin, Brazil. Interpretation 2:1–12. https://doi.org/10.1190/INT-2014-0213.1

Mescolotti PC, Varejão FG, Warren LV, Ladeira FSB, Giannini PCF, Assine ML (2019) The sedimentary record of wet and dry eolian systems in the Cretaceous of Southeast Brazil. Braz J Geol 49(3). https://doi.org/10.1590/2317-4889201920190057

Mestriner G, Marsola JCA, Nesbitt SJ, Da-Rosa AAS, Langer M (2023) Anatomy and phylogenetic affinities of a new silesaurid assemblage from the Carnian beds of south Brazil. J Vertebr Paleontol. https://doi.org/10.1080/02724634.2023.2232426

Milani EJ (1992) Intraplate tectonics and the evolution of the Paraná Basin, Southern Brazil. In: De Wit MJ, Ransome ID (eds) Inversion tectonics of the Cape Fold Belt, Karoo and Cretaceous basins of Southern Africa. A.A. Balkema, Rotterdam, pp 101–130

Milani EJ, De Wit MJ (2008) Correlations between the classic Paraná and Cape-Karoo sequences of South America and southern Africa and their basin in fills flanking the Gondwanides: du Toit revisited. Geol Soc Lond, Spec Publ 294(1):319–341. https://doi.org/10.1144/SP294.17

Milani EJ, Ramos VA (1998) Orogenias paleozóicas no domínio sul-ocidental do Gondwana e os ciclos de subsidência da Bacia do Paraná: Brazilian. J Geol 28(4):473–484

Milani EJ, Thomaz Filho A (2000) Sedimentary basins of South America. In: Cordani UG, Milani, EJ, Thomaz Filho A, Campos DA (eds) Tectonic evolution of South America. 31st. IGC, Rio de Janeiro, pp 389–449

Milani EJ, Zalán PV (1999) An outline of the geology and petroleum systems of the Paleozoic interior basins of South America. Episodes 22(3):199–205

Milani EJ, Melo JHG, Souza PA, Fernandes LA, França AB (2007) Bacia do Paraná. Bol Geociênc Petrobrás 15(2):265–287

Mizusaki AMP, Thomaz Filho A (2004) O magmatismo pós-Paleozóico no Brasil. In: Mantesso-Neto V, Bartorelli A, Carneiro CDR, Brito-Neves BB (eds) Geologia do Continente Sul Americano: Evolução da Obra de Fernando Flávio Marques de Almeida, chap XVII. Beca, São Paulo, pp 280–291

Müller RT, Garcia MS (2023) A new silesaurid from Carnian beds of Brazil fills a gap in the radiation of avian line archosaurs. Sci Rep 13:4981. https://doi.org/10.1038/s41598-023-32057-x

Nascimento DL, Martinez P, Batezelli A, Ladeira F, Corrêa L (2022) From the micromorphology of paleoweathering fronts to paleoenvironmental analysis: a case study of the Cretaceous dune fields of Sanfranciscana Basin, Brazil. Catena 211:106008

Nogueira ACR, Rabelo CEN, Góes AM, Cardoso AR, Bandeira J, Rezende GL, Santos RF, Truckenbrodt W (2021) Evolution of Jurassic intertrap deposits in the Parnaíba Basin, northern Brazil: the last sediment-lava interaction linked to the CAMP in West Gondwana. Palaeogeogr, Palaeoclimatol, Palaeoecol 572:110370. https://doi.org/10.1016/j.palaeo.2021.110370

Oliveira DC, Mohriak WU (2003) Jaibaras Trough: an important element in the early tectonic evolution of the Parnaíba interior sag Basin, Northeastern Brazil. Mar Pet Geol 20:351–383

Olsen P, Sha J, Fang Y, Chang C, Whiteside JH, Kinney S, Sues H-D, Kent D, Schaller M, Vajda V (2022) Arctic ice and the ecological rise of the dinosaurs. Sci Adv 8:eabo6342. https://doi.org/10.1126/sciadv.abo6342

Pacheco C, Müller RT, Langer M, Pretto FA, Kerber L, Silva SD (2019) *Gnathovorax cabreirai*: a new early dinosaur and the origin and initial radiation of predatory dinosaurs. PeerJ 7:e7963. https://doi.org/10.7717/peerj.7963

Pedreira AJ, Bahia RBC (2000) Sedimentary basins of Rondônia State, Brazil: response to the geotectonic evolution of the Amazonic craton. Rev Bras Geociênc 30(3):477–480

Pedreira da Silva AJ, Lopes RC, Vasconcelos AM, Bahia RBC (2003) Bacias Sedimentares Paleozóicas e Meso-Cenozóicas Interiores. In: Bizzi LA, Schobbenhaus C, Vidotti RM, Gonçalves JH (eds) Geologia, Tectônica e Recursos Minerais do Brasil. CPRM, Brasília, pp 55–85

Pereira RM, Nobre JM, Saraiva CAF, d'Ávila RS, Freire AFM (2021) Acquisition of spectral gamma-ray data from cuttings, with application in the paleoenvironmental interpretation of the Cabeças Formation, Parnaíba Basin, Brazil. Braz J Geophys 39(4):585–595. https://doi.org/10.22564/rbgf.v38i4.2120

Petri S (1983) Brazilian Cretaceous paleoclimates: evidence from clay-minerals, sedimentary structures and palynomorphs. Rev Bras Geociênc 13(4):215–222

Petri S (1987) Cretaceous paleogeographic maps of Brazil. Palaeogeogr Palaeoclimatol Palaeoecol 59(1987):117–168

Petri S (1998) Paleoclimas da era Mesozóica no Brasil - evidências paleontológicas e sedimentológicas. Rev Univ Guarulhos 6:22–38

Ponte FC (Coord.) (1972) Análise comparativa da paleogeologia dos litorais atlântico brasileiro e africano. Petrobrás, Curso Projetos Especiais de Geologia, CPEG IV, Salvador, Bahia (unpublished).

Ponte FC (1992) Origem e evolução das pequenas bacias cretácicas do interior do Nordeste do Brasil. In: 2° Simpósio sobre as Bacias Cretácicas Brasileiras, Rio Claro, 1992, São Paulo, Universidade Estadual Paulista, Resumos expandidos, pp 55–58

Popoff M (1988) Du Gondwana à l'Atlantique sud: les connexions du fossé de la Bénoué avec les bassins du Nord-Est brésilien jusqu'à l'ouverture du golfe de Guinée au Crétacé inférieur. J Afr Earth Sci 7:409–431

Porto A, Carvalho C, Lima C, Heilbron M, Caxito F, La Terra E, Fontes SL (2022) The Neoproterozoic basement of the Parnaíba Basin (NE Brazil) from combined geophysical-geological analysis: a missing piece of the western Gondwana puzzle. Precambrian Res 379:106784. https://doi.org/10.1016/j.precamres.2022.106784

Rapozo BF, Córdoba VC, Antunes AF (2021) Tectono-stratigraphic evolution of a cretaceous intracontinental rift: example from Rio do Peixe Basin, north-eastern Brazil. Mar Pet Geol 126:104899. https://doi.org/10.1016/j.marpetgeo.2021.104899

Ribeiro AC, Poyato-Ariza FJ, Bockmann FA, Carvalho MR (2018) Phylogenetic relationships of Chanidae (Teleostei: Gonorynchiformes) as impacted by *Dastilbe moraesi*, from the Sanfranciscana basin, Early Cretaceous of Brazil. Neotrop Ichthyol 16(3):e180059. https://doi.org/10.1590/1982-0224-20180059

Riff D, Souza RG, Carvalho IS (2018) Primeiro Registro Icnológico de Dinosauria na Bacia Bauru. In: 11° Simpósio Brasileiro de Paleontologia de Vertebrados, Teresina 2018, Sociedade Brasileira de Paleontologia, Boletim de Resumos, p 91

Rossetti DF (2001) Arquitetura Deposicional da Bacia de São Luís–Grajaú. In: Rossetti DF, Góes AM, Truckenbrodt W (eds) O Cretáceo na Bacia de São Luís–Grajaú. Museu Paraense Emílio Goeldi, Belém, pp 31–46

Rossetti DF, Truckenbrodt W (1997) Revisão estratigráfica para os depósitos do Albiano-Terciário Inferior (?) na Bacia de São Luís, Maranhão. Belém, Bol Mus Para Emilio Goeldi – Sér Ciênc Terra 9:29–41

Rossetti L, Lima EF, Waichel BL, Hole MJ, Simões MS, Scherer CM (2018) Lithostratigraphy and volcanology of the Serra Geral Group, Paraná-Etendeka Igneous Province in southern Brazil: towards a formal stratigraphical framework. J Volcanol Geoth Res 355:98–114. https://doi.org/10.1016/j.jvolgeores.2017.05.008

Santos EJ, Brito Neves BB (1984) Província Borborema. In: Almeida FFM, Hasui Y (coord) O Pré-Cambriano no Brasil. Editora Edgard Blücher, São Paulo, Brazil, pp 123–186

Salgado-Campos VMC, Carvalho IS, Bertolino LC, Duarte TA, Araújo BC, Borghi L (2021) Clay mineralogy and lithogeochemistry of lutites from the Lower Cretaceous Crato Member, Araripe Basin, NE Brazil: implications for paleoenvironmental, paleoclimatic and provenance reconstructions. J S Am Earth Sci 110:103329

Salgado-Campos VMC, Carvalho IS, Bertolino LC, Borghi L, Rios-Netto AM, Araújo BC, Souza DS, Ferreira LO, Bobco FER (2022) Unraveling an alkaline lake and a climate change in Northeastern Brazil during the Late Aptian. Sed Geol 442:106290

Santos MECM (1971) Um nôvo artrópodo da Formação Areado, Estado de Minas Gerais. An Acad Bras Ciênc 43:415–420

Santos MECM, Carvalho MSS (2009) Paleontologia das bacias do Parnaíba, Grajaú e São Luís. Serviço Geológico do Brasil, CPRM, Ministério das Minas e Energia, 215 p

Scaramuzza dos Santos AC, Guerra-Sommer M, Degani-Schmidt I, Sieloch AM, Carvalho IS, Mendonça Filho JG, Mendonça JO (2020) Fungus-plant interactions in Aptian Tropical Equatorial Hot arid belt: white rot in araucarian wood from Crato fossil Lagerstätte (Araripe Basin, Brazil). Cretac Res 114:104525

Scaramuzza dos Santos AC, Sieloch AM, Guerra-Sommer M, Degani-Schmidt I, Carvalho I (2021) *Agathoxylon santanensis* sp. nov. from the Aptian Crato fossil Lagerstatte, Santana Formation, Araripe Basin, Brazil. J S Am Earth Sci 112:103633

Scaramuzza dos Santos AC, Guerra-Sommer M, Degani-Schmidt I, Sieloch AM, Mendonça JO, Mendonça Filho, JG, Carvalho IS (2022) Record of *Brachyoxylon patagonicum*, a Cheirolepidiaceae wood preserved by gelification in the Aptian Maceió Formation, Sergipe–Alagoas Basin, NE Brazil. J S Am Earth Sci 118:103950

Scaramuzza dos Santos AC, Guerra-Sommer M, Barboza EG, Degani-Schmidt I, Siegloch AM, Vieira CEL, Vieira DT, Bardola TP, Schultz CL (2023) Stressing environmental conditions in the "petrified forest" from the Mata sequence in the Triassic context of the Paraná Basin. J S Am Earth Sci. https://doi.org/10.1016/j.jsames.2023.104415

Scherer CMS (2000) Eolian dunes of the Botucatu formation (Cretaceous) in Southernmost Brazil: morphology and origin. Sed Geol 137:63–84

Scherer CMS (2002) Preservation of aeolian genetic units by lava flows in the Lower Cretaceous of the Paraná Basin, Southern Brazil. Sedimentology 49(1):97–116

Scherer CMS, Lavina EL (2005) Sedimentary cycles and facies architecture of aeolian-fluvial strata of the Upper Jurassic Guará Formation, Southern Brazil. Sedimentology 52:1323–1341

Scherer CMS, Lavina EL (2006) Stratigraphic evolution of a fluvial-eolian succession: the example of the Upper Jurassic-Lower Cretaceous Guará and Botucatu formations, Paraná Basin, Southernmost Brazil. Gondwana Res 9:475–484

Scherer CMS, Faccini UF, Lavina EL (2002) Arcabouço estratigráfico do Mesozóico da Bacia do Paraná. In: Holz M, De Ros LF (eds) Geologia do Rio Grande do Sul. Instituto de Geociências/CIGO, pp 335–354

Sgarbi GNC (2000) The Cretaceous Sanfranciscan Basin, eastern plateau of Brazil. Rev Bras Geociênc 30(3):450–452

Sgarbi GNC, Sgarbi PBA, Campos JEG, Dardenne MA, Penha UC (2001) Bacia Sanfranciscana: o registro Fanerozóico da Bacia do São Francisco In: Pinto CP, Martins-Neto MA (eds) Bacia do São Francisco: Geologia e Recursos Naturais. SBG, Belo Horizonte, pp 93–138

Sgarbi PCB, Heaman LM, Gaspar JC (2004) U-Pb perovskite ages for Brazilian kamafugitic rocks: further support for a temporal link to a mantle plume hotspot track. J S Am Earth Sci 16(8):715–724. https://doi.org/10.1016/j.jsames.2003.12.005

Silva RC, Carvalho IS, Schwanke C (2007) Vertebrate dinoturbation from the Caturrita Formation (Late Triassic, Paraná Basin), Rio Grande do Sul State, Brazil. Gondwana Res 11:303–310

Silva RC, Carvalho IS, Fernandes ACS (2008) Pegadas de Dinossauros do Triássico (Formação Santa Maria) do Brasil. Ameghiniana 45(4):783–790

Simões TR, Kammerer CF, Caldwell MW, Pierce SE (2022) Successive climate crises in the deep past drove the early evolution and radiation of reptiles. Sci Adv 8(33):eabq1898. https://doi.org/10.1126/sciadv.abq1898

Siqueira LP (1989) Bacia dos Parecis. Bol Geociênc Petrobras 3:3–16

Srivastava NK, Carvalho IS (2004) Bacias do Rio do Peixe. Fund Paleontol Phoenix 71:1–4

Spigolon ALD, Alvarenga CJS (2002) Fácies e elementos arquiteturais resultantes de mudancas climáticas em um ambiente desértico: Grupo Urucuia (Neocretáceo), Bacia Sanfranciscana. Rev Bras Geociênc 32(4):579–586

Skelton P (2003) The Cretaceous world. Cambridge University Press, Cambridge, UK, 360 p. ISBN 978-0521831123

Souza-Lima W, Silva RO (2018) Aptian-Albian paleophytogeography and paleoclimatology from Northeastern Brazil sedimentary basins. Rev Palaeobot Palynol 258:163–189. https://doi.org/10.1016/j.revpalbo.2018.08.003

Teramoto EH, Gonçalves RD, Chang HU (2020) Hydrochemistry of the Guarani Aquifer System modulated by mixing with underlying and overlying hydrostratigraphic units. J Hydrol: Reg Stud 30:100713. https://doi.org/10.1016/j.ejrh.2020.100713

Toczeck A, Schmitt RS, Braga MAS, Miranda FP (2019) Tectonic evolution of the Paleozoic Alto Tapajós intracratonic basin—a case study of a fossil rift in the Amazon Craton. J S Am Earth Sci 94:102225. https://doi.org/10.1016/j.jsames.2019.102225

Trompette R, Egydio-Silva M, Tommasi A, Vauchez A, Uhlein A (1993) Amalgamação do Gondwana Ocidental no Panafricano-Brasiliano e o papel da geometria do Cráton do São Francisco na arquitetura da Faixa Ribeira. Rev Bras Geociênc 23:187–193

Tucker ME, Benton MJ (1982) Triassic environments, climates and reptile evolution. Palaeogeogr Palaeoclimatol Palaeoecol 40:361–379. https://doi.org/10.1016/0031-0182(82)90034-7

Tucker M, Dias-Brito D (2017) Petrologia Sedimentar Carbonática: iniciação com base no registro geologico do Brasil. IGCE/UNESP, Rio Claro, 208 p

Valença LMM, Neumann VH, Mabesoone JM (2003) An overview on Callovian-Cenomanian intracratonic basins of Northeast Brazil: onshore stratigraphic record of the opening of the Southern Atlantic. Geol Acta 1(3):261–275. https://doi.org/10.1244/105.000001614

Vaz PT, Rezende NGAM, Wanderley Filho JR, Travassos WAS (2007) Bacia do Parnaíba. Bol Geociênc Petrobras 15:253–263

Viana MSS, Lima Filho MF, Carvalho IS (1993) Borborema Megatracksite: uma base para correlação dos "arenitos inferiores" das bacias intracontinentais do Nordeste do Brasil. Simp Geol Nord, Soc Bras Geol/Núcl Nord, Bol 13:23–25

Whiteside JH, Lindström S, Irmis RB, Glasspool IJ, Schaller MF, Dunlavey M, Nesbitt SJ, Smith ND, Turner AH (2015) Extreme ecosystem instability suppressed tropical dinosaur dominance for 30 million years. Proc Natl Acad Sci USA 112:7909–7913. https://doi.org/10.1073/pnas.1505252112

Zaher H, Pol D, Carvalho AB, Nascimento PM, Riccomini C, Larson P, Juarez-Valieri R, Pires-Domingues R, Silva NJd Jr, Campos DA (2011) A complete skull of an Early Cretaceous sauropod and the evolution of advanced titanosaurians. PlosOne 6:1–10

Zaher H, Pol D, Navarro BA, Delcourt R, Carvalho AB (2020) An Early Cretaceous theropod dinosaur from Brazil sheds light on the cranial evolution of the Abelisauridae. CR Palevol 19(6):101–115. https://doi.org/10.5852/cr-palevol2020v19a6

Chapter 2
Triassic Tracks from Paraná Basin: The First Data on the Origin of Dinosauria

Rafael Costa da Silva ⓘ

2.1 Introduction

The Triassic period stands as a prominent epoch in Earth's geological history, marked by deep global transformations that occurred at the end of the Paleozoic and persisted throughout the Mesozoic, ultimately leading to the establishment of the Cenozoic biotas. One of the most notable paleobiological events during the Triassic was the emergence of a variety of archosaur lineages and the appearance of the first dinosaurs.

These significant events are magnificently preserved in the Triassic sequence of the Rosário do Sul Group, Paraná Basin, a globally renowned deposit that offers invaluable insights into the genesis of dinosaurs. This geological unit is widely acknowledged for its exceptionally diverse fauna, featuring a wealth of osteological remains, including temnospondyl "amphibians", archosauromorphs such as rhynchosaurs, "thecodonts" and dinosaurs, therapsids such as cynodonts and dicynodonts, and small tetrapods as procolophonids and rhynchocephalians (Holz and De Ros 2000). Particularly noteworthy is the exponential growth in the number of basal dinosaur taxa discovered in the Triassic of southern Brazil over the last decade (Cabreira et al. 2011, 2016; Müller et al. 2018; Pretto et al. 2018; Pacheco et al. 2019; Marsola et al. 2019; Müller 2021).

The vast majority of this paleontological record within the Rosário do Sul Group comprises osteological evidence that has been studied for over a century. Only recently vertebrate footprints and tracks been discovered, including dinosaur ichnofossils dating back to the same period as their skeletal remains (Cargnin et al. 2001; Silva et al. 2007a, 2008a, b, c, 2012). While Triassic dinosaur footprints are relatively abundant worldwide, in Brazil, they are exclusively known within this geological unit, given that Triassic-age rocks are somewhat rare in Brazil and, when present, have been

R. C. da Silva (✉)
Museu de Ciências da Terra, Serviço Geológico do Brasil—SGB/CPRM, Av. Pasteur 404, Urca, Rio de Janeiro, RJ 22290-255, Brazil
e-mail: rafael.costa@sgb.gov.br

I. S. Carvalho and G. Leonardi (eds.), *Dinosaur Tracks of Mesozoic Basins in Brazil*,
https://doi.org/10.1007/978-3-031-56355-3_2

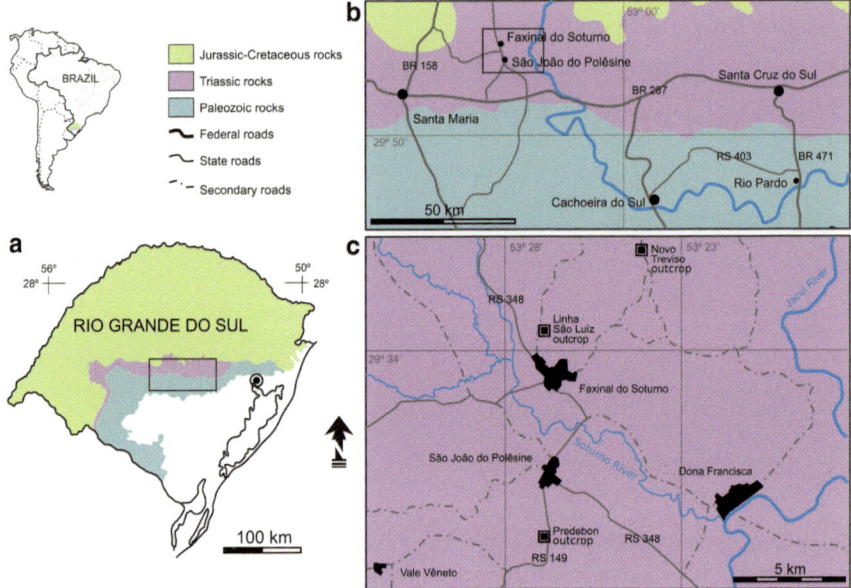

Fig. 2.1 Location of the Triassic dinosaur footprints ichnosites from Brazil (modified from Silva et al. 2008c, 2012). **a** Distribution of Triassic rocks in the Rio Grande do Sul State, Southern Brazil; **b** detailed view of the central region of the state; **c** detailed view of the "Quarta Colônia" region, with the dinosaur footprints ichnosites

insufficiently studied, with no known occurrences of tetrapod fossils. The study of these fossil footprints provides invaluable data on diversity, functional morphology, behavior and ecological relationships of extant animals, as well as sedimentological characteristics of the substrate and paleoenvironmental aspects. Traditionally, fossil tracks have not been employed for biostratigraphic purposes, although they can serve as significant indicators in determining the age of successions when body fossils are absent, or as supplementary aids in refining age estimates when associated with them (Lucas 1998). In the case of the tracks from the Rosário do Sul Group, they offer valuable insights into the chronological context of the rocks (Fig. 2.1).

2.2   Geological Context

The Paraná Basin, one of the world's largest intracratonic basins, extends laterally from the Central-West region of Brazil to Argentina, Uruguay, and Paraguay. This extensive basin is underlain by a crystalline basement and comprises a significant vertical extent, with sedimentary and igneous rocks spanning from the Late Ordovician to the Early Cretaceous (Milani et al. 2007). The geologic strata within the Paraná

Basin represent the superposition of three distinct sequences, corresponding to the Ordovician-Devonian, Carboniferous-Triassic, and Jurassic-Cretaceous periods.

The genesis of sediment deposition during the Triassic, where the earliest dinosaur tracks were preserved, traces back to earlier geological events. During the Permian period, the Paraná Basin was covered by a substantial body of water, characterized by marine systems and carbonate sedimentation. Towards the Permian-Triassic boundary, sedimentation displayed regressive tendencies, indicating a pronounced shift towards continental conditions during this phase. Over time, the extensive water body gradually receded, predominantly influenced by tides (Gama 1979; Milani et al. 2007). Towards the later stages of this period, lacustrine and fluvial depositional systems emerged, followed by the establishment of fluvial-eolian systems in restricted regions of the basin during the Early Triassic (Gama 1979; Milani et al. 2007). The onset of the Triassic was marked by a significant tectonic event, known as the La Ventana Orogeny, which occurred in the southern part of Gondwana. This event led to uplifts in various sectors of the basin, along with intermittent depositional hiatuses.

During the Triassic, fluvial-eolian systems associated with shallow and localized lakes formed in the southern region, leaving a record known as the Rosário do Sul Group (Andreis et al. 1980; Scherer et al. 2000). Subsequently, the initial movements related to the Gondwana fragmentation caused the uplift of some basin's portions, leading to an erosive stage that continued until the mid-Jurassic (Milani et al. 2007).

The Rosário do Sul Group is traditionally divided into the Pirambóia, Sanga do Cabral, Santa Maria, and Caturrita formations (Andreis et al. 1980), with the latter three restricted to the Rio Grande do Sul State in the southern region of Brazil. The status of the Pirambóia Formation is currently under discussion and its age may be younger (Christofoletti et al. 2021). Dinosaur tracks occur in the Santa Maria and Caturrita formations. Traditionally, the Santa Maria Formation is interpreted as deposited by a continental fluvio-lacustrine system (Andreis et al. 1980; Zerfass et al. 2003; Da-Rosa 2005) and is divided into the Passo das Tropas and Alemoa members (Fig. 2.2). The lowermost member (Passo das Tropas) consists of conglomerates and coarse sandstones, representing an interlaced, ephemeral, high-energy fluvial system (Zerfass et al. 2003). The upper member, Alemoa, is characterized by reddish mudstones, massive or finely laminated, interbedded with siltstones and fine sandstones, calcrete levels, and paleosols (Zerfass et al. 2003; Da-Rosa 2005). At the top of the unit, the mudstones are interbedded with tabular to lenticular, fine to medium-grained whitish sandstones, displaying small to medium-sized cross-bedding with intraclasts.

The Caturrita Formation comprises conglomeratic to fine sandstones with planar or trough cross-bedding and horizontal stratification, along with massive or fine laminated siltstones and sandy siltstones (Andreis et al. 1980). These sedimentary rocks are interpreted as meandering river deposits in an alluvial plain, featuring lateral associations of paleosol levels with multiepisodic sandy facies from fluvial channels. The prevailing climate during this period is believed to have been warm and humid (Andreis et al. 1980). The Caturrita Formation has been subdivided into two parts based on distinct fossil assemblages, with the lower part considered Norian

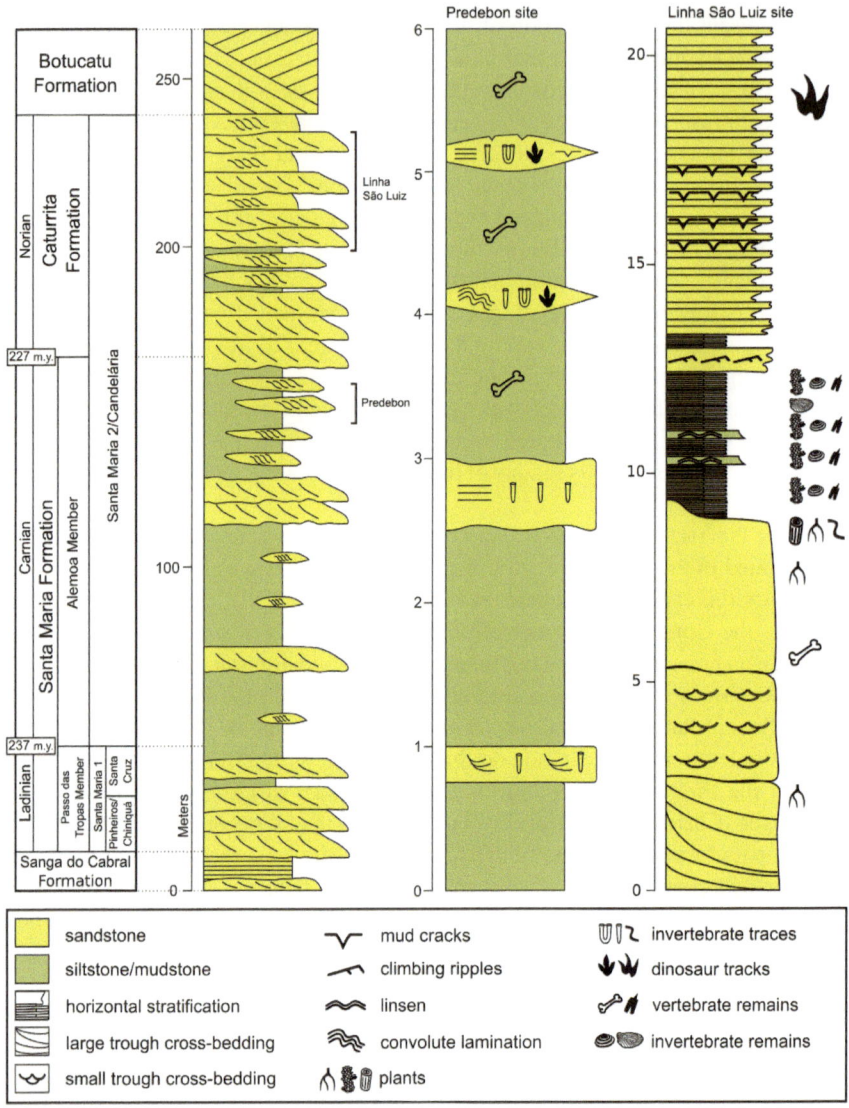

Fig. 2.2 Compound section of the Triassic rocks from southern Brazil and detailed sections of the dinosaur footprints ichnosites. The geochronological data, lithostratigraphic unit's limits and nomenclature were based in Andreis et al. (1980), Scherer et al. (2000), Zerfass et al. (2003), Da-Rosa (2005), Silva et al. (2008c, 2012), and Horn et al. (2014, 2018)

due to its vertebrate fauna content (Barberena 1977; Schultz et al. 2000; Rubert and Schultz 2004) and the uppermost part ("Arenito Mata") likely Rhaetian due to the presence of allochthonous *Araucarioxylon* wood logs (Guerra-Sommer and Cazzulo Klepzig 2000; Faccini 2007).

Triassic sedimentation in the Paraná Basin predominantly occurred within the temporal interval between the Lower Triassic and Upper Triassic, as defined by Milani (2000) within the context of the Gondwana I and Gondwana II supersequences. These deposits encompass fluvio-lacustrine and eolian sediments, including the Sanga do Cabral, Santa Maria, and Caturrita formations. However, the age of the Pirambóia Formation remains controversial, with some researchers suggesting it is Triassic (e.g. Scherer et al. 2000; Milani et al. 2007) or even Jurassic age (Christofoletti et al. 2021).

The Triassic rocks of Rio Grande do Sul have been studied by Zerfass et al. (2003) within the framework of sequence stratigraphy and categorized into two second-order depositional sequences: the Sanga do Cabral supersequence (equivalent to the Sanga do Cabral Formation) and the Santa Maria supersequence (equivalent to the Santa Maria, Caturrita formations, and the Arenito Mata). The former likely encompasses sediments from ephemeral and low-sinuosity fluvial systems during the Neoinduan. The Santa Maria supersequence encompasses low-sinuosity rivers, lakes, and deltas, further subdivided into three third-order sequences: Santa Maria 1 (Ladinian), Santa Maria 2 (Carnian to Norian), and Santa Maria 3 (possibly Rhaetian to Lower Jurassic) (Zerfass et al. 2003).

Zerfass et al. (2003) propose that the Caturrita Formation represents a transgressive tract system, comprising two, or part of them, third-order sequences deposited in a tectonically and structurally disturbed area. According to Zerfass (2007), these deposits originated from a high-energy, low-sinuosity, and ephemeral fluvial system, featuring associated alluvial plains or lakes, within an extensional basin contemporaneous with those in South Africa (Waterberg) and western Argentina (Cuyo and Ischigualasto) during the Triassic. The framework for the Triassic period in Southern Brazil is still evolving, and lithostratigraphic and chronostratigraphic changes are expected. In recent studies, Ladinian deposits have been more precisely delineated, resulting in a new stratigraphic sequence equivalent to the *Santacruzodon* AZ biozone (Horn et al. 2014). According to this proposal, the third-order sequence Santa Maria 1 from Zerfass et al. (2003) would be equivalent to the sequences Pinheiros-Chiniquá (*Dinodontosaurus* AZ, Ladinian-Carnian) and Santa Cruz (*Santacruzodon* AZ, Carnian). The Santa Maria 2 would be equivalent to the Candelária sequence (*Hyperodapedon* AZ, Santa Maria Formation, and *Riograndia* AZ, Caturrita Formation, Carnian-Norian).

Recently, the widespread presence of massive siltstones in the Santa Maria Supersequence has been interpreted as an eolian contribution in the form of loess (Horn et al. 2018), revealing a rather complex sedimentary record. In this scenario, the two lowermost Sequences (Pinheiros-Chiniquá and Santa Cruz) would consist of ephemeral braided river deposits overlain by dry mudflat deposits. In the Candelária Sequence, only the sheetflood delta facies association is present. The Santa Maria

Supersequence is believed to have been deposited under arid to semi-arid conditions, while the base of the Candelária Sequence records a shift from fluvial/eolian depositional systems to fluvial/deltaic systems, along with the disappearance of loess deposits and a transition from predominantly terrestrial to more aquatic faunas (Horn et al. 2018).

The Triassic dinosaur footprints in Brazil originate from only two locations, the Predebon site and the Linha São Luiz site (Silva et al. 2007a, 2008a, b, c, 2012; Figs. 2.1 and 2.2), although other questionable occurrences have been documented in the literature (see Chapter 1). The Predebon site is situated in the municipality of São João do Polêsine, Rio Grande do Sul, with coordinates 29° 38′ 29.14″ S; 53° 26′ 52.14″ W (Fig. 2.1). Covering approximately 100 m in length and six meters in height (Fig. 2.3a–c), the studied section corresponds to the upper portion of the Alemoa Member of the Santa Maria Formation, near the contact with the Caturrita Formation. It can be classified into four distinct facies:

- Facies 1: comprises massive reddish mudstones with calciferous nodules and Rhynchosauria bones.
- Facies 2: reddish or whitish fine tabular sandstones with small scale cross-bedding, calciferous nodules at the top of the layer and invertebrate trace fossils, mainly *Skolithos* isp.
- Facies 3: fine, reddish, tabular massive sandstones with horizontal lamination in the upper layers and a high degree of bioturbation, mostly of *Skolithos* isp.
- Facies 4: red to orangish fine sandstones with horizontal millimeter to centimeter-thick laminations, forming lenses of a few meters in length, with invertebrate trace fossils (*Skolithos* isp. and *Arenicolites* isp.) and vertebrates tracks and trackways, together with desiccation mud-cracks and convolute laminations.

Facies 2 and 3 occur intercalated with Facies 1 in the lower portion of the outcrop, while Facies 4 and 5 occur intercalated with Facies 1 in the upper portion.

The faciological interpretations of the rocks in the upper portion of the Santa Maria Formation have been subject to controversy, and various interpretations can be found in the literature. For example, mudstones are traditionally interpreted as lacustrine deposits, while the lenticular layers represented by Facies 4 could be associated with small channels resulting from subaerial exposure events. A more recent explanation suggests that this sequence could have been formed by a fluvial system with meandering to stable channels, where the mudstones represent floodplain deposits; the tabular sandstones correspond to main channels, while the small sandstone lenses are interpreted as crevasse splay deposits (Fonseca and Scherer 1998; Schultz et al. 2000).

The Linha São Luiz site (29° 33′ 45″ S; 53° 26′ 48″ W, Fig. 2.1) is located in the municipality of Faxinal do Soturno, state of Rio Grande do Sul, Brazil. The outcrop reaches about 20 m thick (Fig. 2.3d–f) and exhibits a stratigraphic succession consisting of basal sandstones, intermediate mudstones, and upper rhythmic alternations of sandstones and mudstones (Fig. 2.2). In the basal portion, there are low angle cross bedded and well sorted fine to medium sandstones as well as short scale trough

Fig. 2.3 Triassic dinosaur footprints ichnosites from southern Brasil. **a** General view of the Predebon site, upper portion of Santa Maria Formation; **b** detailed view of the outcrop, showing the distribution of mudstone and sandstone facies; **c** detailed view from the top, showing the sandstone lens where the footprints were found; **d** general view of the Linha São Luiz site, Caturrita Formation; **e** upper portion of the outcrop, showing the tabular sandstones; **f** detailed view of the tabular layer with the dinosaur footprint

cross-bedded sandstones followed by massive or sigmoidal trough cross-laminated fine sands associated with crevasse splay deposits (Faccini 2007). These sedimentary features are indicative of a fluvio-deltaic depositional system (Zerfass 2007).

The overlying laminated mudstone-siltstone interval is interpreted as representing a lacustrine setting with a more permanent water body. This horizon reveals the presence of fossilized woody remains from both autochthonous and parautochthonous gymnosperms, diverse conifer branches, and rare Equisetales (Pires and Guerra-Sommer 2004; Dutra and Crisafulli 2009). Additionally, it preserves impressions of sterile and reproductive parts of Bennettitales (Barboni and Dutra 2013), accompanied by Conchostraca, insects, and fish scales, providing invaluable paleontological insights (Pires and Guerra-Sommer 2004; Dutra and Crisafulli 2009). In fact, the paleoflora of this locality significantly differs from the traditional associations of *Dicroidium* and *Araucarioxylon* commonly found in the Rosário do Sul Group (Guerra-Sommer et al. 2000).

The uppermost part of the outcrop comprises heterolithic deposits of sandstones and mudstones, with compelling evidence of episodic subaerial exposure. These rocks have been identified as indicative of crevasse splay deposits and host remarkable dinosaur footprints associated with desiccation cracks, occurring on the surface of a thick tabular sandstone layer.

2.3   Footprints: Diversity and Paleobiological Interpretation

The Triassic dinosaur footprints from Brazil belong to three different forms, originating from the two mentioned localities. The first two kinds are from the Santa Maria Formation in the Predebon site and have been identified as "Dinosaur tracks indet." and *?Grallator* isp. (Silva et al. 2008a). The third type comes from the Linha São Luiz site in the Caturrita Formation and has been originally identified as *Eubrontes* isp. (Silva et al. 2012).

The "Dinosaur tracks indet." consist of two samples (Fig. 2.4a, b) housed in the paleontological collection of the Museu de Ciências Naturais (MCN-Museum of Natural Sciences) at the Fundação Zoobotânica do Rio Grande do Sul (FZBRS-Zoobotanical Foundation of Rio Grande do Sul), labeled as MCN-PIC.022 and MCN-PIC.023, respectively. These footprints are isolated impressions preserved as concave epireliefs on finely laminated sandstones with desiccation cracks. The upper layers are broken, while the deeper layers are deformed by the weight of the trackmaker. The footprint producer's laterality (right or left autopodium) cannot be determined due to the angular nature of the posterior margin of the prints. It is not possible to determine whether they were produced by a right or left autopodium. The posterior margin of the footprints is angular and there are no phalangeal or plantar pads present. The MCN-PIC.022 footprint is tridactyl and digitigrade, featuring sharp claws and acute hypexes. The central digit displays a constriction in its proximal portion. The length of the footprint, measured along the axis of the central digit, is 8.5 cm. The right and central digits exhibit a divergence angle of 57°, while the angle between the central and left digits corresponds to 32°. The total divergence angle is 89°. The MCN-PIC.023 footprint preserves two digits but is broken, lacking the right portion. The digits display sharp claws and acute hypexes, suggesting that the trackmaker was possibly a tridactyl dinosaur with bipedal locomotion. According to Thulborn (1989), the height at the pelvic girdle (h) can be estimated based on the taxonomic group and size range of the footprint. In this context, h would correspond to 4.6 times the length of the footprint for small bipedal dinosaurs (with footprints measuring less than 25 cm), resulting in an estimated height of the hip joint with the pelvic girdle of approximately 40 cm for the MCN-PIC.022 footprint.

A third footprint was tentatively attributed to the ichnogenus *Grallator* Hitchcock 1858. The sample MCN-PIC.021 (Fig. 2.4c) is also housed in the paleontological collection of the MCN (FZBRS). *?Grallator* isp. is an isolated, tridactyl track, characterized by its mesaxonic and digitigrade nature, as well as sharp claws and acute

Fig. 2.4 Dinosaur footprints from Santa Maria Formation, Predebon site. **a, b** MCN-PIC.023 and MCN-PIC.022, "Dinosaur tracks indet."; **c** MCN-PIC.021, *?Grallator* isp. (modified from Silva et al. 2008a)

hypices. Its posterior portion is missing. The central digit is longer than the others and bears three preserved phalangeal pads, while the lateral digits exhibit two preserved phalangeal pads. The digit on the right side is slightly shorter, possibly corresponding to digit II. In this case, the footprint would represent an impression of a left foot; the divergence angle between digits II and III would be 21°, and between digits III and IV would be 22°. The total divergence angle is 43°. Despite being incomplete, this footprint exhibits characteristics that allow its attribution, albeit with a certain degree of uncertainty, to the ichnogenus *Grallator*. These features include the presence of three digits separated by low interdigital angles, with digit III being substantially longer than the others, and digits II and IV having similar lengths I (Silva et al. 2008a; Klein and Lucas 2021). It is worth noting that although the ichnogenus *Grallator* is more commonly associated with the Lower Jurassic (Haubold 1986), it has also been identified in Upper Triassic rocks (starting from the Carnian) in regions such as South Africa (Olsen and Galton 1984; Raath et al. 1990), Europe, and North America (e.g., Haubold 1986). Traditionally, *Grallator* is attributed to small theropod dinosaurs (Olsen and Galton 1984).

Dinosaur track records in Triassic rocks are relatively common; however, debates persist regarding the identification of the trackmaker and the age of the oldest ichnites. Tridactyl footprints are frequently ascribed to dinosaurs, with occurrences dating back to the Early Triassic (e.g., Demathieu 1989; Avanzini 2002; Marsicano et al. 2004). Nevertheless, these records often entail uncertainties related to dating, questionable interpretations, or are tracks attributed to other Archosauromorpha (Thulborn 1990; King and Benton 1996; Klein and Lucas 2021). Notably, some footprints attributed to theropod and sauropodomorph dinosaurs have been documented in rocks of Carnian age within the Portezuelo Formation, Argentina (Marsicano and Barredo 2004). Also, tridactyl footprints of quadrupedal and bipedal trackways related to the ichnogenera *Atreipus-Grallator* ("*Coelurosaurichnus*") mark the Anisian-Ladinian transition (Klein and Lucas 2021).

The earliest indisputable occurrences of dinosaur footprints can be tracked back to the Carnian period, becoming more common from the Norian stage onwards (Lockley and Meyer 2000; Klein and Lucas 2021). As the upper section of the Santa Maria Formation was deposited during the Carnian stage (Scherer et al. 2000; Rubert and Schultz 2004; Lucas 1998, 2001; Langer 2005; Langer et al. 2018), the tracks from this formation are likely attributed to basal dinosaurs rather than other Triassic archosaurian groups. Morphological characteristics, such as the presence of a longer digit III compared to the others and digitigrade locomotion, lend support to this hypothesis (Thulborn 1990).

The sequence comprising the Santa Maria Formation yields fossils of some of the world's oldest dinosaurs, including *Staurikosaurus pricei* Colbert 1970 and other recently discovered species such as *Saturnalia tupiniquim* Langer et al. 1999, *Pampadromaeus barberenai* Cabreira et al. 2011, *Buriolestes schultzi* Cabreira et al. 2016, *Bagualosaurus agudoensis* Pretto et al. 2018, *Gnathovorax cabreirai* Pacheco et al. 2019, *Nhandumirim waldsangae* Marsola et al. 2019, and *Erythrovenator jacuiensis* Müller 2021. The foot bones were not found preserved in the fossils of the dinosaurs *Staurikosaurus, Pampadromaeus*, and *Erythrovenator*. In *Nhandumirim*, only a few isolated bone elements were discovered. Although establishing direct associations between fossil tracks and known taxa through skeletons is challenging, a brief comparison with some Triassic dinosaurs of Brazil may provide insights into the trackmakers. Notably, *Staurikosaurus, Saturnalia, Gnathovorax*, and potentially *Buriolestes* exhibit dimensions consistent with the discovered footprints. However, due to limited knowledge regarding foot anatomy in most early dinosaurs, their morphological features tend to be more generic, making it challenging to conclusively link footprints to specific species. *Staurikosaurus*, a theropod resembling *Herrerasaurus*, is often reconstructed with functionally tridactyl feet, with digit III being longer than the others. Similarly, Langer (2003) observed that *Saturnalia* possessed functionally tridactyl feet, with the three central metatarsals forming a slender unit, distinguishing it from other prosauropods and *Herrerasaurus*, with metatarsal III being longer. *Gnathovorax* also demonstrates size compatibility with the footprints (Pacheco et al. 2019), while *Buriolestes* and *Bagualosaurus* potentially align with the required dimensions (Cabreira et al. 2016; Pretto et al. 2018). Conversely, the genera *Pampadromaeus, Nhandumirim*, and *Erythrovenator* are too small to be plausible trackmakers (Cabreira et al. 2011; Marsola et al. 2019). Hence, it is plausible that *Staurikosaurus, Saturnalia, Gnathovorax*, and potentially *Buriolestes* could have produced footprints akin to the studied material. However, further research is necessary to conclusively ascertain which species best matches the morphological characteristics of the trackmakers.

The occurrences of dinosaur footprints from the Predebon outcrop constitute the only securely identified ones from the Brazilian Triassic. Among the sites with known fossil footprint records in the Paraná Basin, the Predebon Site stands out due to its superior preservation quality and remarkable diversity (Silva et al. 2007a, 2008a, b, c), containing nine morphotypes of lacertoid, mammaloid, and dinosauroid tracks. The footprints attributed to *Eubrontes* Hitchcock 1845, from the Caturrita Formation, correspond to one sample deposited in the paleontological collection of MCN (FZBRS) under the number MCN-PIC.030 (Fig. 2.5a), while the other remains in situ, described under the field number FSSL-02 (Fig. 2.5b, Silva et al. 2012). Casts of these footprints are available in MCN (FZBRS) and the Museu de Ciências da Terra, Serviço Geológico do Brasil—SGB (Museum of Earth Sciences, Geological Survey of Brazil) in Rio de Janeiro. These footprints are digitigrade, tridactyl, and mesaxonic, with acute hypices and sharp digital ends. The posterior margin exhibits a pronounced and evident posteromedial notch. The MCN-PIC.030 sample is incomplete due to erosive effects and represents the impression of a left foot, with the plantar portion, digit IV, and the proximal part of digits II and III preserved. Digit IV is the most complete, featuring three phalangeal pads and a claw mark. On the other hand, the FSSL-02 footprint represents the complete impression of a right foot, with the digits curving medially. The footprint is longer than wide, and there are no phalangeal or plantar pads. Digits II and IV show nearly equal length, while digit III is longer.

The footprints clearly correspond to impressions made by dinosaurs and are distinguishable from chirotheroid tracks of archosaurs due to their strong mesaxonic and

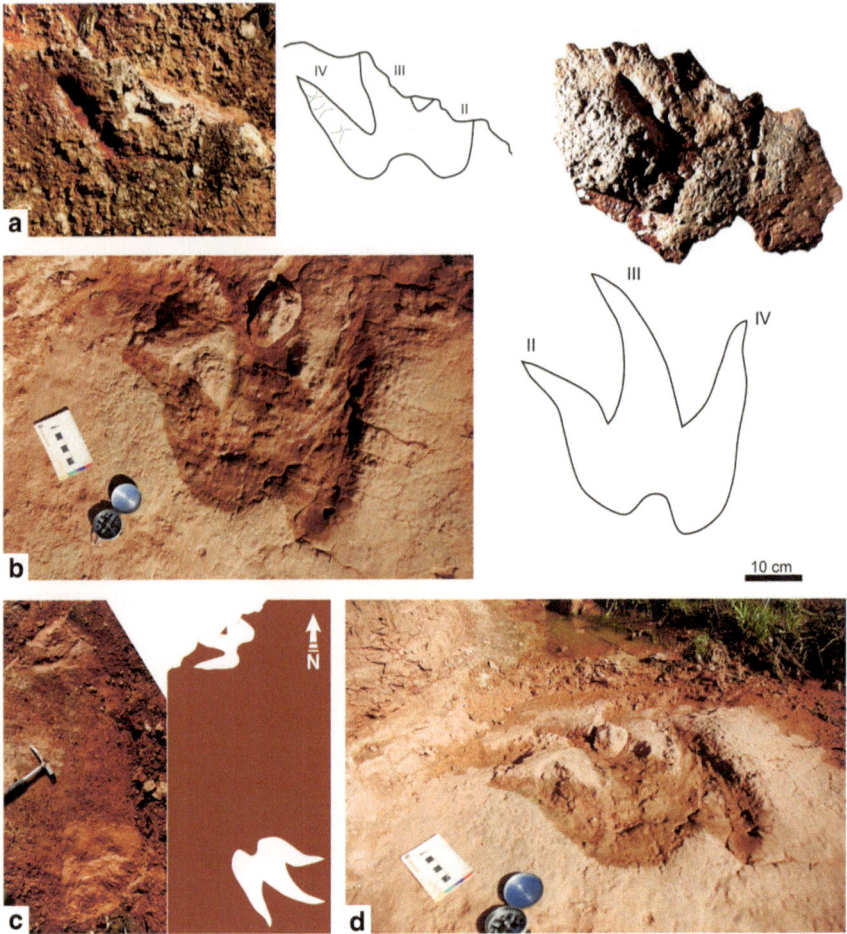

Fig. 2.5 "Theropod tracks indet." from Caturrita Formation, Linha São Luiz site. **a** Sample MCN-PIC.030, incomplete footprint; **b** footprint FSSL-02, which remains in situ; **c** composite photograph illustrating the original position of the two footprints; **d** specimen FSSL-02 in perspective, showing the deformation of the substrate (modified from Silva et al. 2012)

digitigrade pattern and because of their tridactily (Fig. 2.5c). The icnotaxonomic determination, however, may be reconsidered, as there are certain difficulties arising from the poor preservation of more detailed morphological features such as digital pads or the precise delineation of their contours. The footprints were attributed to *Eubrontes* by Silva et al. (2012), considering their fully digitigrade stance, typical theropod tridactyl morphology, and large size, exceeding 28 cm in length (as defined by Olsen et al. 1998). The material was then compared to *Eubrontes veillonensis* Lapparent and Montenat 1967, and the ichnogenera *Gigandipus* Hitchcock 1855 (also considered a behavioral variant of *Eubrontes* sensu Rainforth 2004) and *Tyrannosauropus* Haubold 1971. A comparison could also be drawn with *Columbosauripus*

Sternberg 1932, in which the toes exhibit tapering (gradually narrowing), and the total divergence (II–IV) is typically wider than in *Eubrontes*. In *Eubrontes* tracks, following the tendency for large animals to have stouter toes, the toes are thick up to the distal end of the toe (Farlow 2018), from which a slender claw protrudes, and the total divergence is usually less than 51°. Therefore, specimens from the Caturrita Formation might be more appropriately classified as "Theropod tracks indet.", a designation that will be consistently used henceforth, and their classification as *Eubrontes* should be abandoned.

Based on the FSSL-02 footprint, the size of the trackmaker was estimated using the equations proposed by Thulborn (1989) through morphometric ratios and allometric equations. In both methods, the height (h) was estimated to be 2.10 m, corresponding to an animal up to 8 m in length, similar in size to a large *Allosaurus*. All this information suggests that a large theropod dinosaur produced these footprints. The known paleofauna of the Linha São Luiz outcrop mainly includes small-sized animals such as sphenodonts, procolophonids, lepidosauriformes, cynodonts, and possible pterosaurs (Bonaparte and Sues 2006; Cisneros and Schultz 2003; Bonaparte et al. 2010a, b; Bonaparte et al. 2001, 2003), and one small dinosaur (Rubert and Schultz 2004; Bonaparte et al. 2006). The paleofauna of the Caturrita Formation also includes dicynodonts, archosaurs, and the dinosaurs *Guaibasaurus candelariensis* Bonaparte et al. 1999, *Unaysaurus tolentinoi* Leal et al. 2004, *Sacisaurus agudoensis* Ferigolo and Langer 2006, and *Macrocollum itaquii* Müller et al. 2018 (e.g. Holz and De Ros 2000). Among the previously described forms, none could be responsible for the footprints studied here. In fact, the *Grallator* footprint from the Santa Maria Formation (Silva et al. 2008a) is representative of what would be expected for the Brazilian Triassic dinosaurs.

2.4 On the Age of the Earliest Large Theropod Tracks

The age of the earliest occurrences of large dinosaur tracks has been a subject of much debate. Haubold (1971) considered the presence of *Eubrontes* both at the end of the Triassic and the beginning of the Jurassic. Olsen and Galton (1984) proposed that the appearance of *Eubrontes* coincides with the base of the Jurassic. Subsequently, the ichnogenus was regarded as diagnostic of the Jurassic, and its sudden appearance indicated a significant increase in the size of theropod dinosaurs at the Triassic-Jurassic boundary (e.g., Lockley and Hunt 1994; Olsen et al. 2002). However, Triassic footprints larger than 25 cm in length were reevaluated by Lucas et al. (2006) and attributed to *Eubrontes*, with occurrences ranging between 26 and 50 cm and ages from the Norian to the Rhaetian, and possibly the Carnian. Nevertheless, size criteria alone cannot provide a solid ichnotaxonomic basis; morphological elements must be considered for proper identification.

It now seems evident that the appearance of large dinosaur tracks occurred before the beginning of the Jurassic, and there were Triassic animals capable of producing footprints at least of medium *Eubrontes* size. However, theropod footprints over 40 cm in length are more characteristic of the Jurassic.

The Norian age of the Caturrita Formation was initially established through correlations with faunas from the Los Colorados Formation in Argentina (Rubert and Schultz 2004; Soares and Schultz 2006). However, earlier studies considered the Argentine deposits to be of Rhaetian age (Lucas 1998; Heckert and Lucas 1998). Recently, high-precision U–Pb zircon geochronology has dated the maximum age of the Caturrita Formation to 225 Ma (Langer et al. 2018). Nevertheless, some elements of the paleofauna and paleoflora found in the Linha São Luiz outcrop within the Caturrita Formation exhibit affinities with the Jurassic. These include sphenodontids, Brasilodontidae cynodonts, small-sized dinosaurs, the procolophonid *Soturnia caliodon*, and a probable primitive pterosaur (Bonaparte et al. 1999, 2001, 2003, 2006, 2010b; Cisneros and Schultz 2003; Martinelli et al. 2005; Bonaparte and Sues 2006; Soares et al. 2011). Regarding the paleofloristic content, Barboni and Dutra (2013) identified a new species of Bennettitales, *Williamsonia potyporanae*, at the Linha São Luiz site. The advanced morphology of this species suggests a Rhaetian or even younger age.

The challenges in correlating the paleobiota of the Linha São Luiz site arise from its location in a depressed block, sometimes placing it at the same topographic level as the upper beds of the Santa Maria Formation (Da-Rosa and Faccini 2005). Additionally, the endemism of most of its faunal components further complicates the correlation. According to Silva et al. (2012), the occurrence of fossil footprints and the paleofloristic and paleofaunistic data suggest a Rhaetian age for the upper portion of the Caturrita Formation, or at least for the top of the Linha São Luiz site, which might even correspond to a new geological unit. Further studies in the region are warranted.

A third set of purportedly Triassic tracks was previously described (Cargnin et al. 2001; Silva et al. 2007b). They studied rounded contour structures exposed on a horizontal outcrop at a locality known as Novo Treviso, also in Rio Grande do Sul, which was previously attributed to the Caturrita Formation. They concluded that these structures corresponded to undertracks produced by large vertebrates, possibly prosauropod dinosaurs. Identifying and interpreting vertebrate-generated bioturbations can be challenging, often relying more on sedimentological criteria than morphological features. Large vertebrates, particularly, can create metric-scale deformations in relation to the trampled surface due to the significant vertical pressure exerted on the sediment.

The rocks from the Novo Treviso locality were subsequently studied in a geological mapping conducted by the Geological Survey of Brazil (SGB–CPRM) and identified, along with other outcrops in the region, as belonging to the Guará Formation (Zerfass 2007). The Guará Formation predominantly crops out in the western region of the State of Rio Grande do Sul (Scherer et al. 2000) and consists of a succession of fine to conglomeratic sandstones with small to large-scale cross-beddings and planar lamination, interspersed with centimeter-thick pelite layers (Scherer et al. 2000). These deposits were formed by interlaced fluvial systems associated with lakes and aeolian dunes (Scherer et al. 2000). Based on the fossil content, which mainly includes fish, conchostracans, mollusks, and dinosaur footprints (Scherer and Lavina 2005; Dentzien-Dias et al. 2007; Zerfass 2007), the age of the Guará Formation was estimated to range from the Late Jurassic to the Early Cretaceous. Consequently, the Novo Treviso locality corresponds to Jurassic-Cretaceous deposits rather than Triassic, as previously believed.

The Guará Formation in the western region of Rio Grande do Sul contains occurrences of Sauropoda and Theropoda dinosaur footprints (Dentzien-Dias et al. 2007). The circular structures exposed in Novo Treviso resemble descriptions of Sauropoda footprints in the literature (e.g., Thulborn 1990; Lockley 1991) and are likely associated with these animals. Additionally, they share similarities in shape and dimensions with those found in the western part of the state, where footprints show limited morphological preservation, lacking digital structures, but exhibiting well-marked deformation features. These deformations are visible even in longitudinal cross-sections of vertical exposures, suggesting relatively deep impressions (Dentzien-Dias et al. 2007).

2.5   Paleogeographical Distribution of the Footprints

Once considered a relatively rare group in the Triassic period, dinosaurs rose to become the dominant terrestrial animals throughout the Mesozoic era. In the last few years, our understanding of this scenario has evolved, revealing a more diverse and widespread presence of dinosaurs in Triassic faunas than previously imagined. Additionally, there is mounting evidence supporting the hypothesis that dinosaurs originated in South America, between Argentina and southern Brazil (Irmis et al. 2007; Langer et al. 2018; Garcia et al. 2019; Novas et al. 2021).

This context holds significant implications for comprehending the dinosaur footprints found in the Brazilian Triassic. In recent decades, the Brazilian record of dinosauriforms has substantially expanded, encompassing a broad range of sauropodomorphs, herrerasaurids, silesaurids, and potential theropods (Ferigolo and Langer 2006; Langer and Ferigolo 2013; Cabreira et al. 2016; Marsola et al. 2019; Müller et al. 2018; Pacheco et al. 2019; Novas et al. 2021).

Particularly noteworthy is the Quarta Colônia region, situated in the central area of Rio Grande do Sul State, which has recently gained increasing importance in paleontological research. This relevance stems from frequent discoveries of Triassic fossils within the region. The fossils found in the Quarta Colônia area exhibit remarkable diversity, including some of the oldest known dinosaurs, as well as tracks and traces of vertebrates (Silva et al. 2007a, 2008a, b, c, 2012). Through ongoing exploration and scientific investigation, these findings contribute significantly to our understanding of the early evolutionary history and ecological dynamics of dinosaurs, shedding light on their pivotal role in shaping ecosystems during the Triassic period.

The Triassic formations in Brazil and Argentina present a remarkably diverse and relatively abundant record of dinosauriforms. Together, the Brazilian and Argentine sequences provide a comprehensive stratigraphic succession of the Carnian and

Norian stages (Novas et al. 2021), documenting a representative assemblage of vertebrate faunas and the early radiation of dinosaurs, with a growing number of taxa
being discovered annually. Each new finding changes the phylogenetic proposal and
engenders novel evolutionary relationships (Ferigolo and Langer 2006; Langer and
Ferigolo 2013; Cabreira et al. 2016; Marsola et al. 2019; Müller et al. 2018; Pacheco
et al. 2019; Novas et al. 2021). Nevertheless, a consensus on the phylogeny of the
early dinosaurs and their precursors has not yet been achieved (Novas et al. 2021).

In general, the Triassic dinosaurian fauna consists of dinosauromorphs, encompassing three clades: Lagerpetidae, Silesauridae, and Dinosauria (Garcia et al. 2019).
Although Lagerpetidae and Silesauridae precede Dinosauria in age, all three groups
coexisted for at least 21 million years (Irmis et al. 2007; Langer et al. 2018;
Garcia et al. 2019), from the middle Carnian, around 233 million years ago. The
strata containing the oldest Dinosauria, represented by the lower portion of the
Ischigualasto Formation in Argentina and the lower segment of the Candelária
sequence in Brazil (corresponding to the uppermost part of the Santa Maria Formation), are dated between the middle and late Carnian, approximately 231 to 233
million years ago (Martinez et al. 2011, 2013; Langer et al. 2018). Other occurrences
of dinosauromorphs, such as the Maleri Formation in India and the Pebbly Arkose
Formation in Zimbabwe, lack precise dating but are considered Carnian based on
biostratigraphic correlation. These occurrences testify to the broader geographical
distribution of these animals during this interval, extending to the eastern side of
Pangaea (Langer et al. 2010, 2018; Novas et al. 2011, 2021). The taxonomic ascrip-
tions of pre-Norian dinosaurs to Theropoda, Sauropodomorpha, and Ornithischia
remain tentative (Novas et al. 2021) and should be interpreted with caution.

In North America, the earliest occurrences of dinosaurs are found in the Chinle
Formation, dating to the Norian age. However, the faunas in this region are taxo-
nomically less diverse and numerically less abundant compared to their counterparts
in South America (Irmis et al. 2011). Notably, the North American faunas occupied
paleolatitudes closer to the equator, whereas the South American assemblages were
situated at higher latitudes, approximately 30° S to 35° S. This latitudinal distinction
highlights the significant influence of latitude in shaping the geographical distribution
of early dinosaurs (Langer et al. 2010; Novas et al. 2011). The delayed appearance
of the first dinosaurs in the northern region may be attributed to climatic changes
that reduced the contrasts between temperate and tropical climate zones, thereby
lessening the barriers that impeded the dispersal of these animals (Kent et al. 2014).

2.6   Paleoenvironmental and Paleoclimatic Contexts

Regarding the preservation of tracks from the Santa Maria Formation (Predebon
outcrop), detailed morphological analysis has shed light on behavioral characteristics of the animals responsible for these traces. Notably, evidence suggests activities
such as swimming, occasional bipedalism, and potential climbing behavior in sphen-
odontian tracks (Silva et al. 2008c). In contrast, non-mammalian cynodonts exhibit a

gait consistent with basal amniotes, including tail dragging (Silva et al. 2007a, 2008a, b, c). Preservation of these tracks is influenced by the presence and thickness of a water film during their formation, as well as subsequent subaerial exposure. This categorization has resulted in five distinct types: underwater tracks, semi-aquatic tracks, semi-terrestrial tracks, wet substrate tracks, and damp substrate tracks (Silva et al. 2007a). Of these, wet and damp substrate tracks have demonstrated the best preservation conditions. The presence of temporary channels in seasonal climates, within the floodplain system where these footprints were produced, indicates regions with varying water depth. Some footprints near the center of the channels could be subaquatic or semi-aquatic, while those closer to the margins and more exposed to subaerial conditions likely resulted in the other preservation types. These findings contribute to our understanding of the anatomy and behavior of the trackmakers and their interactions with the surrounding environment. Regarding the dinosaur footprints from Santa Maria Formation, they were likely created on a damp and relatively non-plastic substrate, followed by prolonged subaerial exposure. This suggests that some of these tracks may be undertracks, given the absence of certain surface features, such as well-defined claw and digital pad impressions. The presence of various morphological types of footprints within a single outcrop of the Santa Maria Formation has revealed a complex ichnocoenosis, comprising different kinds of sphenodontians, cynodonts, and dinosaurs. This is in contrast to the known skeleton record from the top of the Alemoa Member. Interestingly, this ichnocoenosis hints at a paleofauna similar to that found in the Caturrita Formation, indicating that ichnofossils may precede the record of groups represented by body fossils (Lockley 1991), providing valuable complementary information.

The two footprints attributed to "Theropod tracks indet." clearly exhibit characteristics typical of large theropod dinosaurs (Haubold 1971; Thulborn 1990; Lockley 1991). Certain features of the footprints can provide valuable insights into the behavior of the track-makers, substrate conditions during deposition, and the preservation of the strata. The footprints preserved in a tabular sandstone layer show signs of plastic deformation (Fig. 2.5c). During field preparation (Silva et al. 2012), it was observed that the sandy siltstone layer covering the footprints was also deformed and churned above them, indicating that the trackmaker stepped on the fresh silt sediment, penetrating it, and reaching the sand below (Brand and Kramer 1996; Milàn and Bromley 2006). While the impressions are not recognizable in the siltstone layer due to the fluid nature of the original substrate, they are well-preserved in the sandstone layer below (Fig. 2.5d). This preservation process enhances the potential for footprints' survival by eliminating exposure to air after their formation. The original footprint should have been larger than the preserved one, with its size controlled by the depth of the impression. This kind of preservation can be understood as a variation of the classic undertracks and ghost prints, similar to what has been termed "cut undertracks" by Goldring and Seilacher (1971). Furthermore, the total divergence of the studied footprints is slightly high for their size, suggesting that they were produced in softer and wetter soil. Footprints formed in such substrates tend to exhibit higher divergence angles than those on firm ground (Currie and Sarjeant 1979; Thulborn 1990).

The preservation of Triassic dinosaur footprints in Brazil occurred in fluvial envi-ronments associated with the deposition of overbank sediments in floodplains (Santa Maria Formation) and crevasse splay deposits (Caturrita Formation) (Fig. 2.6). Both facies are similar components of fluvial systems, where unconsolidated sediments are deposited due to river overflow, followed by progressive drying and compaction of the deposited layer, creating a suitable substrate for new sedimentation. This process results in a continuous gradation of sediment moisture and viscosity condi-tions, eventually reaching the optimal composition for the preservation of different kinds of fossil tracks. Thus, fluvial systems hold significant potential for vertebrate ichnology and play a crucial role in preserving Triassic dinosaur footprints in Brazil.

The establishment of fluvial systems and the radiation of dinosauromorphs during the Carnian in the Brazilian Triassic rocks appear to be associated with profound

Fig. 2.6 Art conception of the Triassic dinosaur trackmakers in their respective environments. Left, reconstruction of the *?Grallator* trackmaker in the floodplains of the Santa Maria Formation during the Carnian Pluvial Episode. Right, "Theropod tracks indet." trackmaker walks on the newly deposited sands of crevasse splay deposits of the younger Caturrita Formation (Art by Guilherme Gehr)

global climatic changes that occurred during this period. The Triassic was typically dominated by arid climates. However, during the Carnian, this dominance was disrupted by several episodes of significant increase in precipitation, resulting in a warmer and more humid global climate and causing extensive environmental changes. This series of events is referred to as the "Carnian Pluvial Episode," which occurred around 234 to 232 million years ago, likely triggered by massive volcanic eruptions in the Wrangellia province in western Canada, resulting in considerable basalt emissions (Dal Corso et al. 2020). These eruptions released significant amounts of greenhouse gases into the atmosphere, leading to peaks of global warming and inducing the pluvial episodes. This interval coincides with the deposition of the early Candelária sequence and the appearance of dinosauromorph fossils in Brazil.

The Carnian Pluvial Episode resulted in significant environmental changes, leading to disturbances in both marine and terrestrial ecosystems, followed by substantial aridification. These alterations caused the extinction of several species and facilitated the emergence of new faunal and floral radiations during the remainder of the Triassic period. In the Triassic rocks of Brazil, there is a noticeable decline in dicynodonts, rhynchosaurs, and pseudosuchians during this interval (Dal Corso et al. 2020). In this context, dinosaurs likely appeared and diversified as opportunistic and generalist fauna, in contrast to more specialized synapsids and archosaurs, experiencing rapid expansion across Pangaea (Benton et al. 2018). The transition is also evident in the replacement of the *Dicroidium* flora with conifer-dominated forests.

The record of Triassic fossil tracks seems to correspond to this pattern, demonstrating the appearance of successive dinosaurian groups and the replacement of the non-mammalian Synapsida and non-dinosaur Archosauria fauna with dinosauriforms. As observed by Bernardi et al. (2018), footprints of the ichnogenus *Grallator*, associated with small bipedal dinosaurs, emerge in the second half of the Carnian, while *Eubrontes* or large theropod footprints, associated with medium to large-sized carnivores, appear during the Norian. Although the paleoenvironmental and paleobiogeographic conditions are not the same, and the genera and species of the trackmakers are different as well, a similar pattern of replacement of smaller and more generalist dinosaurs by larger, more specialized ones seems to emerge in the discoveries in Brazilian Triassic associations.

2.7 Conclusions

The tridactyl footprints from the Santa Maria Formation have been identified as "Dinosaur tracks indet." and *?Grallator* isp., attributed to dinosauriform animals. The genera *Staurikosaurus*, *Saturnalia*, and *Sacisaurus*, known from rocks of the Alemoa-Caturrita sequence, show morphological similarities consistent with the animals that produced these tracks. The occurrences of dinosaur footprints from the Santa Maria Formation represent the oldest in Brazil.

The footprints from the Caturrita Formation were identified as "Theropod tracks indet.", produced by large theropod dinosaurs. The dimensions of these footprints are more advanced than those commonly found in the Carnian/Norian, aligning them with those found after the Rhaetian/Jurassic.

Furthermore, previously registered footprints in the Caturrita Formation at the Novo Treviso site, initially attributed to prosauropod dinosaurs, have been reinterpreted as belonging to the Guará Formation. This reevaluation places their age within the Late Jurassic to Early Cretaceous period.

The depositional sequence of Santa Maria-Caturrita formations, including both its osteological and ichnological records, provides an almost continuous record of the interval between the emergence of dinosaurs and their establishment as ecologically dominant elements in continental faunas. These events appear to be closely associated with the Carnian Pluvial Episode, a series of environmental changes linked to global warming, resulting in significant extinctions and the replacement of flora and fauna. The Triassic dinosaur footprints in Brazil, although currently limited to few layers, serve as valuable complements to the osteological record and contribute to the corroboration of these events.

Nonetheless, several questions remain unanswered, and there are still gaps that need to be filled. New research and the discovery of new specimens are essential to a more comprehensive understanding of the Triassic ichnocoenoses in Southern Brazil.

References

Andreis RR, Bossi GE, Montardo DK (1980) O Grupo Rosário do Sul (Triássico) no Rio Grande do Sul—Brasil. In: 31° Congresso Brasileiro de Geologia, Sociedade Brasileira de Geologia, Camboriú, Anais, vol 2, pp 659–673

Avanzini M (2002) Dinosauromorph tracks from the Middle Triassic (Anisian) of the Southern Alps (Valle di Non—Italy). Boll Soc Paleontol Ital 41:37–40

Barberena MC (1977) Bioestratigrafia preliminar da Formação Santa Maria. Pesqui Geociênc 7:111–129

Barboni R, Dutra TL (2013) New "flower" and leaves of Bennettitales from Southern Brazil and their implication in the age of the Lower Mesozoic deposits. Ameghiniana 50(1):14–32

Benton MJ, Bernardi M, Kinsella C (2018) The Carnian Pluvial Episode and the origin of dinosaurs. J Geol Soc 175:1019–1026. https://doi.org/10.1144/jgs2018-049

Bernardi M, Gianolla P, Petti FM, Mietto P, Benton MJ (2018) Dinosaur diversification linked with the Carnian Pluvial Episode. Nat Commun 9:1499. https://doi.org/10.1038/s41467-018-03996-1

Bonaparte JF, Ferigolo J, Ribeiro AM (1999) A new early Late Triassic saurischian dinosaur from Rio Grande do Sul State, Brazil. In: Proceedings of the second Gondwanan dinosaur symposium 15, Tokyo, National Science Museum monographs, vol 15, pp 89–109

Bonaparte JF, Ferigolo J, Ribeiro AM (2001) A primitive Late Triassic "Ictidosaur" from Rio Grande do Sul, Brazil. Palaeontology 44:623–635

Bonaparte JF, Martinelli AG, Schultz CL, Rubert R (2003) The sister group of mammals: small cynodonts from the Late Triassic of southern Brazil. Rev Bras Paleontol 5:5–27

Bonaparte JF, Sues HD (2006) A new species of Clevosaurus (Lepidosauria: Rhynchocephalia) from the Upper Triassic of Rio Grande do Sul, Brazil. Palaeontology 49:917–923

Bonaparte JF, Brea G, Schultz CL, Martinelli AG (2006) A new specimen of *Guaibasaurus cande-lariensis* (basal Saurischia) from the Late Triassic Caturrita Formation of southern Brazil. Hist Biol 19:73–82

Bonaparte JF, Schultz CL, Soares MB, Martinelli AG (2010a) The Faxinal do Soturno local fauna, Late Triassic of Rio Grande do Sul, Brazil. Rev Bras Paleontol 13:233–246

Bonaparte JF, Schultz CL, Soares MB (2010b) Pterosauria from the Late Triassic of Southern Brazil. In: Bandyopadhyay S (ed) New aspects of Mesozoic biodiversity. Springer, Berlin, pp 63–72

Brand LR, Kramer J (1996) Underprints of vertebrate and invertebrate trackways in the Permian Coconino Sandstone in Arizona. Ichnos 4:225–230

Cabreira SF, Schultz CL, Bittencourt JS, Soares MB, Fortier DC, Silva LR, Langer MC (2011) New stem-sauropodomorph (Dinosauria, Saurischia) from the Triassic of Brazil. Naturwissenschaften 98:1035–1040. https://doi.org/10.1007/s00114-011-0858-0

Cabreira SF, Kellner AWA, Dias-da-Silva S, Silva LR, Bronzati M, Marsola JCA, Müller RT, Bittencourt JS, Batista BJ, Raugust T, Carrilho R, Brodt A, Langer MC (2016) A unique Late Triassic dinosauromorph assemblage reveals dinosaur ancestral anatomy and diet. Curr Biol 26:3090–3095

Cargnin D, Ferigolo J, Ribeiro AM, Negri FR, Carvalho IS (2001) Pegadas Fósseis do Triássico da Bacia do Paraná (Grupo Rosário do Sul), Rio Grande do Sul, Brasil. Rev Bras Paleontol 2:71

Christofoletti B, Peixoto BCPM, Warren LVW, Inglez L, Fernandes MA, Alessandretti L, Perinotto JAJ, Simões MG, Assine ML (2021) Dinos among the dunes: dinoturbation in the Pirambóia Formation (Paraná Basin), São Paulo State and comments on cross-section tracks. J S Am Earth Sci 109:103252. https://doi.org/10.1016/j.jsames.2021.103252

Cisneros JC, Schultz CL (2003) *Soturnia caliodon* n.g. y n.sp., a procolophonid reptile from the Upper Triassic of Southern Brazil. Neues Jahrb Geol Palaeontol Abh 227:365–380

Colbert EH (1970) A saurischian dinosaur from the Triassic of Brazil. Am Mus Novit 2405:1–39

Currie PJ, Sarjeant WAS (1979) Lower Cretaceous dinosaur footprints from the Peace River Canyon, British Columbia, Canada. Palaeogeogr, Palaeoclim, Palaeoecol 28:103–115

Dal Corso J, Bernardi M, Sun Y, Song H, Seyfullah LJ, Preto N, Gianolla P, Ruffell A, Kustatscher E, Roghi G, Merico A, Hohn S, Schmidt AR, Marzoli A, Newton RJ, Wignall PB, Benton MJ (2020) Extinction and dawn of the modern world in the Carnian (Late Triassic). Sci Adv 6(38):eaba0099. https://doi.org/10.1126/sciadv.aba0099

Da-Rosa AAS (2005) Paleoalterações em depósitos sedimentares de planícies aluviais do Triássico Médio a Superior do sul do Brasil: caracterização, análise estratigráfica e preservação fossilífera. PhD thesis, Universidade do Vale do Rio dos Sinos, São Leopoldo, Brasil

Da-Rosa AAS, Faccini U (2005) Delimitação de blocos estruturais de diferentes escalas em seqüências mesozóicas do Estado do Rio Grande do Sul: implicações bioestratigraficas. GAEA 1:16–23

Demathieu G (1989) Appearance of the first dinosaur tracks in the french Middle Triassic and their probable significance. In: Gillette DD, Lockley MG (eds) Dinosaur tracks and traces. Cambridge University Press, Cambridge, pp 201–207

Dentzien-Dias P, Schultz CL, Scherer CMS, Lavina EC (2007) The trace fossil record from the Guará Formation (Upper Jurassic?), southern Brazil. Arq Mus Nac 65:585–600

Dutra TL, Crisafulli A (2009) *Kaokoxylon zalesskyi* (Sahni) Maheswari em en los niveles superiores de la Secuencia Santa Maria 2 (Formacion Caturrita), Cuenca de Paraná, Brasil. GAEA 5:61–69

Faccini UF (2007) Tectonic and climatic induced changes in depositional styles of the Mesozoic sedimentary record of southern Paraná Basin, Brazil. In: Iannuzzi R, Boardman D (eds) Problems in western Gondwana geology - I workshop - "South America - Africa correlations: du Toit revisited", extended abstracts, pp 42–45

Farlow JO (2018) Noah's Ravens: interpreting the makers of tridactyl dinosaur footprints. Indiana University Press

Ferigolo J, Langer MC (2006) A Late Triassic dinosauriform from South Brazil and the origin of the ornithischian predentary bone. Hist Biol 19:23–33

Fonseca MM, Scherer CMS (1998) The Meso and Late Triassic of South brazilian Gondwanaland: a process-oriented analysis and the fluvial deposits. Hallesches Jahrb Geowiss 5:51–52

Gama E Jr (1979) A sedimentação do Grupo Passa Dois (Exclusive Formação Irati): um modelo geomórfico. Rev Bras Geociênc 9(1):1–16

Garcia MS, Müller RT, Da-Rosa AAS, Dias-da-Silva S (2019) The oldest known co-occurrence of dinosaurs and their closest relatives: a new lagerpetid from a Carnian (Upper Triassic) bed of Brazil with implications for dinosauromorph biostratigraphy, early diversification and biogeography. J S Am Earth Sci 91:302–319. https://doi.org/10.1016/j.jsames.2019.02.005

Goldring R, Seilacher A (1971) Limulid undertracks and their sedimentological implications. Neues Jahrb Geol Palaontol Abh 137:422–442

Guerra-Sommer M, Cazzulo-Klepzig M (2000) The Triassic taphoflora from Paraná Basin, southern Brazil: an overview. Rev Bras Geociênc 30:481–485

Guerra-Sommer M, Cazzulo-Klepzig M, Bolzon RT, Alves LSR, Ianuzzi R (2000) As floras triássicas do Rio Grande do Sul: Flora *Dicroidium* e Flora *Araucarioxylon*. In: Holz M, De Ros LF (eds) Paleontologia do Rio Grande do Sul. CIGO/UFRGS, pp 85–106

Haubold H (1986) Archosaur footprints at the terrestrial Triassic-Jurassic transition. In: Padian K (ed) The begginning of the age of dinosaurs. Cambridge University Press, Cambridge, pp 189–201

Haubold H (1971) Ichnia Amphibiorum et Reptiliorum fossilium. In: Encyclopedia of Paleoherpetology, vol 18, pp 1–124

Heckert A, Lucas S (1998) First occurrence of *Aetosaurus* (Reptilia: Archosauria) in the Upper Triassic Chinle Group (USA) and its biochronological significance. Neues Jahrbuch Geol Paläontologie Abh 10:604–612

Hitchcock E (1845) An attempt to name, classify, and describe the animals that made the fossil footmarks of New England. Ann Assoc Am Geogr 6:23–25

Hitchcock E (1855) Impressions (chiefly tracks) on alluvial clay, in Hadlay, Mass. Am J Sci 2:391–396

Hitchcock E (1858) Ichnology of New England: a report on the sandstone of the Connecticut valley and its fossil footmarks. Commonwealth of Massachusetts, Boston, William White, 220 p

Holz M, De Ros LF (eds) (2000) Paleontologia do Rio Grande do Sul. CIGO/UFRGS, Porto Alegre, 398 p

Horn BLD, Melo TM, Schultz CL, Philipp RP, Kloss HP, Goldberg K (2014) A new third-order sequence stratigraphic framework applied to the Triassic of the Paraná Basin, Rio Grande do Sul, Brazil, based on structural, stratigraphic and paleontological data. J S Am Earth Sci 55:123–132. https://doi.org/10.1016/j.jsames.2014.07.007

Horn BLD, Goldberg K, Schultz CL (2018) A loess deposit in the Late Triassic of southern Gondwana, and its significance to global paleoclimate. J S Am Earth Sci 81:189–203. https://doi.org/10.1016/j.jsames.2017.11.017

Irmis RB, Nesbitt SJ, Padian K, Smith ND, Turner AH, Woody D, Downs A (2007) A Late Triassic dinosauromorph assemblage from New Mexico and the rise of dinosaurs. Science 317:358–361

Irmis RB, Mundil R, Martz JW, Parker WG (2011) High-resolution U-Pb ages from the Upper Triassic Chinle Formation (New Mexico, USA) support a diachronous rise of dinosaurs. Earth Planet Sci Lett 309(3–4):258–267. https://doi.org/10.1016/j.epsl.2011.07.015

Kent DV, Malnis PS, Colombi CE, Alcober OA, Martínez RN (2014) Age constraints on the dispersal of dinosaurs in the Late Triassic from magnetochronology of the Los Colorados Formation (Argentina). PNAS 111:7958–7963. https://doi.org/10.1073/pnas.1402369111

King MJ, Benton MJ (1996) Dinosaurs in the Early and Mid Triassic?—the footprint evidence from Britain. Palaeogeogr Palaeoclimatol Palaeoecol 122:213–225

Klein H, Lucas SG (2021) The Triassic tetrapod footprint record. N M Mus Nat Hist Sci Bull 83:1–194

Langer MC (2003) The pelvic and hind limb anatomy of the stem-sauropodomorph *Saturnalia tupiniquim* (Late Triassic, Brazil). PaleoBios 23:1–30

Langer MC (2005) Studies on continental Late Triassic tetrapod biochronology. I. The type locality of *Saturnalia tupiniquim* and the faunal succession in south Brazil. J S Am Earth Sci 19:205–218

Langer MC, Abdala F, Richter M, Benton M (1999) A sauropodomorph dinosaur from the Upper Triassic (Carnian) of southern Brazil. C R Acad Sci 329:511–517

Langer MC, Ezcurra MD, Bittencourt JS, Novas FE (2010) The origin and early evolution of dinosaurs. Biol Rev 85:55–110

Langer MC, Ferigolo J (2013) The Late Triassic dinosauromorph *Sacisaurus agudoensis* (Caturrita Formation; Rio Grande do Sul, Brazil): anatomy and affinities. Geol Soc, Lond, Spec Publ 379:353–392

Langer MC, Ramezani J, Da-Rosa AAS (2018) Upb age constraints on dinosaur rise from south Brazil. Gondwana Res 57:133–140. https://doi.org/10.1016/j.gr.2018.01.005

Lapparent AF, Montenat C (1967) Les empreintes des pas de reptiles de l'Infralias du Veillon, Vendée. Mem Soc Géol Fr 46:1–44

Leal LA, Azevedo SA, Kellner AWA, Da-Rosa AAS (2004) A new early dinosaur (Sauropodomorpha) from the Caturrita Formation (Late Triassic), Paraná Basin, Brazil. Zootaxa 690:1–24

Lockley MG (1991) Tracking dinosaurs. University Press, Cambridge

Lockley MG, Meyer C (2000) Dinosaurs tracks and other fossil footprints of Europe. Columbia University Press, New York

Lockley MG, Hunt AP (1994) A review of Mesozoic vertebrate ichnofaunas of the western Interior United States: evidence and implications of a superior track record. In: Caputo MV et al (eds) Mesozoic systems of the Rocky Mountain Region, USA. SEPM, Denver, pp 95–108

Lucas SG (1998) Global Triassic tetrapod biostratigraphy and biochronology. Palaeogeogr Palaeoclimatol Palaeoecol 143:347–384

Lucas SG (2001) Age and correlation of Triassic tetrapod assemblages from Brazil. Albertiana 26:13–20

Lucas SG, Klein H, Lockley MG, Spielmann JA, Gierlinski G, Hunt AP, Tanner LH (2006) Triassic-Jurassic stratigraphic distribution of the theropod footprint ichnogenus *Eubrontes*. N M Mus Nat Hist Sci Bull 37:86–93

Marsicano CA, Barredo SP (2004) A Triassic tetrapod footprint assemblage from southern South America: palaeobiogeographical and evolutionary implications. Palaeogeogr Palaeoclimatol Palaeoecol 203:313–335

Marsicano CA, Arcucci AB, Mancuso A, Caselli AT (2004) Middle Triassic tetrapod footprints of southern South America. Ameghiniana 41:171–184

Marsola JCA, Bittencourt JS, Butler RJ, Da-Rosa ASS, Sayão JM, Langer MC (2019) A new dinosaur with theropod affinities from the Late Triassic Santa Maria Formation, South Brazil. J Vertebr Paleontol e1531878. https://doi.org/10.1080/02724634.2018.1531878

Martinelli AG, Bonaparte JF, Schultz CL, Rubert RR (2005) A new Tritheledontid (Therapsida, Eucynodontia) from the Late Triassic of Rio Grande do Sul (Brazil) and its phylogenetic relationships among carnivorous non-mammalian eucynodonts. Ameghiniana 42:91–208

Martinez RN, Sereno P, Alcober OA, Colombi C, Renne PR, Montañez IP, Currie BS (2011) A basal dinosaur from the dawn of the dinosaur era in southwestern Pangea. Science 331:206–210. https://doi.org/10.1126/science.1198467

Martinez RN, Apaldetti C, Alcober OA, Colombi CE, Sereno PC, Fernandez E, Malnis PC, Correa GA, Abelin D (2013) Vertebrate succession in the Ischigualasto Formation. J Vertebr Paleontol 32(sup1):10–30

Milàn J, Bromley RG (2006) True tracks, undertracks and eroded tracks, experimental work with tetrapod tracks in laboratory and field. Palaeogeogr Palaeoclimatol Palaeoecol 231:253–264

Milani EJ (2000) Geodinâmica Fanerozóica do Gondwana sul-ocidental e a evolução geológica da Bacia do Paraná. In: Holz M, De Ros LF (eds) Geologia do Rio Grande do Sul. UFRGS/Instituto de Geociências – CIGO, Porto Alegre, pp 275–302

Milani EJ, Melo JHG, Souza PA, Fernandes LA, França AB (2007) Bacia do Paraná. Bol Geociênc PETROBRÁS 15:265–287

Müller RT (2021) A new theropod dinosaur from a peculiar Late Triassic assemblage of southern Brazil. J S Am Earth Sci 107:103026. https://doi.org/10.1016/j.jsames.2020.103026

Müller RT, Langer MC, Dias-da-Silva S (2018) An exceptionally preserved association of complete dinosaur skeletons reveals the oldest long-necked sauropodomorphs. Biol Lett 14(11):1744–9561. https://doi.org/10.1098/rsbl.2018.0633

Novas FE, Ezcurra MD, Chatterjee S Kutty TS (2011) New dinosaur species from the Upper Triassic Upper Maleri and Lower Dharmaram formations of Central India. Earth Environ Sci Trans R Soc Edinb 101:333–349

Novas FE, Agnolin FL, Ezcurra MD, Müller RT, Martinelli AG, Langer MC (2021) Review of the fossil record of early dinosaurs from South America, and its phylogenetic implications. J S Am Earth Sci 110:103341. https://doi.org/10.1016/j.jsames.2021.103341

Olsen PE, Galton PM (1984) A review of the Reptile and Amphibian Assemblages from the Stormberg of Southern Africa, with special emphasis on the footprints and age of the Stormberg. Palaeontol Afr 25:87–110

Olsen PE, Kent DV, Sues HD, Koeberl C, Huber H, Montanari A, Rainforth EC, Powell SJ, Szajna MJ, Hartline BW (2002) Response to comment on "Ascent of dinosaurs linked to an iridium anomaly at the Triassic-Jurassic boundary. Science 296:1305–1307

Olsen PE, Smith JB, McDonald NG (1998) Type material of the type species of the classic theropod footprint genera *Eubrontes*, *Anchisauripus*, and *Grallator* (Early Jurassic, Hartford and Deerfield Basins, Connecticut and Massachusetts, USA). J Vertebr Paleontol 18:586–601

Pacheco C, Müller RT, Langer MC, Pretto FA, Kerber L, Dias-da-Silva S (2019) *Gnathovorax cabreirai*: a new early dinosaur and the origin and initial radiation of predatory dinosaurs. PeerJ 7:e7963. https://doi.org/10.7717/peerj.7963

Pires EF, Guerra-Sommer M (2004) *Sommerxylon spiralosus* from Upper Triassic in southernmost Paraná Basin (Brazil): a new taxon with taxacean affinity. Acad Bras Ciências 76:595–609

Pretto FA, Langer MC, Schultz CL (2018) A new dinosaur (Saurischia: Sauropodomorpha) from the Late Triassic of Brazil provides insights on the evolution of sauropodomorph body plan. Zool J Linn Soc 1–29. https://doi.org/10.1093/zoolinnean/zly028.

Raath MA, Kitching JW, Shone RW, Rossouw GJ (1990) Dinosaur tracks in Triassic Molteno sediments: the earliest evidence of dinosaurs in South Africa. Palaeontol Afr 27:89–95

Rainforth EC (2004) The footprint record of Early Jurassic dinosaurs in the Connecticut Valley: status of the taxon formerly known as *Brontozoum*. In: Geological Society of America, abstracts with programs, vol 36, p 96

Rubert RR, Schultz CL (2004) Um novo horizonte de correlação para o Triássico Superior do Rio Grande do Sul. Pesqui Geociênc 31:71–88

Scherer CMS, Lavina ELC (2005) Sedimentary cycles and facies architecture of aeolian–fluvial strata of the Upper Jurassic Guará Formation, southern Brazil. Sedimentology 52:1323–1341

Scherer CMS, Faccini UF, Lavina EL (2000) Arcabouço estratigráfico do Mesozóico da Bacia do Paraná. In: Holz M, De Ros LF (eds) Geologia do Rio Grande do Sul. UFRGS/Instituto de Geociências - CIGO, Porto Alegre, pp 335–354

Schultz CL, Scherer CMS, Barberena MC (2000) Bioestratigraphy of Southern Brazilian Middle-Upper Triassic. Rev Bras Geociênc 30:495–498

Silva RC, Carvalho IS, Fernandes ACS, Ferigolo J (2007a) Preservação e contexto paleoambiental das pegadas de tetrápodes da Formação Santa Maria (Triássico Superior) do Sul do Brasil. In: Carvalho IS et al (eds) Paleontologia: Cenários da Vida 1. Interciência, Rio de Janeiro, pp 525–532

Silva RC, Carvalho IS, Schwanke C (2007b) Vertebrate dinoturbation from the Caturrita Formation (Late Triassic, Paraná Basin), Rio Grande do Sul State, Brazil. Gondwana Res 11:303–310

Silva RC, Carvalho IS, Fernandes ACS (2008a) Pegadas de dinossauros do Triássico (Formação Santa Maria) do Brasil. Ameghiniana 45:783–790

Silva RC, Carvalho IS, Fernandes ACS, Ferigolo J (2008b) Pegadas teromorfóides do Triássico Superior (Formação Santa Maria) do Sul do Brasil. Rev Bras Geociênc 38(1):100–115

Silva RC, Ferigolo J, Carvalho IS, Fernandes ACS (2008c) Lacertoid footprints from the Upper Triassic (Santa Maria Formation) of Southern Brazil. Palaeogeogr Palaeoclimatol Palaeoecol 262(3–4):140–156

Silva RC, Barboni R, Dutra T, Godoy MM, Binotto RN (2012) Footprints of large theropod dinosaurs and implications on the age of Triassic biotas from Southern Brazil. J S Am Earth Sci 39:16–23. https://doi.org/10.1016/j.jsames.2012.06.017

Soares MB, Schultz CL (2006) Proposta de nova denominação para a Cenozona de Ictidosauria, do Triássico Superior (Formação Caturrita) do Rio Grande do Sul. In: Boletim de Resumos do V Simpósio Brasileiro de Paleontologia de Vertebrados, Santa Maria, p 41

Soares MB, Schultz CL, Horn BLD (2011) New information on *Riograndia guaibensis* Bonaparte, Ferigolo & Ribeiro, 2001 (Eucynodontia, Tritheledontidae) from the Late Triassic of southern Brazil: anatomical and biostratigraphic implications. An Acad Bras Ciênc 83:329–354

Sternberg CM (1932) Dinosaur tracks from Peace River, British Columbia. Natl Mus Can Bull 68:59–85

Thulborn T (1989) The gaits of dinosaurs. In: Gillette DD, Lockley MG (eds) Dinosaur tracks and traces. Cambridge University Press, Cambridge, pp 39–50

Thulborn T (1990) Dinosaur tracks. Chapman & Hall, London

Zerfass H (2007) Geologia da Folha Agudo, SH.22-V-C-V, escala 1:100.000. Serviço Geológico do Brasil – CPRM. https://rigeo.sgb.gov.br/handle/doc/10287

Zerfass H, Lavina EL, Schultz CL, Garcia AGV, Faccini UF, Chemale F Jr (2003) Sequence stratigraphy of continental Triassic strata of southernmost Brazil: a contribution to Southwestern Gondwana palaeogeography and palaeoclimate. Sediment Geol 161:85–105.

Chapter 3
Dinosaur Tracks and Trackways from the Upper Jurassic Guará Formation, Paraná Basin, Brazil

Heitor Francischini[ID], **Denner Deiques Cardoso**[ID], and **Paula Dentzien-Dias**[ID]

3.1 Introduction

The Late Jurassic, known as "the Age of Brontosaurs" (Lockley and Meyer 2000), is widely known for its giant dinosaurs, which are well represented in several tracksites around the world. In Europe, there are more than 30 sites with well-preserved dinosaur tracks, coming mainly from the Iberian Peninsula and the Jura Mountains (e.g., Lockley and Meyer 2000; Belvedere et al. 2019). In North America, the Upper Jurassic Morrison Formation provides impressive long trackways and evidence of sauropod gregariousness at the Purgatoire Valley ichnosite of Colorado (Lockley et al. 1986). In other hand, with the exception of the High Atlas Mountains in Morocco (Dutuit and Ouazzou 1980; Belvedere et al. 2010, 2019), the Late Jurassic dinosaur track record is very scarce and fragmentary in Gondwana, with punctual occurrences in Australia (Romilio et al. 2021), Chile (Moreno and Pino 2002; Moreno et al. 2004; Moreno and Benton 2005), Colombia (Moreno-Sánchez et al. 2011), Guyana (Leonardi 1994), Uruguay (Mesa and Perea 2015), and Brazil (see below; Table 3.1).

Contrasting to Triassic and Cretaceous systems, the Jurassic of South America is constrained, with few continental sedimentary units occurring along the continent (Table 3.1; Bonaparte 1981; Leonardi 1994). As a consequence, the biota of such period is poorly known (e.g., Colbert 1977). In Brazil, Jurassic fossils follow the

H. Francischini (✉) · P. Dentzien-Dias
Departamento de Paleontologia e Estratigrafia, Instituto de Geociências, Universidade Federal do Rio Grande do Sul, Estado do Rio Grande do Sul, Avenida Bento Gonçalves, Porto Alegre 9500, 91501-970, Brazil
e-mail: heitor.francischini@ufrgs.br

D. D. Cardoso
Programa de Pós-Graduação em Geociências, Instituto de Geociências, Universidade Federal do Rio Grande do Sul, Avenida Bento Gonçalves, Porto Alegre, Estado do Rio Grande do Sul 9500, 91501-970, Brazil

Table 3.1 Summary of the Jurassic dinosaur track-bearing geological units of South America (ordered by age)

Unit	Age	Country	References
Marifil Volcanic Complex	Early Jurassic (pre-Pliensbachian)	Argentina	Díaz-Martínez et al. (2017)
La Matilde Formation	Middle Jurassic (Bajocian-Callovian)	Argentina	Casamiquela (1964); Coria and Carabajal (2004)
Baños del Flaco Formation	Late Jurassic (Tithonian)	Chile	Casamiquela and Fasola (1968); Moreno and Pino (2002); Moreno and Benton (2005)
Estación Member, San Salvador Formation	Late Jurassic (Kimmeridgian)–Lower Cretaceous	Chile	Moreno et al. (2004)
Batoví Member, Tacuarembó Formation	Late Jurassic (Kimmeridgian-Tithonian)	Uruguay	Mesa and Perea (2015)
Guará Formation	Late Jurassic (Kimmeridgian-Tithonian)	Brazil	Dentzien-Dias et al. (2007) and further references in this chapter
Takutu Formation	Late Jurassic-Early Cretaceous	Guyana	Leonardi (1994)
Chacarilla Formation	Late Jurassic-Early Cretaceous	Chile	Salinas et al. (1991); Rubilar et al. (2000)
Arcabuco Formation	Late Jurassic-Early Cretaceous	Colombia	Moreno-Sánchez et al. (2011)
Pirambóia Formation	Jurassic? (see text)	Brazil	Christofoletti et al. (2021)
Indeterminate unit at Faxinal do Soturno	Jurassic? (see text)	Brazil	Silva et al. (2012)

same pattern, with sparse vertebrate body fossils found in the Pastos Bons (Alparcata Basin), Sergi, Aliança (both Recôncavo-Tucano-Jatobá Basins), Missão Velha, and Brejo Santo (both Araripe Basin) formations. Dinosaur remains comprise few records from the Sergi and Aliança formations (Upper Jurassic; Dom João local stage): one isolated caudal vertebral centrum of an allosauroid of carcharodontosaurian affinity (Bandeira et al. 2021) and one Neotheropoda caudal vertebra (Oliveira et al. 2022), respectively.

The Brazilian Jurassic dinosaur track record is also very incomplete and is restricted to the Paraná Basin. Recently, Christofoletti et al. (2021) described large-sized (about 50 cm long) tracks preserved in cross section from the Pirambóia Formation sandstones of the São Paulo state, considering this unit Late Jurassic in age and chronocorrelated to the Guará Formation. Historically, the age of the Piram-bóia Formation has been debated and recent propositions (such as Silva et al. 2023) have considered it Triassic based on subsurface data. A second record comprises two theropod tracks (*Eubrontes* isp.) described by Silva et al. (2012) in the upper beds (here assigned to an indeterminate geological unit, different from the orig-inal Caturrita Formation assignation) of the Linha São Luiz site, at the Faxinal do Soturno municipality, Rio Grande do Sul state. Although Silva et al. (2012) regarded this occurrence as Late Triassic, plants (conifer wood: *Agathoxylon africanum*, *Chapmanoxylon jamuriense*, *Kaokoxylon zalesskyi*, *Megaporoxylon kaokense*, and *Sommerxylon spiralosus*; branches and leaves: *Brachyphyllum* sp., *Pagiophyllum* sp. and *Pterophyllum*?; Benettitales reproductive structure: *Williamsonia potyporaneae*) and conchostracans (Eosestheriidae: *Nothocarapacestheria soturnensis*; Fushuno-graptidae: *Australestheria* sp.) found in the levels below to the *Eubrontes*-bearing level point to a (Early to Middle?) Jurassic age for the dinosaur tracks (Pires and Guerra-Sommer 2004; Dutra and Crisafulli 2002; Barboni and Dutra 2013; Rohn et al. 2014; Jenisch et al. 2017). Lastly, the Guará Formation contains a more complete and abundant dinosaur track record (Fig. 3.1). The early reports of dinosaur tracks from this unit come from the early 2000's (Schultz et al. 2002; Scherer and Lavina 2005) and, since then, systematic fieldworks performed by our team have shed light on the composition and diversity of this ichnofauna (Fig. 3.2). This work brings a summary of the state of the art of the Guará Formation dinosaur ichnology based on the successive fieldworks made by our team in the last 20 years and discusses the next steps on the study of Jurassic dinosaurs in Brazil.

3.2 Geological Context

The Guará Formation is a unit that crops out in the southern Brazil (Rio Grande do Sul and Paraná states) and in the northern Uruguay (mainly in the Rivera and Tacuarembó departments), where it is known as the Batoví Member of the Tacuarembó Formation (Lavina et al. 1985; Perea and Martínez 2004; Scherer and Lavina 2005; Reis et al. 2019). Because of the lithological continuity between these units along the Brazil-Uruguay border, the authors often refer to them as representing a single depositional environment and respective lithostratigraphic unit, the Guará-Batoví. Additionally, the Guará-Batoví also occurs in subsurface in Mato Grosso do Sul, Santa Catarina and São Paulo states (Silva et al. 2023). From a lithological point of view, the Guará Formation is composed of fine- to coarse-grained sandstones (quartzarenites) and mudstones and represents a large distributive fluvial system characterized by amalga-mated perennial braided rivers, where fluvial currents deposited alloctonous remains in association to diverse parautoctonous elements (Lavina et al. 1985; Scherer and Lavina 2005; Amarante et al. 2019; Reis et al. 2019). Eolian dunes and sandsheets also occur, but seem to be restricted to southwestern Rio Grande do Sul state and northern Uruguay. The eolian paleocurrents have a W-E direction, while the fluvial ones are NNE-SSW (Scherer and Lavina 2005; Reis et al. 2019).

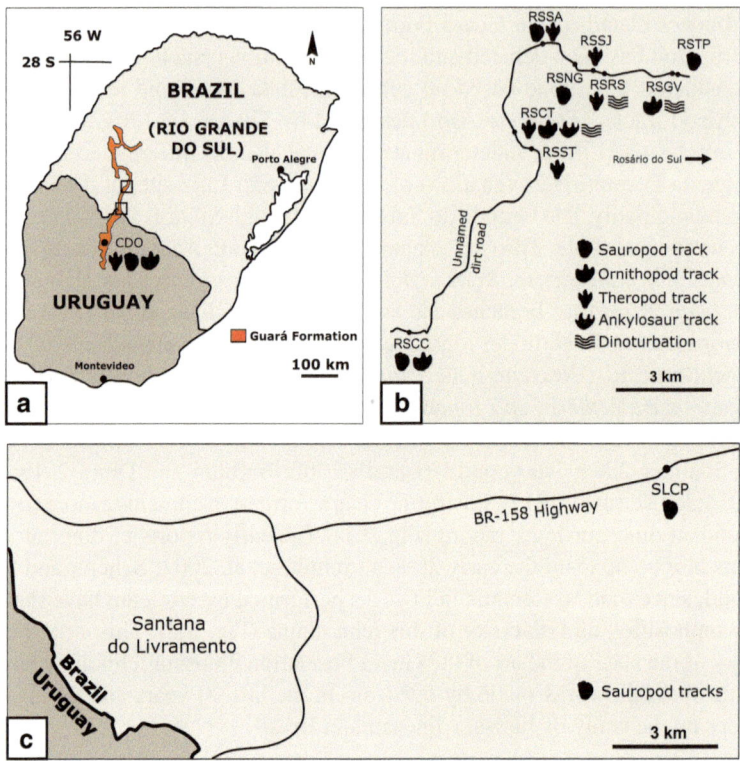

Fig. 3.1 Summary of the occurrence of dinosaur tracks in the Upper Jurassic Guará-Batoví unit of Brazil (Rio Grande do Sul state) and Uruguay. Localities: Cañada del Ombú (CDO), Cerro Caverá (RSCC), Cerro Torneado (RSCT), Granja Santa Vitória (RSGV), Norival Gonçalves (RSNG), Rancho Sossego (RSRS), "Saurópodes" (RSSA), Sanga do Jacaré (RSSJ), Sanga do Torneado (RSST), Touro Passo (RSTP), and Cerro Palomas (SLCP). The Juventina Rosa (RSJR) locality is not depicted

In Uruguay, the Batoví Member of the Tacuarembó Formation is considered Kimmeridgian–Tithonian in age, as estimated by the presence of the hybodontid shark *Priohybodus* cf. *P. arambourgi*, the fushunograptid conchostracan *Orthestheria (Migransia) ferrandoi* and the theropod dinosaurs *Ceratosaurus* sp. and *Torvosaurus* sp. (Perea and Martínez 2004; Soto and Perea 2008; Perea et al. 2009; Soto et al. 2020a, b). Other vertebrates of the Batoví paleofauna include the coelacanth *Mawsonia* sp., the lungfishes "*Ceratodus*" *tiguidiensis* and *Ceratodus africanus,* the pholidosaur crocodyliform *Meridiosaurus vallisparadisi*, the turtle *Tacuarembemys kusterae*, and the ctenochasmatid pterosaur *Tacuadactylus luciae,* besides indeterminate abelisaurid theropods (Soto and Perea 2010; Fortier et al. 2011; Perea et al. 2014; Soto et al. 2021, 2022; Toriño et al. 2021). Sauropod, theropod and ornithopod tracks also occur in the Batoví Member of the Tacuarembó Formation (Mesa and Perea 2015). The lithostratigraphic continuity between the Guará and

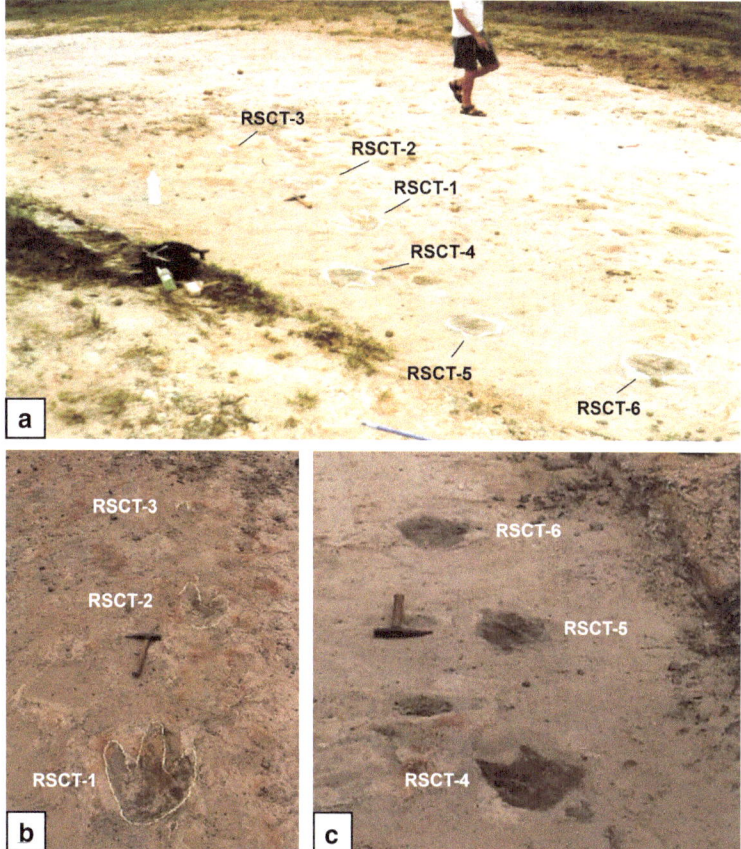

Fig. 3.2 Historical record of the first findings of dinosaur tracks in the Guará Formation at the Cerro Torneado (RSCT) locality

Tacuarembó formations allows the extrapolation of the same age to the Brazilian territory.

The fossil record of the Guará Formation (in Brazil) does not include body fossils besides conchostracans (Fig. 3.3a–d), but invertebrate trace fossils (*Arenicolites* isp., *Beaconites coronus* and *Taenidium barretti;* Fig. 3.3e), tetrapod paleoburrows and dinosaur tracks are commonly found (Netto 1989; Dentzien-Dias et al. 2007, 2008, 2012; Francischini et al. 2015, 2018). These traces occur mainly in sand interdune deposits, while bioturbation structures are rare in the eolian dunes (Dentzien-Dias et al. 2008).

Up to this point, eleven fossiliferous outcrops of the Guará Formation provide ichnological evidence of dinosaurs, ten of them in the Rosário do Sul municipality and one in the Santana do Livramento municipality (Rio Grande do Sul state), Brazil (Fig. 3.4).

Fig. 3.3 **a–d** Conchostracans from the Arroio do Beco locality, Santana do Livramento municipality. Scale bars: 1 mm; **e** *Taenidium barretti* from the Cerro Palomas (SLCP) locality

Fig. 3.4 **a–b** Mode of occurrence of the Guará Formation outcrops in Brazil. **a** aerial view from
the Cerro Caverá (RSCC) locality; **b** Overview of the Sanga do Torneado (RSST) locality. The
theropod track UFRGS-PV-0207-G can be seen inside the ravine; **c, d** Signs of the occurrence of
dinosaur tracks and trackways of the Cerro Torneado (RSCT) and Cerro Palomas (SLCP) localities,
Rosário do Sul and Santana do Livramento municipalities, respectively

RSCC (Rosário do Sul, Cerro Caverá; UTM 21 J 0675706/6645873): this outcrop is
composed of sandstones originated in a large sheet of bioturbated fine sand, where
a wide-gauge sauropod trackway with five defined footprints and at least four other

isolated, less defined tracks, were found (Fig. 3.4a; Deiques 2023). The tracks remain *in situ*;

RSCT (Rosário do Sul, Cerro Torneado; UTM 21 J 0679148/6652013): this outcrop presents basal layers composed of sandstones clearly coming from paleodunes with paleowinds directed to the north, while the upper layers reveal a sequence of fully bioturbated eolian sand sheets (Fig. 3.2; Dentzien-Dias et al. 2007). The tracks FURG-H-480, UFRGS-PV-0204-G (formerly UFRGS-PV-0004-J/K), UFRGS-PV-0205-G (formerly UFRGS-PV-0005-J/K), UFRGS-PV-0206-G, and UFRGS-PV-0208-G were collected on an unnamed local road, but several (RSCT-1 to RSCT-14) still remains *in situ* (see Francischini et al. 2018, for example);

RSGV (Rosário do Sul, Granja Santa Vitória; UTM 21 J 684,118/6653718) is composed of a layer of sandstones from eolian sand sheets, about 30 cm thick, totally dinoturbed (Dentzien-Dias et al. 2007);

RSJR (Rosário do Sul, Juventina Rosa; UTM 22 J 068203/6653736): the base of this outcrop is composed of sandstones from fossil eolian dunes and the top of sandstone from eolian sand sheets. Only one isolated track was found and it remains *in situ* (Dentzien-Dias et al. 2007);

RSNG (Rosário do Sul, Norival Gonçalves; UTM 21 J 0681104/6654026): the sandstones of this outcrop come both from eolian dunes and sand sheets. Only one isolated sauropod track was found and it remains *in situ* (Dentzien-Dias et al. 2007);

RSRS (Rosário do Sul, Rancho Sossego; UTM 21 J 681913/6653733): composed, from the base to the top, of fine sandstones of fluvial origin, other sandstones from eolian sand sheets, a layer of massive siltstone, and more sandstone from sand sheets (Deiques 2023). All tracks from RSRS remain *in situ*;

RSSA (Rosário do Sul, "Saurópodes"; UTM 21 J 680026/6654689): the outcrop is composed of, from the base to the top, fine sandstones of fluvial origin, conglomerates with siltstone intraclasts, sandstone from sand sheet, fine and conglomeratic sandstones, and again layers of fine sandstones (Deiques 2023). The sauropod tracks from RSSA still remain *in situ*;

RSSJ (Rosário do Sul, Sanga do Jacaré; UTM 21 J 0681689/6653839): composed only of sandstones produced in paleodune environments, whose paleowinds were directed eastward. The track UFRGS-PV-0203-G (formerly UFRGS-PV-0003-J/K of Dentzien-Dias et al. 2007) was collected from the dunes of this site;

RSST (Rosário do Sul, Sanga do Torneado; UTM 21 J 0679626/6651467): this outcrop is composed of a succession of conglomeratic and fine sandstones of fluvial origin and sandstones of eolian origin (Deiques 2023). The only track found here is UFRGS-PV-0207-G, which was collected from inside the main ravine (see Fig. 3.4b);

RSTP (Rosário do Sul, Touro Passo; UTM 21 J 0684252/6653733): composed of a succession of eolian and lacustrine sediments. All tracks remain *in situ* (Dentzien-Dias et al. 2007);

SLCP (Santana do Livramento, Cerro Palomas; UTM 21 J 0656439/6588052): outcrop adjacent to the BR-158 highway, km 549. From base to top there is a succession of sandstones correspondent to eolian dunes, sheets of eolian sand, lacustrine layers and a new succession of dunes (Dentzien-Dias et al. 2007). Sauropod tracks occur in the sand sheet bed and remain in situ. Invertebrate traces occur in the top of the outcrop (Fig. 3.3e).

Besides the sites described above, all from the southwestern area of Rio Grande do Sul, there is an enigmatic occurrence at Novo Treviso (Faxinal do Soturno municipality, central zone of the Rio Grande do Sul State; UTM 22 J 267188/6731040). This occurrence is discussed in Sect. 3.5, under the light of new data.

In Uruguay, sauropod, theropod and ornithopod tracks occur in the Cañada del Ombú I and II localities, respectively at the km 262.5 and 262.4 of the Ruta 26 at the Tacuarembó department (Mesa and Perea 2015).

3.3 Footprints: Diversity and Paleobiological Interpretation

3.3.1 Sauropod Tracks

Sauropod tracks (Fig. 3.5) occur in the following sites: RSCC, RSNG, RSSA, and SLCP in Brazil (Dentzien-Dias et al. 2007; Deiques 2023) and at Cañada del Ombú I in Uruguay (Mesa and Perea 2015). Apparently, only the tracks in SLCP and Cañada del Ombú I represent true tracks, because of their tridimensional morphology and the presence of marginal ridges following the track outline and sediment filling the track shaft. All remaining are composed of rounded, digitless and shallow undertracks, with a mean diameter of 50 cm and up to 45 cm depth (Dentzien-Dias et al. 2007, 2008; Francischini et al. 2015). All of the known tracks are pes prints, except by a single manus track found in the SLCP site (Fig. 3.5b, c). It is D-shaped and is placed just anteriorly to the pes imprint of the same side, evidencing a heteropod producer (Dentzien-Dias et al. 2012). According to Francischini et al. (2015), the mean pelvic girdle height estimation (measurement following Thulborn 1989) of the sauropod trackmakers of the Guará Formation is 2.92 m, but they could reach the maximum height of 3.12 m.

Two morphotypes of sauropod trackways were recognized in the Guará Formation by Dentzien-Dias et al. (2007), following the proposition of Farlow (1992) and Lockley et al. (1994a): in the wide-gauge trackways the autopodia are placed away from the trackway midline; contrasting, in the narrow-gauge trackways manus and pes tracks are closer to the midline. The occurrence of wide- and narrow-gauge trackways in the Guará Formation was, then, followed by the subsequent papers (Dentzien-Dias et al. 2008, 2012; Francischini et al. 2015). Nevertheless, quantitative methods for sauropod trackway classification using their gauge were proposed by Romano et al. (2007) and Marty et al. (2010), where the trackway gauge can also be accessed by calculating parameters, such as the trackway ratio (TR) and the

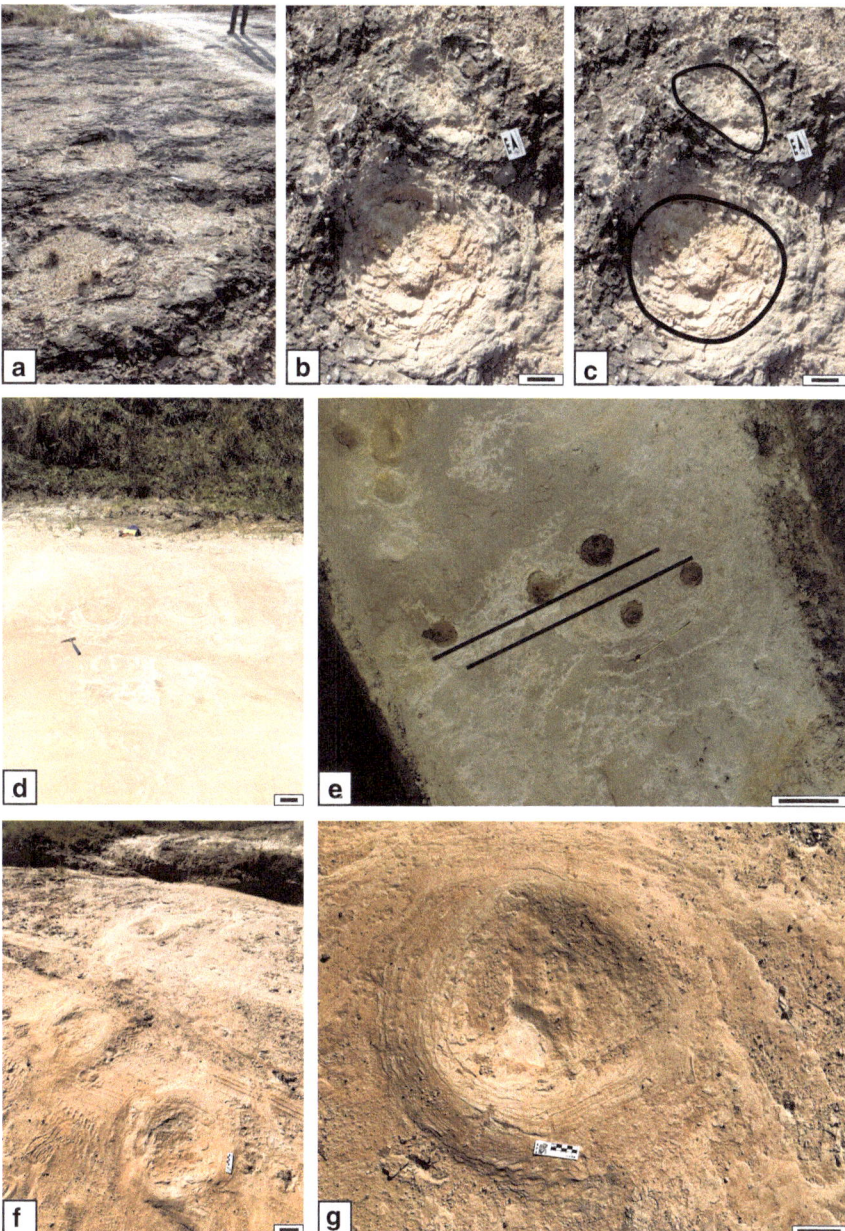

Fig. 3.5 Sauropod tracks from the Guará Formation, Upper Jurassic of Brazil. **a–c** Trackway ("south trackway" of Dentzien-Dias et al., 2007) from the Cerro Palomas (SLCP) locality. **a** A manus-pes couple is depicted in **b** and **c**; **d–e** Very wide-gauge trackway from the Cerro Caverá (RSCC) locality, seen in oblique (**d**) and perpendicular (**e**) views. **f** Wide-gauge trackway from the "Saurópodes" (RSSA) locality in oblique view; **g** Detail of the track RSSA-5 (in perpendicular view), showing the extensive zone of disturbed sediment around it. Scale bars: 15 cm **b, c**, 30 cm **d**, 1 m **e**, and 12 cm **f, g**

ratio between the width of the angulation pattern and the corresponding track length (WAP/PL for pes tracks and WAM/ML for manus tracks). TR and WAP/PL values for the Guará-Batoví sauropod trackways are provided in Table 3.2 and indicate that they rather represent narrow- to very wide-gauges. According to Dentzien-Dias et al. (2007, 2012), a narrow-gauge trackway also occurs in the SLCP site (named by them as the "south trackway"; Fig. 3.5a–c). However, new fieldwork data suggest that it can be actually composed by two independent rows of tracks, positioned near parallelly. Therefore, for this instance, the narrow-gauge morphotype remains restricted to the Cañada del Ombú I trackway B.

These morphotypes of trackways were recognized in several localities around the world (Farlow 1992; Lockley et al. 1994a, 1994b; Wilson and Carrano 1999; Marty et al. 2010), being ichnotaxonomically correspondent to the ichnogenera *Brontopodus* (wide-gauge trackways), *Parabrontopodus* and *Breviparopus* (these two latter narrow-gauge trackways) (Dutuit and Ouazzou 1980; Farlow et al. 1989; Lockley et al. 1994a).

Although the Guará-Batoví outcrops have preserved narrow- to very wide-gauge trackways, the ichnotaxonomical assignation of them are still to be done. The lack of any type of digit impression in these tracks precludes a proper assignment and, therefore, the access to the true ichnodiversity of sauropod trackmakers. Some digit-less sauropod ichnotaxa (e.g., *Titanopodus* and *Rotundichnus*) have been proposed (Hendricks 1981; González-Riga and Calvo 2009), but we believe that the lack of digit impressions in the Guará-Batoví tracks is rather an ichnotaphonomic artifact than a true morphological feature. In addition, pes tracks from the Guará-Batoví are nearly rounded, differing from the longer than wide pattern of *Brontopodus* and *Parabrontopodus* pedes. The predominance of pes tracks also precludes the estimation of the heteropody ratio, which is a main ichnotaxobasis for sauropod ichnotaxonomy.

Wide-gauge trackways and their respective ichnotaxa are usually assigned to titanosauriform sauropods (Brachiosauridae, Euhelopodidae, Titanosauria and

Table 3.2 Estimated parameters for the Guará/Batoví sauropod tracks and the main sauropod ichnotaxa. The trackway ratio (TR) follows Romano et al. (2007) and the ratio between the width of the angulation pattern and the pes length (WAP/PL) follows Marty et al. (2010). *Breviparopus*, *Brontopodus* and *Parabrontopodus* are represented by their type materials (Marty et al. 2010). The data from the Cañada del Ombú localities were taken from Mesa and Perea (2015)

Sauropod trackway	TR (%)	WAP/PL	Gauge
Breviparopus	50.2	0.9	Narrow
Parabrontopodus	51.9	0.87	Narrow
Brontopodus	36.04	1.46	Wide
RSCC	25.97	2.97	Very wide
RSSA	40.30	1.52	Wide
SLCP (north trackway)	29.06	2.44	Very wide
Cañada del Ombú A	45.86	1.14	Medium
Cañada del Ombú B	52.06	0.92	Narrow

related forms; see Lockley et al. 1994a and Wilson and Carrano 1999), but this direct correlation is contested by some authors (e.g., Santos et al. 1994). Indeed, Titanosauriformes were not recorded in Late Jurassic of South America so far, being rare in the entire Gondwana in this time interval. Although there are several anatomical features on the limbs and girdles of Titanosauria species that can be directly related to the posture of the group (see García et al. 2015), D'Emic (2012) argued that some non-titanosaurian sauropods (such as the macronarian *Tehuelchesaurus* and the brachiosaurid *Giraffatitan*) also have the architecture required to produce wide-gauge trackways. For Henderson (2006), the ability to produce wide-gauge trackways is widespread among Sauropoda and evolved independently in every lineage in which species exceeded 12 tons. Based on these data and on the fact that the Guará-Batoví trackways do not preserve any other diagnostic feature apart from the gauge, we are not able to assign them to any particular sauropod taxon. Even though the track-trackmaker correlation for the Guará-Batoví trackways is still pending on better preserved materials, the disparity in posture recorded in the narrow- to very wide-gauge trackways could reflect a diverse sauropod fauna.

3.3.2 Theropod Tracks

Theropod tracks are tridactyl, mesaxonic and present digits terminating into claws and, usually, a V-shaped or bilobated "heel" (Fig. 3.6). The Guará Formation theropod tracks are 17–51 cm long and 15–30 cm wide and, except by the trackways composed of RSCT-1–RSCT-3 and RSCT-4–RSCT-6, they are found mainly isolated (Dentzien-Dias et al. 2007; Francischini et al. 2015). Theropod tracks occur in the RSCT, RSRS, RSSJ, and RSST sites in Brazil (Dentzien-Dias et al. 2007; Deiques 2023) and at the Cañada del Ombú II locality of Uruguay (Mesa and Perea 2015), where they have a presumed affinity to *Therangospodus*. This affinity, however, is not followed here because of the lack of fine details. Indeed, except for UFRGS-PV-0207-G (found at RSST; see below), all the theropod tracks from the Guará-Batoví unit are composed of undertracks.

The true track UFRGS-PV-0207-G (Fig. 3.6c, d) was found inside a ravine at the RSST site. It is an isolated left track of 17.8 cm in length and 16.7 cm in width (length/width ratio: 1.08) which has short and blunt digits. Despite its size, this track is quite robust and presents a weak mesaxony (anterior triangle length/width ratio of 0.368; sensu Lockley 2009). These latter characters approximate the track UFRGS-PV-0207-G to the ichnogenera *Iberosauripus* Cobos et al. 2014 and *Jurabrontes* Marty et al. 2017 from the Upper Jurassic of Europe. A detailed ichnotaxonomical analysis of this track is in progress (Deiques, 2023) and will be published elsewhere in the future. Using the methods proposed by Thulborn (1989), the trackmaker hip height can be estimated at 81.5 cm.

Because there are no completely preserved feet of Late Jurassic large ceratosaurians or non-coelurosaurian tetanurans besides *Allosaurus* (Rauhut et al. 2018), a synapomorphy-based track-trackmaker correlation (sensu Carrano and

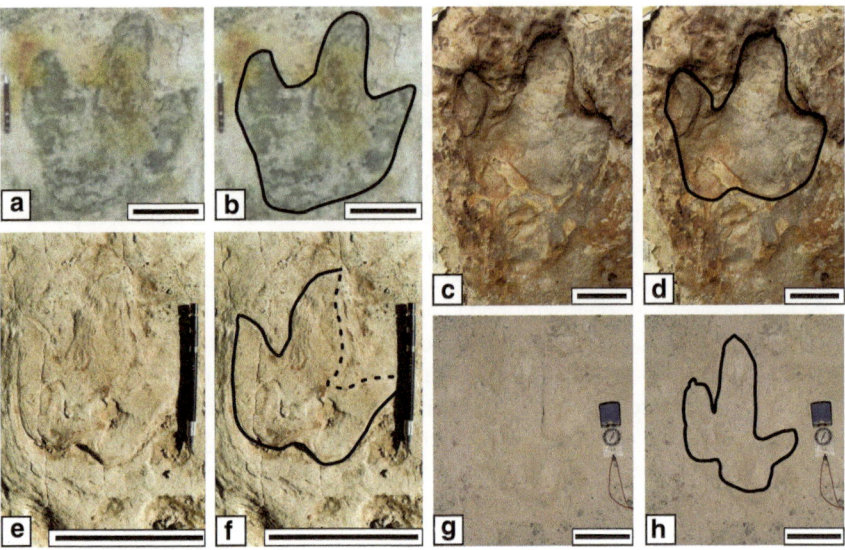

Fig. 3.6 Theropod tracks from the Guará Formation, Upper Jurassic of Brazil. **a, b** track RSCT-1 from the Cerro Torneado (RSCT) locality; **c, d** track UFRGS-PV-0207-G, from the Sanga do Torneado (RSST) locality; **e, f** unnamed track from the Cerro Torneado (RSCT) locality; **g, h** unnamed track from the Cerro Torneado (RSCT) locality. Scale bars: 12 cm (**a, b** and **e, f**), 5 cm (**c, d**) and 16 cm (**g, h**)

Wilson 2001) of the Guará-Batoví tracks is debilitated. However, the similarity of UFRGS-PV-0207-G with *Iberosauripus* and *Jurabrontes* led us to speculate about its trackmaker. Cobos et al. (2014) correlated robust tracks with weak mesaxony (their Ichno-Group 2, which includes *Iberosauripus* and similar tracks) from the Tithonian–Berriasian of Spain to robust megalosaurids, such as *Torvosaurus*. More slender and elongated tracks (i.e., the Ichno-Group 1 of Cobos et al. 2014, which includes *Bueckeburgichnus*, *Hispanosauripus* and *Megalosauripus*) are assigned to allosaurids. Marty et al. (2017) agreed with this proposition and, based on the weak mesaxony and wide "heel", included the ichnogenus *Jurabrontes* in the Cobos' et al. (2014) ichno-group 2. Indeed, *Torvosaurus* sp. teeth were found in the Batoví Member of the Tacuarembó Formation (Soto et al. 2020b), reinforcing the possibility of the trackmaker of UFRGS-PV-0207-G being a megalosaurid. Considering the size of UFRGS-PV-0207-G (in comparison to other Late Jurassic robust tracks and to the size of the teeth found in Uruguay; Belvedere et al. 2019; Soto et al. 2020b), it could have been produced by a smaller taxon or even a juvenile individual. Other tracks with relatively longer digit III also occur in the Guará-Batoví unit (see, for example, Fig. 3.6e–h), probably represent different ichnotaxa and trackmakers (possibly ceratosaurids; Francischini et al. 2015). A deeper discussion about these tracks and their supposed trackmakers can be seen in Deiques (2023).

3.3.3 Ankylosaur Tracks

The first record of ankylosaurian tracks in the Guará Formation was published by Francischini et al. (2018), who described a trackway composed of four consecutive pes (RSCT-10, RSCT-11, RSCT-13 and RSCT-14) and one manus (RSCT-12) from the Cerro Torneado locality (Fig. 3.7a, b). In spite of being undertracks, the morphology of these tracks are worth noting, mainly because of their clear tetradactyly and paraxony, contrasting to all previous known tracks from both the Guará Formation and the Batoví Member of the Tacuarembó Formation until that moment. The paraxonic pes presents four short and blunt digits and a broad heel. The manus is considerably smaller (with evident heteropody) and is positioned antero-laterally to the pes. Similar to the pes tracks, the manus is tetradactyl and paraxonic, but is comparatively much more anteroposteriorly short and is medio-laterally symmetric. All of the preserved digits lack claw marks. A new material (UFRGS-PV-0208-G; Fig. 3.7c, d) was recently found in the same locality and its proper description will be published elsewhere in the future (Deiques 2023).

Ankylosaur tracks are scarce around the world, with few Cretaceous ichnosites producing well preserved vestigia of these animals (see McCrea et al. 2001 for a review). Among them, three ichnospecies can be unambiguously assigned to ankylosaurs due to their morphology and age: *Ligabueichnium bolivianum* Leonardi 1984, *Metatetrapous valdensis* (Nopcsa 1923), and *Tetrapodosaurus borealis* (Sternberg 1932). *Ligabueichnium*, from the Upper Cretaceous of Bolivia, differs from the Guará ankylosaurian tracks by presenting pentadactyl pes, which completely overprints the manus (Leonardi 1984). *Tetrapodosaurus borealis* is the most ubiquitous ichnotaxon, occurring in several Aptian–Cenomanian sites of North America (McCrea et al. 2001; Lockley et al. 2021), with a putative Late Jurassic record (Hups et al. 2008) pending confirmation. Although there are many similarities between *Tetrapodosaurus* and the tracks from the Guará Formation, they differ mainly in: (i) the number of digits in the manus (*Tetrapodosaurus* is pentadactyl, while RSCT-12 is tetradactyl); (ii) the shape of the proximal margin of the manus (concave in *Tetrapodosaurus* and convex in RSCT-12); and (iii) the proportions of the pes (*Tetrapodosaurus* have longer than wide pes and the RSCT tracks have the inverse proportion) (Sternberg 1932; McCrea 2000). Regarding *Metatetrapous valdensis*, from the Lower Cretaceous Bückeberg Formation of Germany, it is very similar to the RSCT tracks described here (Francischini et al. 2018). Indeed, the assignment of the Brazilian tracks to this ichnospecies was pending on the discovery of new and better-preserved materials, which was partially solved by the discovery and collection of UFRGS-PV-0208-G.

During the Kimmeridgian–Tithonian interval, there are few dinosaur taxa that share the combination of quadrupedity, heteropody and tetradactyly, characters present in the RSCT tracks described by Francischini et al. (2018). Late Jurassic stegosaurs and ornithopods, for example, are mainly tridactyl during that period and some species from the latter group can walk bipedally (Lockley and Hunt 1998; Whyte and Romano 2001; Díaz-Martínez et al. 2015). Ceratopsid pes tracks are also tetradactyl and paraxonic pes tracks (McCrea et al. 2001), but the occurrence of

Fig. 3.7 Ankylosaur tracks from the Guará Formation, Upper Jurassic of Brazil. **a, b** pes (RSCT-11) and manus (RSCT-12) tracks from the Cerro Torneado (RSCT) locality; **c, d** Track UFRGS-PV-0207-G, also from the Cerro Torneado (RSCT) locality. Scale bars: 30 cm (**a, b**) and 5 cm (**c, d**)

these dinosaurs in the Guará Formation is unlikely because they are chronostratigra-phycally restricted to the Cretaceous. Based on these differences, Francischini et al. (2018) regarded the RSCT tracks as ankylosaur-made, with the osteological record, the geographical distribution and the temporal range of the group suggesting that the trackmakers possibly were nodosaurids.

The South American record of Ankylosauria is scarce, with few osteological and ichnological materials coming from the Cretaceous of Argentina and Bolivia (see Francischini et al., 2018 for a synthesis). Some trackways from the purported Upper Jurassic-Lower Cretaceous La Puerta Formation of Bolivia (Tunasniyoj locality) were described by Apesteguía and Gallina (2011) and, along with the

Guará record, they could represent the oldest occurrence of ankylosaurs in South America. However, the Bolivian site was reconsidered as belonging to the Upper Triassic Ipaguazú Formation and the tracks assigned to cf. *Brachychirotherium* isp. (Apesteguía et al. 2021). Consequently, the ankylosaurian tracks from the RSCT locality represent the only Jurassic occurrence of these armored dinosaurs in South America and the oldest record of the group for the entire Gondwana.

3.3.4 Ornithopod Tracks

Ornithopods tracks are underrepresented in the Guará Formation (Fig. 3.8). They are small- to large-sized tridactyl tracks, generally almost as wide as long (length/ width ratio: 0.96–1.26) with low mesaxony. Different from theropod tracks, they are medio-laterally subsymmetrical. One trackway from the RSCT locality, composed of the tracks RSCT-4–RSCT-6 (Fig. 3.2c), and three isolated tracks (from the RSCT, RSGV and RSJR localities) were previously attributed to this group (Dentzien-Dias et al. 2007; 2008). The mentioned trackway is composed of relatively large footprints (43 cm long and 34 cm wide). The size of the isolated tracks is 15–25 cm long and 13–26 cm wide (Dentzien-Dias et al. 2007).

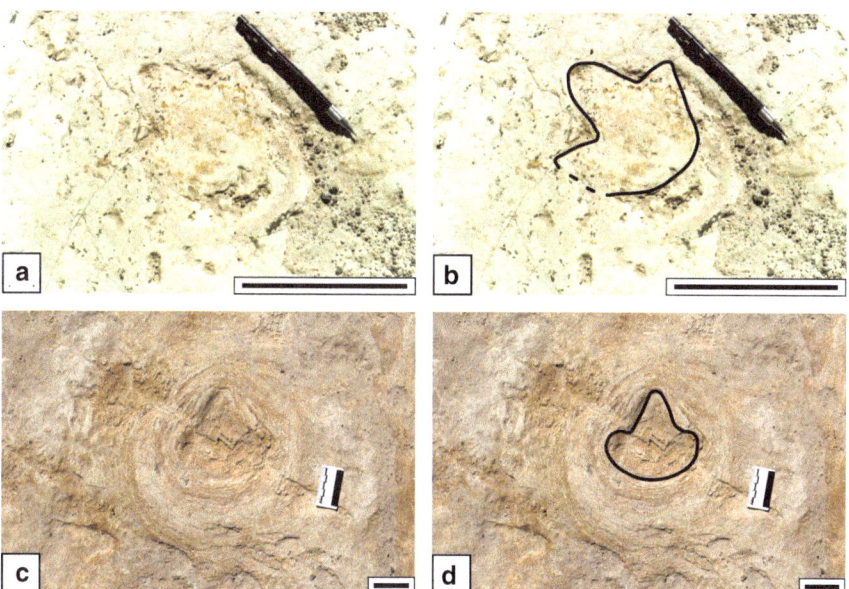

Fig. 3.8 Ornithopod tracks from the Guará Formation, Upper Jurassic of Brazil. **a, b** Track RSCT-3 from the Cerro Torneado (RSCT) locality; **c, d** Track RSCC-6 from the Cerro Caverá (RSCC) locality. Scale bars: 12 cm **a, b** and 5 cm **c, d**

A new ornithopod track (RSCC-6) was recently found in the Cerro Caverá locality (Fig. 3.8c, d). This track is tridactyl, mesaxonic, with rounded toes and a broad, U-shaped "heel". It is 13 cm long by 13 cm wide (length/width ratio: 1) and no claw marks can be seen in this track. The digit III is relatively longer than the other and the divarication angle is high, but it is not clear if these features can represent preservational artifacts or not (all ornithopod tracks found so far in the Guará-Batoví are interpreted as undertracks). In the Uruguayan Cañada del Ombú I locality, only one isolated ornithopod track was described (their Track C; Mesa and Perea 2015). It differs in many morphological aspects from the ornithopod tracks that occur in the Guará Formation of Brazil. The footprint described by Mesa and Perea (2015) is wider (20 cm) than longer (18 cm) (length/width ratio: 0.9), with a U-shaped heel. The digits are rounded and blunt, without claw marks (Mesa and Perea 2015). The Brazilian ornithopod tracks, on the other hand, have relatively well-developed digit III (Fig. 3.8), contrasting to the near rounded overall morphology of the Uruguayan one. Manus tracks were never found in this formation, neither in Brazil, nor in Uruguay.

The rounded morphology and relatively small size of the "heel" impression of the RSCT-3 and RSCC-6 tracks (Fig. 3.8) are very similar to the expected for the Díaz-Martínez' et al. (2015) Group 1 of ornithopod tracks, which includes the ichnogenus *Iguanodontipus* Sarjeant et al. (1998) and other indeterminate materials within the ichnofamily Iguanodontipodidae. However, as mentioned above, the relatively long digit III is not in accordance with the diagnosis of *Iguanodontipus* (which have digits II-IV of similar length). Any of the tracks known so far present the nearly parallel-oriented digits or the quadripartite configuration in the pes tracks, which are typical of some ornithopod ichnotaxa (such as *"Amblydactylus"*, *Caririchnium* and *Hadrosauropodus*, for example; Currie and Sarjeant 1979; Lockley et al. 2014) or the Groups 2 and 3 of Díaz-Martínez et al. (2015). New and better-preserved material is required for a more precise ichnotaxonomy. If confirmed, the Guará tracks would represent the oldest record of Iguanodontipodidae, filling the gap noticed by Díaz-Martínez et al. (2015).

The body fossil record of Ornithopoda in South America is still restricted to the Cretaceous (Coria 2016), precluding a proper synapomorphy-based or coincident track-trackmaker correlation. The rarity of ornithopod tracks in the Guará Formation is one more evidence that these ornithischians were not common elements in the South American Late Jurassic ecosystems.

3.3.5   The Enigmatic Structures from Novo Treviso

Originally, Cargnin et al. (2001) reported the presence of a "set of rounded structures of different diameters" at an outcrop of the Caturrita Formation in Novo Treviso (Fig. 3.9a–c). Some of these structures are cylindrical, with vertical walls, while others were described as 30 cm wide and 15 cm long tridactyl tracks (Cargnin et al. 2001). The structures occur in massive fine sandstones, often filled with grayish

Fig. 3.9 Cylindrical structures found in several geological units in Rio Grande do Sul state, Brazil. **a–c** structures found in the sandstones (Guará Formation) from Novo Treviso (Faxinal do Soturno municipality) and described by Cargnin et al. (2001) and Silva et al. (2007) as tetrapod tracks. Their genesis could be related to physical and/or chemical weathering of the matrix; **d** Structures found in sandstones (uncertain unit and age) from Santa Maria municipality; **e, f** structures found in sandstones (Botucatu Formation) from Rio da Ilha (Taquara municipality); **g, h** Structures found in weathered gneiss or granite from Canguçu municipality. Arrow in **h** indicates the concordant disposition between the cylindrical structure and the matrix plain of fault (white intermittent lines). Images **a–c** and **d** kindly ceded by Dr. Tânia L. Dutra and Dr. Cesar L. Schultz, respectively. Map legend: 1. Santa Maria, 2. Faxinal do Soturno, 3. Taquara, 4. Canguçu

muddy sandstones with high amounts of iron oxide. According to Cargnin et al. (2001), the most probable trackmakers would be synapsids because of their abundance in the presumed Caturrita Formation (Late Triassic in age). Posteriorly, these structures were restudied by Silva et al. (2007), who described them as rounded (in planview) or semi-cylindrical tubes (in cross section) with diameter of 20 to 50 cm; "finger-like projection" structures width between 16 and 30 cm and length between 12 and 27 cm; and structures composed of the combinations of the referred circular structures and kidney-shaped, with mean size of 35 cm. Most of these structures are more than 40 cm deep deformations in the strata. Silva et al. (2007) interpreted these as large tetrapod tracks which disrupted the substrate homogeneity by foot pression during the track registration. The finger-shaped projections would correspond to undertracks produced by the same trackmakers. Although morphological details were lacking, Silva et al. (2007) assigned these tracks to early-diverging sauropodomorphs.

Subsequently to the publication of the paper by Silva et al. (2007), Zerfass (2007) published a geological map of Novo Treviso region, indicating that the referred sandstones are part of the Guará Formation, leading Silva (2008) to attribute the structures to sauropod and theropod dinosaurs. On the other hand, a new analysis made by our team in Novo Treviso and other localities allowed us to propose a new hypothesis on the nature of these structures (Francischini et al. 2017). Approximately 50 structures similar to those from Novo Treviso were found in a locality at the Rio da Ilha district, Taquara municipality, northeast Rio Grande do Sul state (UTM 22 J 528374.02/6721031.64; Fig. 3.9e, f). The sandstones of Rio da Ilha site present large-sized cross-stratification and inverse grading, suggesting that the matrix is part of the eolian deposits of the Botucatu Formation (Lower Cretaceous). Similar cylindrical structures were also found in weathered gneiss or granites of the crystalline shield at Canguçu (UTM22J 346617.72/6514793.84; Fig. 3.9g, h). The presence of these structures in non-sedimentary rocks is strong evidence of their non-biogenic nature. The disposition of some of these structures along the fault plane corroborates this hypothesis. Besides, the lack of evidence of perturbation in the subjacent strata suggest that their genesis could be related to physical and/or chemical weathering of the matrix, producing depressions (weathering pits or pans) that are filled posteriorly or concomitantly to their generation. Weathering pits or pans are erosional features produced by differential erosion in outcrops of large horizontal exposition of rocks (e.g., Paradise 2022). Although it is not under the scope of this work to elucidate this issue, our observation indicates a non-biogenic nature for the Novo Treviso structures (Francischini et al. 2017). Systematic research is still needed for better understanding these enigmatic structures and, for this instance, we do not regard them as dinosaur footprints.

3.4 Paleogeographical Distribution of the Footprints

According to Bonaparte (1980) there are some affinities among the South American sauropods and theropods with those found on the Morrison Formation (Kimmeridgian–Tithonian of the USA). Consequently, Gondwana was not totally isolated during the Late Jurassic, and some degree of faunal interchange with Laurasia could perhaps occur. The Guará-Batoví would be, in this case, a witness of this paleobiogeographic framework, yielding both the occurrence of widespread dinosaur genera (such as *Torvosaurus* and *Ceratosaurus*; Soto et al. 2020a, b), and ichnological evidence of the presence of nodosaurid ankylosaurs (Francischini et al. 2018).

The absence or rarity of Jurassic faunas in latitudes lower than 35°S in South America was noticed by Bonaparte (1979, 1996), who interpreted this situation as a limitation imposed by a desert environment that restricted geographically and chronologically the occurrence of tetrapods. Even though, for Bonaparte (1979), such conditions are represented by the Botucatu desert (currently regarded to as Early Cretaceous; see Francischini et al. 2015), other geological units (e.g., the Pedreira Sandstone and possibly the Pirambóia Formation in Brazil) can attest to the remarkably arid conditions present in the low latitude regions of South America during the Jurassic. The possibility that this arid belt would work as a biogeographic barrier for fauna dispersal is still to be tested, but apparently it does not prevent some degree of faunal interchange. Actually, the occurrence of the Ichno-Group 2 (i.e., *Iberosauripus* or similar forms) of Cobos et al. (2014), represented by UFRGS-PV-0207-G, in the Guará-Batoví enlarges the geographical distribution of such group of theropod tracks and, consequently, of their trackmakers. Except for some occurrences in Morocco (Belvedere et al. 2019), this robust theropod track morphotype was so far restricted to the European Archipelago.

Regarding the temporal distribution of sauropod trackways, there is a dominance of narrow-gauge trackways in Jurassic strata, while wide-gauge trackways are more common in Cretaceous (Lockley et al. 1994a, 1994b; Wilson and Carrano 1999). According to Wilson and Carrano (1999), this transition of dominance occurred during the Kimmeridgian–Tithonian interval, suggesting an evolutionary trend on the increasing in size and changing in the position of the gravity center (migrating anteriorly), which implies in a larger participation of the forelimbs in the body support and in a more columnar posture. The presence of narrow- to very wide-gauge trackways in the Guará-Batoví unit reinforces the interpretation of its age as Kimmeridgian–Tithonian, interval in which a wide spectrum of trackway morphotypes are contemporaneous, suggesting that different groups of sauropod trackmakers could have inhabited the region in the Late Jurassic. This is in accordance with the high diversity of sauropod lineages found in some Upper Jurassic localities of Gondwana (e.g., Bonaparte 1996; Mannion et al. 2019).

The Guará-Batoví unit shares some dinosaur taxa (at least in a more inclusive perspective) with other Upper Jurassic units around the world (see a discussion in Francischini et al. 2018). Notably, megalosauroid theropods are well represented in the Guará-Batoví by *Torvosaurus* sp. teeth and the robust track UFRGS-PV-0207-G, but are also important components of the Kimmeridgian–Tithonian fauna recovered from the Tendaguru (Tanzania; Janensch 1920, but see Soto et al. 2020b), Morrison (USA; Galton and Jensen 1979), Villar del Arzobispo (Spain; Cobos et al. 2014; Malafaia et al. 2017), Alcobaça and Lourinhã (Portugal; e.g., Hendrickx and Mateus 2014) formations. This faunal similarity is reinforced by the occurrence of *Ceratosaurus* sp. and other high-level taxa in the Uruguayan Batoví Member, as well as *Metatetrapous*-like ankylosaur tracks in the Brazilian RSCT locality (Francischini et al. 2018).

3.5 Paleoenvironmental and Paleoclimatic Contexts

Although some of the sauropod tracks (likely those from the SLCP and the Cañada del Ombú I localities) from the Guará-Batoví unit are interpreted as true tracks, it is noteworthy the predominance of pedes in relation to manus. Several localities around the world are known by having sauropod pes-only or pes-dominated trackways (see Falkingham et al. 2012 for examples) and, among all hypotheses proposed to explain this phenomenon, the 'Goldilocks Effect' (Falkingham et al. 2011) is the most widely accepted. Accordingly, manus- or pes-only trackways would be results of the specific interaction of the position of the center of mass of the animal, its relative autopodium surface areas and the substrate shear strength. For Falkingham et al. (2011), sauropod pes-only or pes-dominated trackways would be produced by animals with a more posterior center of mass position (a plesiomorphic feature within Sauropodomorpha). Considering the Guará-Batoví record, it seems contradictory that the wide- to very wide-gauge sauropod trackways from RSCC, RSSA and SLCP are pes-dominated, because it would be expected that sauropods with massive construction of pectoral girdles and forelimbs (i.e., those closely associated with wide-gauge trackways) present an anteriorly-positioned center of mass (Henderson 2006) and, therefore, be more able to produce manus-only or manus-dominated trackways. On the other hand, pes-dominated trackways seem to be more commonly found in non-cohesive sandy substrates (Falkingham et al. 2012). Considering this and the difficulty in ichnotaxonomic assignation for sauropod trackways (but also for tracks produced by other groups), Francischini (2018) considered that the ichnological record of the Guará-Batoví unit is deeply affected by the 'Goldilocks Effect'.

Invertebrate traces can be observed in different outcrops and are assigned to *Taenidium barretti, Beaconites coronus* and *Arenicolites* isp. that are known to occur in wet environments (Nascimento et al. 2023). Conchostracans are also found, both in the near the Arroio do Beco locality of Brazil and in two localities of Uruguay (Ruta 5, km 396 and Cerro Batoví Dorado), where they are identified as the species *Orthestheria (Migransia) ferrandoi* (Yanbin et al. 2004). These occurrences suggest that the eolian deposits of the Guará-Batoví unit would present some degree of moisture and, consequently, the environment does not represent an arid and dry desert. The presence of unionoid bivalve (Martínez and Figueiras 1991; Martínez et al. 1993), fish (e.g., Soto and Perea 2010; Toriño et al. 2021), turtle (Perea et al. 2014) and pholidosaurid crocodylian (Fortier et al. 2011) remains in the (perennial?) fluvial beds of Uruguay reinforces this assumption.

Mesozoic laminated eolian interdune or sand-sheet are known to preserve many dinosaur tracks (Loope and Milàn 2016). In these laminated beds the undertracks often go deep in the sediment. According to Loope and Milàn (2016), the surface of damp interdunes can be lightly cemented by salt, so when the dinosaur stepped in the sand not only the sediment directly under the feet was deformed, but also a large area around the registered track. This preservation can be seen in the dinosaur tracks from the Middle Jurassic Entrada Formation (USA; Loope and Milàn 2016 and references therein), but also in the tracks from the Guará-Batoví (see for example, Figs. 3.5d–g, 3.7a–d, 3.8c, d). Even though the North and South American units were deposited under similar contexts, the presence of salts in the Guará-Batoví is still unknown. Modern erosion (most of which anthropic due to intense traffic of agricultural machinery or the activity of bulldozers on the dirty roads) of such track-bearing sandstones reveal the complex and somewhat distorted undertracks described here (Fig. 3.10). More data on the substrate properties of the Guará-Batoví tracksites are still needed for better understanding its impact on the preservation of the tracks and, consequently, the composition and diversity of the dinosaur ichnofauna of this Upper Jurassic unit (Fig. 3.11).

3.6 Conclusions

The Guará-Batoví unit from Brazil and Uruguay is one of the most important windows for the study of Late Jurassic environments and their biota in South America. For more than 20 years, our team has been prospecting, collecting and studying the fossil record of this unit in the Brazilian territory, with many findings being published. Here we provided a concise outlook about the dinosaur tracks and trackways from the Guará-Batoví unit, focusing on our findings in Rosário do Sul and Santana do Livramento municipalities of the Rio Grande do Sul state. Saurischian (Sauropoda and Theropoda) footprints are dominant in the eolian sand sheet deposits of the Guará-Batoví, with the sauropod lineage represented only by its tracks and trackways. Ornithischian tracks are rarer, but at least two lineages (Ornithopoda and Ankylosauria) occur in the Guará-Batoví strata.

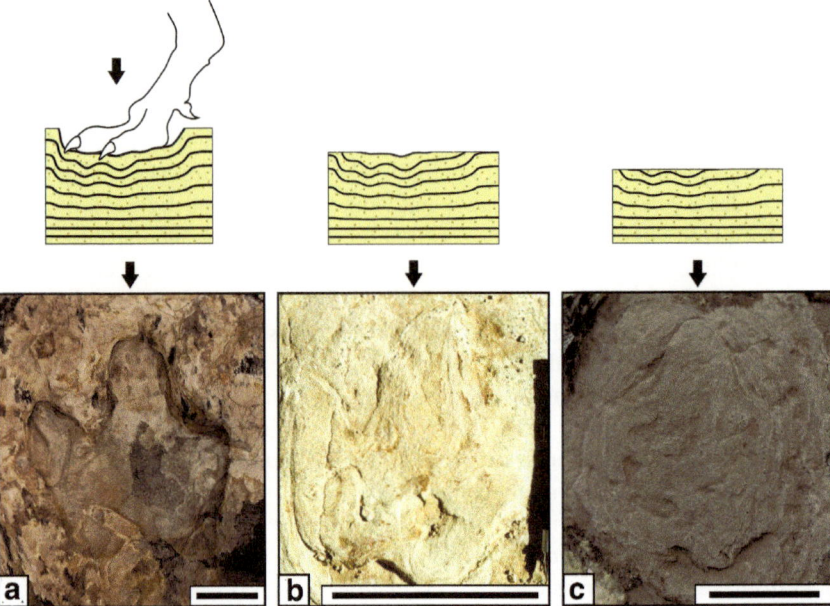

Fig. 3.10 Registration and erosion of the dinosaur undertracks of the Guará-Batoví unit, evidenced by theropod tracks. **a** track registration and the true track UFRGS-PV-0207-G, from the Sanga do Torneado locality. Scale bar: 5 cm; **b** partially eroded track with some details still preserved (unlabeled track from the Cerro Torneado locality). Scale bar: 12 cm; **c** very eroded track with the lost of morphological detail, represented by UFRGS-PV-0206-G, from the Cerro Torneado locality. Scale bar: 7 cm

The Guará Formation was formalized in the early 2000's and since then, the knowledge about the Late Jurassic dinosaur ichnofauna was notably expanded. Up to this point, sauropod, theropod, ankylosaur and ornithopod tracks were found in Brazil and, together with the body fossils from the Uruguayan territory, comprise an important dinosaur record for South America, often compared to the so-called Morrison, Lourinhã and Tendaguru strata. New data are still to be published and future contributions will shed more light on the composition, diversity and preservation of the Guará-Batoví dinosaurs. Meanwhile, activities on divulgation and popularization of the Dinosaur Ichnology are at full swing.

Fig. 3.11 The Late Jurassic Guará/Batoví paleoenvironment and the probable producers of the theropod and ankylosaur dinosaur tracks. In this scene, a megalosaurid is defending its meal (a nodosaurid carcass) from the assault of a ceratosaurid. Artwork by Guilherme Gehr

References

Amarante FB, Scherer CMS, Aguilar CAG, Reis ADR, Mesa V, Soto M (2019) Fluvial-eolian deposits of the Tacuarembó Formation (Norte Basin - Uruguay): depositional models and stratigraphic succession. J South Am Earth Sci 90:355–376

Apesteguía S, Gallina PA (2011) Tunasniyoj, a dinosaur tracksite from the Jurassic-Cretaceous boundary of Bolivia. An Acad Bras Ciênc 83(1):267–277

Apesteguía S, Riguetti F, Citton P, Veiga GD, Poiré DG, de Valais S, Zacarías GG (2021) The Ruditayoj-Tunasniyoj fossil area (Chuquisaca, Bolivia): a triassic chirotheriid megatracksite and reinterpretation of purported thyreophoran tracks. Hist Biol 33(11):2883–2896

Bandeira KL, Brum AS, Pêgas RV, Souza LG, Pereira PV, Pinheiro AE (2021) The first Jurassic theropod from the Sergi formation, Jatobá Basin, Brazil. An Acad Bras Ciênc 93:e20201557

Barboni R, Dutra TL (2013) New "flower" and leaves of Benettitales from southern Brazil and their implication in the age of the lower Mesozoic deposits. Ameghiniana 50(1):14–32

Belvedere M, Mietto P, Ishigaki S (2010) A Late Jurassic diverse ichnocoenosis from the siliciclastic Iouaridène Formation (Central High Atlas, Morocco). Geological Quarterly 54(3):367–380

Belvedere M, Castanera D, Meyer CA, Marty D, Mateus O, Silva BC, Santos VF, Cobos A (2019) Late Jurassic globetrotters compared: a closer look at large and giant theropod tracks of North Africa and Europe. J Afr Earth Sc 158:103547

Bonaparte JF (1979) Faunas y paleobiogeografía de los tetrápodes mesozoicos de America del Sur. Ameghiniana 16(3–4):217–238

Bonaparte JF (1996) Late Jurassic vertebrate communities of Eastern and Western Gondwana. GeoRes For 1–2:427–432

Bonaparte JF (1980) Jurassic tetrapods from South America and dispersal routes. In: Jacobs L (ed) Aspects of vertebrate history. Museum of Northern Arizona Press, pp 73–98.

Bonaparte JF (1981) Inventario de los vertebrados jurássicos de América del Sur. In: Volkheimer W, Musacchio EA (eds) Cuencas sedimentarias del Jurásico y Cretácico de América del Sur, Buenos Aires: Comité Sudamericano del Jurásico y Cretácico, p 661–683.

Cargnin D, Ferigolo J, Ribeiro AM, Negri FR, Carvalho IS (2001) Pegadas fósseis do Triássico da Bacia do Paraná (Grupo Rosário do Sul), Rio Grande do Sul, Brasil. Rev Brasil de Paleontol 2:71–72

Carrano MT, Wilson JA (2001) Taxon distribution and the tetrapod track record. Paleobiology 27(3):564–582

Casamiquela RM, Fasola A (1968) Sobre pisadas de dinosaurios del Cretácico Inferior de Colchagua (Chile). Publ del Depart de Geol Univ de Chile 30:1–24

Casamiquela RM (1964). Estudios icnológicos. problemas y métodos de la icnología con aplicación al estudio de pisadas mesozoicas (Reptilia, Mammalia) de la Patagonia, Buenos Aires: Colegio Industrial Pío IX

Christofoletti B, Peixoto BCPM, Warren LV, Inglez L, Fernandes MA, Alessandretti L, Perinotto JAJ, Simões MG, Assine ML (2021) Dinos among dunes: dinoturbation in the Pirambóia Formation (Paraná Basin), São Paulo State and comments on cross-section tracks. J S Am Earth Sci 109:103252

Cobos A, Lockley MG, Gascó F, Royo-Torres R, Alcalá L (2014) Megatheropods as apex predators in the typically Jurassic ecosystems of the Villar del Arzobispo Formation (Iberian Range, Spain). Palaeogeogr Palaeoclimatol Palaeoecol 399:31–41

Colbert EH (1977) Gondwana vertebrates. In: Laskar B, Raja Rao CS (eds) Fourth International Gondwana Symposium: papers. Delhi, India, pp 1–18

Coria RA (2016) An overview of the ornithischian dinosaurs from Argentina. Contribuciones Del MACN 6:109–117

Coria RA, Carabajal AP (2004) Nuevas huellas de theropoda (Dinosauria: Saurischia) del Jurásico de Patagonia, Argentina. Ameghiniana 41(3):393–398

Currie PJ, Sarjeant WAS (1979) Lower Cretaceous dinosaur footprints from the Peace River Canyon, British Columbia, Canada. Palaeogeogr Palaeoclimatol Palaeoecol 28:103–115

Deiques D (2023) Novos registros de pegadas de dinossauros para a Formação Guará (Jurássico Superior da Bacia do Paraná), Rio Grande do Sul, Brasil. Unpublished Master thesis, Universidade Federal do Rio Grande do Sul, Porto Alegre, Brazil

D'Emic MD (2012) The early evolution of titanosauriform sauropod dinosaurs. Zool J Linn Soc 166:624–671

Dentzien-Dias PC, Schultz CL, Scherer CM, Lavina ELC (2007) The trace fossil record from the Guará Formation (Upper Jurassic?), southern Brazil. Arquivos do Museu Nacional 65(4):585–600

Dentzien-Dias PC, Schultz CL, Bertoni-Machado C (2008) Taphonomy and paleoecology inferences of vertebrate ichnofossils from Guará Formation (Upper Jurassic), southern Brazil. J S Am Earth Sci 25:196–202

Dentzien-Dias P, Figueiredo AEQ, Mesa V, Perea D, Schultz CL (2012) Vertebrate footprints and burrows from the Upper Jurassic of Brazil and Uruguay. In Netto RG, Carmona NB, Tognoli FMW (eds) Ichnology of Latin America: selected papers, Porto Alegre: Monografias da Sociedade Brasileira de Paleontologia, vol 2, pp 129–140

Díaz-Martínez I, Pereda-Suberbiola X, Pérez-Lorente F, Canudo JI (2015) Ichnotaxonomic review of large ornithopod dinosaur tracks: temporal and geographical implications. PLoS ONE 10(2):e0115477

Díaz-Martínez I, González SN, de Valais S (2017) Dinosaur footprints in the Early Jurassic of Patagonia (Marifil Volcanic Complex, Argentina): biochronological and palaeobiogeographical inferences. Geol Mag 154(4):914–922

Dutra TL, Crisafulli A (2002) Primeiro registro de uma associação de lenhos e ramos de coníferas em níveis do final do Triássico Superior no sul do Brasil (Bacia do Paraná, Formação Caturrita). In: VIII Congreso Argentino de Paleontología y Bioestratigrafía, Corrientes: Resúmenes, p 32

Dutuit J-M, Ouazzou A (1980) Découverte d'une piste de dinosaure sauropode sur le site d'empreintes de Demnat (Haut-Atlas marocain). Mémoires de la Société Géologique de France, Nouvelle Série 139:95–102

Falkingham PL, Bates KT, Margetts L, Manning PL (2011) The 'Goldilocks' effect: preservation bias in vertebrate track assemblages. J R Soc Interface 8:1142–1154

Falkingham PL, Bates KT, Mannion PD (2012) Temporal and palaeoenvironmental distribution of manus- and pes-dominated sauropod trackways. J Geol Soc Lond 169:365–370

Farlow JO (1992) Sauropod tracks and trackmakers: integrating the ichnological and skeletal records. Zubía 10:89–138

Farlow JO, Pittman JG, Hawthorne JM (1989) *Brontopodus birdi*, Lower Cretaceous sauropod footprints from the US Gulf Coastal Plain. In: Gillette DD, Lockley GM (eds) Dinosaur tracks and traces. Cambridge University Press, Cambridge, pp 371–394

Fortier D, Perea D, Schultz C (2011) Redescription and phylogenetic relationships of *Meridiosaurus vallisparadisi*, a pholidosaurid from the Late Jurassic of Uruguay. Zool J Linn Soc 163:S257–S272

Francischini H, Dentzien-Dias PC, Fernandes MA, Schultz CL (2015) Dinosaur ichnofauna of the Upper Jurassic/Lower Cretaceous of the Paraná Basin (Brazil and Uruguay). J S Am Earth Sci 63:180–190

Francischini H, Frank HT, Dentzien-Dias P, Schultz CL (2017) Pegadas de dinossauros ou pias de sacrifício? Sobre a origem das estruturas enigmáticas de Novo Treviso (Rio Grande do Sul). Paleontologia Em Destaque 33:116

Francischini H, Sales MAF, Dentzien-Dias P, Schultz CL (2018) The presence of ankylosaur tracks in the Guará Formation (Brazil) and remarks on the spatial and temporal distribution of Late Jurassic dinosaurs. Ichnos 25:177–191

Francischini H (2018) Paleoicnologia de vertebrados da Bacia do Paraná, com ênfase nos registros dos depósitos continentais. Unpublished PhD thesis, Universidade Federal do Rio Grande do Sul, Porto Alegre, Brazil

Galton PM, Jensen JA (1979) A new large theropod dinosaur from the Upper Jurassic of Colorado. Brigham Young Univ Geol Stud 26:1–12

García RA, Salgado L, Fernández MS, Cerda IA, Carabajal AP, Otero A, Coria RA, Fiorelli LE (2015) Paleobiology of titanosaurs: reproduction, development, histology, pneumaticity, locomotion and neuroanatomy from the South American fossil record. Ameghiniana 52(1):29–68

González-Riga BJ, Calvo JO (2009) A new wide-gauge sauropod track site from the Late Cretaceous of Mendoza, Neuquén Basin, Argentina. Palaeontology 52(3):631–640

Henderson DM (2006) Burly gaits: centers of mass, stability, and the trackways of sauropod dinosaurs. J Vertebr Paleontol 26(4):907–921

Hendricks A (1981) Die Saurierfährte von Münchehagen bei Rehburg-Loccum (NW-Deutschland). Abhandlung Landesmuseum Naturkunde Münster 43:1–22

Hendrickx C, Mateus O (2014) *Torvosaurus gurneyi* n. sp., the largest terrestrial predator from Europe, and a proposed terminology of the maxilla anatomy in nonavian theropods. PLoS ONE 9(3):e88905

Hups K, Lockley M, Foster J, Gierlinski G (2008) The first ankylosaur track from the Jurassic. J Vertebr Paleontol 28(3):94A

Janensch W (1920) Über *Elaphrosaurus bambergi* und die megalosaurier aus den Tendaguru-Schichten Deutsch-Ostafrikas. Sitzungsberichte der Gesellschaft Naturforschender Freunde zu Berlin. pp 225–235

Jenisch AG, Lehn I, Gallego OF, Monferran MD, Horodyski RS, Faccini UF (2017) Stratigraphic distribution, taphonomy and paleoenvironments of Spinicaudata in the Triassic and Jurassic of the Paraná Basin. J S Am Earth Sci 80:569–588

Lavina EL, Azevedo SAK, Barberena MC, Ferrando L (1985) Contribuição à estratigrafia e paleoambiente da Formação Tacuarembó no noroeste do Uruguai. Pesquisas 17:5–23

Leonardi G (1984) Le impronte fossili di dinosauri. In Bonaparte JF, Kielan-Jaworowska Z, Leonardi G, Taquet P, Casini G et al. (eds) Sulle Orme dei Dinosauri. Venice, Erizzo, pp 165–186

Leonardi G (1994) Annotated Atlas of South America tetrapod footprints (Devonian to Holocene) with an appendix on Mexico and Central America. Companhia de Pesquisa de Recursos Minerais, Brasília

Lockley MG (2009) New perspectives on morphological variation in tridactyl footprints: clues to widespread convergence in developmental dynamics. Geol Quart 53(4):415–432

Lockley MG, Hunt AP (1998) A probable stegosaur track from the Morrison formation of Utah. Mod Geol 23:331–342

Lockley MG, Meyer C (2000) Dinosaur tracks and other fossil footprints of Europe. Columbia University Press, New York

Lockley MG, Houck KJ, Prince NK (1986) North America's largest dinosaur trackway site: implications for Morrison Formation paleoecology. Geol Soc Am Bull 97:1163–1176

Lockley MG, Farlow JO, Meyer CA (1994a) *Brontopodus* and *Parabrontopodus* ichnogen. nov. and the significance of wide- and narrow-gauge sauropod trackways. Gaia 10:135–145

Lockley MG, Hunt AP, Meyer CA (1994b) Vertebrate tracks and the ichnofacies concept: implications for paleoecology and palichnostratigraphy. In: Donovan S (ed) The paleobiology of trace fossils. John Wiley & Sons, Chichester, pp 241–268

Lockley MG, Xing L, Lockwood JAF, Pond S (2014) A review of large Cretaceous ornithopod tracks, with special reference to their ichnotaxonomy. Biol J Lin Soc 113:721–736

Lockley MG, Kim KS, Lim JD (2021) *Tetrapodosaurus* trackways from the Cretaceous of Colorado and Canada: observations on variability in ankylosaurian gaits. N M Mus Nat Hist Sci Bull 82:227–235

Loope DB, Milàn J (2016) Dinosaur tracks in eolian strata: New insights into track formation, walking kinetics and trackmaker behavior. In: Falkingham PL, Marty D, Richter A (eds) Dinosaur tracks: the next steps. Indiana University Press, Bloomington, pp 358–364.

Malafaia E, Mocho P, Escaso F, Ortega F (2017) New data on the anatomy of *Torvosaurus* and other remains of megalosauroid (Dinosauria, Theropoda) from the Upper Jurassic of Portugal. J Iber Geol 43:33–59

Mannion PD, Upchurch P, Schwarz D, Wings O (2019) Taxonomic affinities of the putative titanosaurs from the Late Jurassic Tendaguru Formation of Tanzania: phylogenetic and
biogeographic implications for eusauropod dinosaur evolution. Zool J Linn Soc 185:784–909

Martínez S, Figueiras A (1991) Two new Mesozoic species of *Diplodon* (Bivalvia: Hyriidae) from
Uruguay. Walkerana 5(14):217–223

Martínez S, Figueiras A, Silva JS (1993) A new unionoid (Mollusca, Bivalvia) from the Tacuarembó
Formation (Upper Triassic-Upper Jurassic), Uruguay. J Paleontol 67(6):962–965

Marty D, Belvedere M, Meyer CA, Mietto P, Paratte G, Lovis C, Thüring B (2010) Comparative
analysis of Late Jurassic sauropod trackways from the Jura Mountains (NW Switzerland) and
the central High Atlas Mountains (Morocco): implications for sauropod ichnotaxonomy. Hist
Biol 22(1–3):109–133

Marty D, Belvedere M, Razzolini NL, Lockley MG, Paratte G, Cattin M, Lovis C, Meyer CA (2017)
The tracks of giant theropods (Jurabrontes curtedulensis ichnogen. & ichnosp. nov.) from the
Late Jurassic of NW Switzerland: palaeoecological & palaeogeographical implications. Hist
Biol 30(7):928–956

McCrea RT, Lockley MG, Meyer CA (2001) Global distribution of purported ankylosaur track occurrences. In: Carpenter K (ed) The armored dinosaurs. Indiana University Press, Bloomington, pp
413–454

McCrea RT (2000) Vertebrate palaeoichnology of the Lower Cretaceous (lower Albian) Gates
Formation of Alberta. PhD thesis, University of Saskatchewan, Saskatoon, Canada

Mesa V, Perea D (2015) First record of theropod and ornithopod tracks and detailed description
of sauropod trackways from the Tacuarembó Formation (Late Jurassic-?Early Cretaceous) of
Uruguay. Ichnos 22:109–121

Moreno K, Benton MJ (2005) Occurrence of sauropod dinosaur tracks in the Upper Jurassic of
Chile (redescription of *Iguanodonichnus frenki*). J S Am Earth Sci 20:253–257

Moreno K, Pino M (2002) Huellas de dinosaurios en la Formación Baños del Flaco (Titoniano-
Jurásico Superior), VI Región, Chile: paleoetología y paleoambiente. Rev Geol Chile 29(2):191–
206

Moreno K, Blanco N, Tomlinson A (2004) Nuevas huellas de dinosaurios del Jurásico Superior en
el norte de Chile. Ameghiniana 41(4):535–544

Moreno-Sánchez M, Cruz AG, Tapias JG (2011) Reporte de huellas de dinosaurios en el Santuario
de Fauna y Flora de Iguaque, en cercanías de Chíquiza (Boyacá, Colombia). Boletín De Geol
33(2):107–118

Nascimento DL, Netto RG, Batezelli A, Ladeira FSB, Sedorko D (2023) *Taenidium barretti* ichno-
fabric and rainfall seasonality: insights into dryland suites of Scoyenia ichnofacies. J Palaeogeogr
12:28–49

Netto RG (1989) Paleoicnologia das sequências eólicas sotopostas à Formação Botucatu no Rio
Grande do Sul. Acta Geol Leopoldensia 12:31–44

Nopcsa F (1923) Die familien der reptilien. Fortschritte der Geol Palaeontol 2:1–210

Oliveira LM, Oliveira EV, Fambrini GL (2022) The first dinosaur from the Jurassic Aliança
Formation of northeastern Brazil, west Gondwana: a basal Neotheropoda and its age and
paleobiogeographical significance. J South Am Earth Sci 116:103835

Paradise TR (2022) chap. 3 Tafoni and other rock basins. In: Shroder JF (ed), Treatise on
geomorphology, vol 12. Elsevier, pp 204–220

Perea D, Soto M, Veroslavsky G, Martínez S, Ubilla M (2009) A Late Jurassic fossil assemblage
in Gondwana: Biostratigraphy and correlations of the Tacuarembó Formation, Parana Basin,
Uruguay. J S Am Earth Sci 28:168–179

Perea D, Martínez S (2004) La Formación Tacuarembó. Su fauna y su edad Jurásico Tardío-Cretácico
Temprano. In: Veroslavsky G, Ubilla M, Martínez S (eds) Cuencas Sedimentarias de Uruguay.
DIRAC, Montevideo, pp 101–112

Perea D, Soto M, Sterli J, Mesa V, Toriño P, Roland G, Silva J (2014) *Tacuarembemys kusterae*, gen.
et sp. nov., a new Late Jurassic-?earliest Cretaceous continental turtle from Western Gondwana.
J Vertebrate Paleontol 34(6):1329–1341

90 H. Francischini et al.

Pires EF, Guerra-Sommer M (2004) *Sommerxylon spiralosus* from Upper Triassic in southernmost Paraná Basin (Brazil): a new taxon with taxacean affinity. An Acad Bras Ciênc 76(3):595–609
Rauhut OWM, Piñuela L, Castanera D, García-Ramos J-C, Cela IS (2018) The largest European theropod dinosaurs: remains of a gigantic megalosaurid and giant theropod tracks from the Kimmeridgian of Asturias, Spain. Peerj 6:e4963
Reis AD, Scherer CMS, Amarante FB, Rossetti MMM, Kifumbi C, Souza EG, Ferronatto JPF, Owen A (2019) Sedimentology of the proximal portion of a large-scale, Upper Jurassic fluvial-aeolian system in Paraná Basin, southwestern Gondwana. J S Am Earth Sci 95:102248
Rohn R, Dutra TL, Cabral MVB (2014) Conchostráceos como evidência de níveis jurássicos na Formação Caturrita, Faxinal do Soturno, Rio Grande do Sul, Brasil, geologia USP. Série Científ 14(1):3–20
Romano M, Whyte MA, Jackson SJ (2007) Trackway ratio: a new look at trackway gauge in the analysis of quadrupedal dinosaur trackways and its implications for ichnotaxonomy. Ichnos 14:257–270
Romilio A, Salisbury SW, Jannel A (2021) Footprints of large theropod dinosaurs in the Middle-Upper Jurassic (lower Callovian–lower Tithonian) Walloon Coal Measures of southern Queensland, Australia. Hist Biol 33(10):2135–2146
Rubilar D, Moreno L, Blanco N (2000) Huellas de dinosaurios ornitópodos en la Formación Chacarilla (Jurásico Superior-Cretácico Inferior), I Región de Tarapacá, Chile. In: IX Congreso Geológico Chileno, Actas, pp 550–554
Salinas P, Marshall LG, Sepúlveda P (1991) Vertebrados continentales del Paleozoico y Mesozoico de Chile. In: VI Congreso Geológico Chileno, Antofagasta, Actas, pp 310–313
Santos VF, Lockley MG, Meyer CA, Carvalho J, Galopim de Carvalho AM, Moratalla JJ (1994) A new sauropod tracksite from the Middle Jurassic of Portugal. Gaia 10:5–13
Sarjeant WAS, Delair JB, Lockley MG (1998) The footprints of Iguanodon: a history and taxonomic study. Ichnos 6(3):183–202
Scherer CMS, Lavina EL (2005) Sedimentary cycles and facies architecture of aeolian-fluvial strata of the Upper Jurassic Guará Formation, southern Brazil. Sedimentology 52:1323–1341
Schultz CL, Scherer CMS, Lavina ELC (2002) Dinosaur's footprints from Guará Formation (Upper Jurassic), Paraná Basin, Southern Brazil. In: VIII Congreso Argentino de Paleontología y Bioestratigrafía, Corrientes: Resúmenes, p 64
Silva RC, Carvalho IS, Schwanke C (2007) Vertebrate dinoturbation from the Caturrita Formation (Late Triassic, Paraná Basin), Rio Grande do Sul State, Brazil. Gondwana Res 11:303–310
Silva RC, Barboni R, Dutra T, Godoy MM, Binotto RB (2012) Footprints of large theropod dinosaurs and implications on the age of Triassic biotas from Southern Brazil. J S Am Earth Sci 39:16–23
Silva FP, Caetano-Chang MR, Chang HK (2023) Stratigraphy of the middle Triassic–lower Cretaceous succession in the Paraná and Uruguayan Chaco-Paraná Basins–an overview based on subsurface data. J S Am Earth Sci 130:104567
Silva RC (2008) Análise das pegadas de tetrápodes do Grupo Rosário do Sul (Triássico, Bacia do Paraná). Unpublished PhD thesis, Universidade Federal do Rio de Janeiro, Rio de Janeiro, Brazil
Soto M, Perea D (2008) A ceratosaurid (Dinosauria, Theropoda) from the Late Jurassic-Early Cretaceous of Uruguay. J Vertebr Paleontol 28(2):439–444
Soto M, Perea D (2010) Late Jurassic lungfishes (Dipnoi) from Uruguay, with comments on the systematics of Gondwanan ceratodontiforms. J Vertebr Paleontol 30(4):1049–1058
Soto M, Toriño P, Perea D (2020a) *Ceratosaurus* (Theropoda, Ceratosauria) teeth from the Tacuarembó Formation (Late Jurassic, Uruguay). J S Am Earth Sci 103:102781
Soto M, Toriño P, Perea D (2020b) A large sized megalosaurid (Theropoda, Tetanurae) from the Late Jurassic of Uruguay and Tanzania. J S Am Earth Sci 98:102458
Soto M, Montenegro F, Toriño P, Mesa V, Perea D (2021) A new ctenochasmatid (Pterosauria, Pterodactyloidea) from the late Jurassic of Uruguay. J S Am Earth Sci 111:103472

Soto M, Delcourt R, Langer MC, Perea D (2022) The first record of abelisaurid (Theropoda: Ceratosauria) from Uruguay (Late Jurassic, Tacuarembó Formation). Hist Biol https://doi.org/10.1080/08912963.2022.2140425

Sternberg CM (1932) Dinosaur tracks from Peace River, British Columbia. Natl Museum Canada Annu Rep 1930:59–85

Thulborn RA (1989) The gaits of dinosaurs. In: Gillette DD, Lockley MG (eds) Dinosaur tracks and traces. Cambridge University Press, Cambridge, pp 39–50

Toriño P, Soto M, Perea D, Carvalho MSS (2021) New findings of the coelacanth *Mawsonia* Woodward (Actinistia, Latimerioidei) from the Late Jurassic–Early Cretaceous of Uruguay: novel anatomical and taxonomic considerations and an emended diagnosis for the genus. J S Am Earth Sci 107:103054

Whyte MA, Romano M (2001) Probable stegosaurian dinosaur tracks from the Saltwick Formation (Middle Jurassic) of Yorkshire, England. Proc Geol Assoc 112:45–54

Wilson JA, Carrano MT (1999) Titanosaurs and the origin of "wide-gauge" trackways: a biomechanical and systematic perspective on sauropod locomotion. Paleobiology 25(2):252–267

Yanbin S, Gallego OF, Martinez S (2004) The conchostracan subgenus Orthestheria (Migransia) from the Tacuarembó Formation (Late Jurassic–?Early Cretaceous, Uruguay) with notes on its geological age. J S Am Earth Sci 16:615–622

Zerfass H (2007) Geologia da Folha Agudo, SH.22-V-C-V, escala 1:100.000. Serviço Geológico do Brasil—CPRM. https://rigeo.sgb.gov.br/handle/doc/10287

Chapter 4
Desert Cretaceous Dinosaurs: The Botucatu Paleodesert and the Footprints Across the Dunes

Marcelo Adorna Fernandes⑩, Luciana Bueno dos Reis Fernandes⑩, and Júlia Beatrice Schutzer⑩

4.1 Introduction

In Brazil, there are many ichnofossils-bearing sites with tetrapod footprints, including sometimes even tracks of dinosaurs. Among the main localities with dinosaur footprints, one stands out, the region of the Araraquara municipality, located in the state of São Paulo. In this locality, during the Mesozoic Era, there was a paleodesert where today, alongside invertebrate, mammal and probably lacertoid tracks, there can be found small- and medium-sized theropod and ornithopod dinosaur footprints. Another important area with dinosaur tracks is around Nioaque, in the Mato Grosso do Sul State.

The dinosaur footprints are common in the sandstones of the Botucatu Formation (Paraná Basin). In situ, levels with dinosaur tracks in this formation were found in some tracksites, mainly in the São Paulo State. Ex-situ, in the sidewalks and other artificial surfaces, dinosaur trackways from the Botucatu Formation were observed and often collected and housed in institutions, from many other towns and other locations, in southeastern and southern Brazil and Paraguay (Leonardi 1994; Leonardi et al. 2007).

The Araraquara outcrops comprise one of the richest tetrapod ichnosites of South America's Lower Cretaceous, in the Botucatu Formation. This unit is represented by reddish sandstones with low-angle (~30°) cross-stratification, medium to large, is interpreted as an extensive dune field implanted on the ancient Gondwanan continent (Scherer 2000). Its age is Early Cretaceous (Berriasian-Barremian; Brückmann et al. 2014).

M. A. Fernandes (✉) · L. B. dos Reis Fernandes · J. B. Schutzer
Departamento de Ecologia E Biologia Evolutiva, Universidade Federal de São Carlos, Estado de São Paulo, 13565-905, Rodovia Washington Luís, Km 235, São Carlos SP-310, Brazil
e-mail: mafernandes@ufscar.br; lucianafernandes@ufscar.br

The first record of a tetrapod trackway for the Botucatu Formation was observed in 1911, by Joviano Augusto do Amaral Pacheco, a mining engineer (Pacheco 1913; Leonardi and Sarjeant 1986). Friedrich Von Huene (1931) cited this record as the first evidence of tetrapod tracks discovered in Latin America.

Giuseppe Leonardi, in 1976, reconnoitered the origin of the flagstone of Pacheco in the quarry Santa Águeda, in the district of Ouro (Araraquara county) and then he discovered in this region numerous other quarries with fossil footprints (mainly mammal and dinosaur tracks), and also in the paving of streets of the city (Leonardi 1980). Later, he and others discovered a score of other ichnosites in the same formation, all around the Paraná Basin (Leonardi 1977, 1980, 1994). Leonardi (1981) described the first ichnogenus denominated *Brasilichnium elusivum* for the Botucatu Formation and attributed those footprints to a small mammal. Fernandes and Carvalho (2008) performed a diagnostic revision of this same ichnospecies, altering the number of functional digits from five to four.

Two new forms of Mammaliaformes footprints for the Botucatu Formation, *Brasilichnium saltatorium* and *Aracoaraichnium leonardii* were described by Buck et al. (2017a, b). Lacertoid tracks were described by Buck et al. (2022) and attributed to possibly Squamata and/or Sphenodontia, suggesting the occurrence of a group hitherto unknown to the Botucatu Formation.

Another ichnotaxon, *Brasilichnium anaiti* D'Orazi Porchetti et al. 2017, was established again for tracks of early mammals. Numerous other locations with the fossil footprints of mammals and especially of bipedal dinosaur tracks in the Botucatu sandstones were discovered in Brazil, on the western flank of the Paraná basin, by Manes et al. (2021), in Mato Grosso do Sul State (MS), in the locality Nioaque and surrounding region. There, sediments previously assigned to the glacial Aquidauana Formation (Permian) were reinterpreted, also and mainly because of the presence of dinosaur tracks, as part of the basal section of the Botucatu Formation. Floodplain and residual channel deposits are among the successions that are believed to represent the lower half of this formation's sequence in the studied area. The ichnofossils that have been found include both isolated tracks of Theropoda and Ornithopoda. This is one of the cases in which the discovery of clear dinosaur footprints led to the raising, in the stratigraphical column, stratigraphic units that were considered to be Paleozoic, at the Mesozoic.

Regarding invertebrate ichnofossils registered for the Botucatu Formation, there is the occurrence of *Taenidium* type tracks attributed to Coleoptera (Fernandes et al. 1988) and others attributed to epistratal invertebrates, possibly insects of the ichnospecies *Paleohelcura araraquarensis* (Peixoto et al. 2020).

During the recovering of sandstone slabs at the São Bento quarry (21° 49′ 03.4″ S e 48° 04′ 22.9″ W), in Araraquara (and in other similar ichnosites), between the years 1976 to 2005, it was possible to distinguish at least four morphotypes for dinosaur footprints, with occurrences of Theropoda and Ornithopoda. One of these was attributed to a medium-sized ornithopod (Fernandes and Carvalho 2007). This trackway, with 5 hind-footprints (see Fig. 1.15A in Chap. 1), corresponds to the largest footprints encountered till this day for this geological unit. More recently, Leonardi et al. (2024) described the ichnogenus and ichnospecies *Farlowichnus rapidus*, attributed to a small Theropoda of which a score of trackways, with up to 7 footprints were found. The trackmaker of this kind of track was considered by Leonardi et al. (2024) related to a Gondwanan theropod family, possibly the ancestral theropods of clades such as noasaurids or velocisaurids. There were also registered two singular occurrences of liquid extrusion associated with dinosaurs (Fernandes et al. 2004). These were denominated, at first, as urolites and later renaming them to micturalites was suggested by Fernandes (2020).

Francischini et al. (2015) suggested that, for the Botucatu Formation, due to the aridity, the dinosaur fauna would be represented basically by Theropoda and Ornithopoda footprints of reduced sizes, when compared to other Mesozoic units of the world. The same climatic condition would be related to the lack of Sauropod footprints.

In the face of the diversity of footprint forms found in the sandstones of the Botucatu Formation, the desert paleoenvironment provided conditions to sustain a community of organisms of different trophic levels, from detritivores to apex predators, such as Theropoda. This ichnofauna of diverse dinosaurs and other animals is very well represented by the footprints and other traces found at the São Bento quarry in Araraquara.

4.2 Geological Context

The Late Jurassic to Early Cretaceous Sequence (Supersequence Gondwana III, Milani 1997; Milani et al. 2007) includes in the Paraná Basin the aeolian deposits of the Botucatu Formation and the basalt flows of the Serra Geral Formation. These formations correspond to the São Bento Group (Schneider et al. 1974). While the basalts of the Serra Geral Formation cover much of the basin, the Botucatu Formation crops out (about 4,500 km; Leonardi et al. 2024) mainly on the basin's borders (Fig. 4.1) and in the central region of the basin it is in subsurface. This unit occupies an area of about 1,300,000 km^2, covering much of southern and southeastern Brazil and marginally, portions of the territory of Argentina, Uruguay, and Paraguay (Milani et al. 2007). Desert conditions have existed since the beginning of fissural volcanism (approximately 134 Ma) associated with the rupture of Gondwana, sometimes covering the pre-existing aeolian landscape (Assine et al. 2004; Carvalho 2022). This volcanism is related to the Paraná-Etendeka province, with its magmatic rocks with its flows or volcanic (basic and, secondarily, acidic) outburst (Turner et al. 1994; Rios et al. 2023). It corresponds to one of the larger igneous provinces (LIP) and events, such as the Central Atlantic Magmatic Province (CAMP), with voluminous lava flows and outpourings of correspondent pyroclastic sediments. This event is related to the initial moments of rupture of Gondwana and the opening of the Atlantic Ocean (Nogueira et al. 2021; Carvalho 2022).

The lower succession of Botucatu Formation (outcrops at Mato Grosso do Sul State) presents medium-grained sandstones, in lenticular bodies exhibiting gradational cycles, with associated conglomeratic sandstones. They are interpreted as

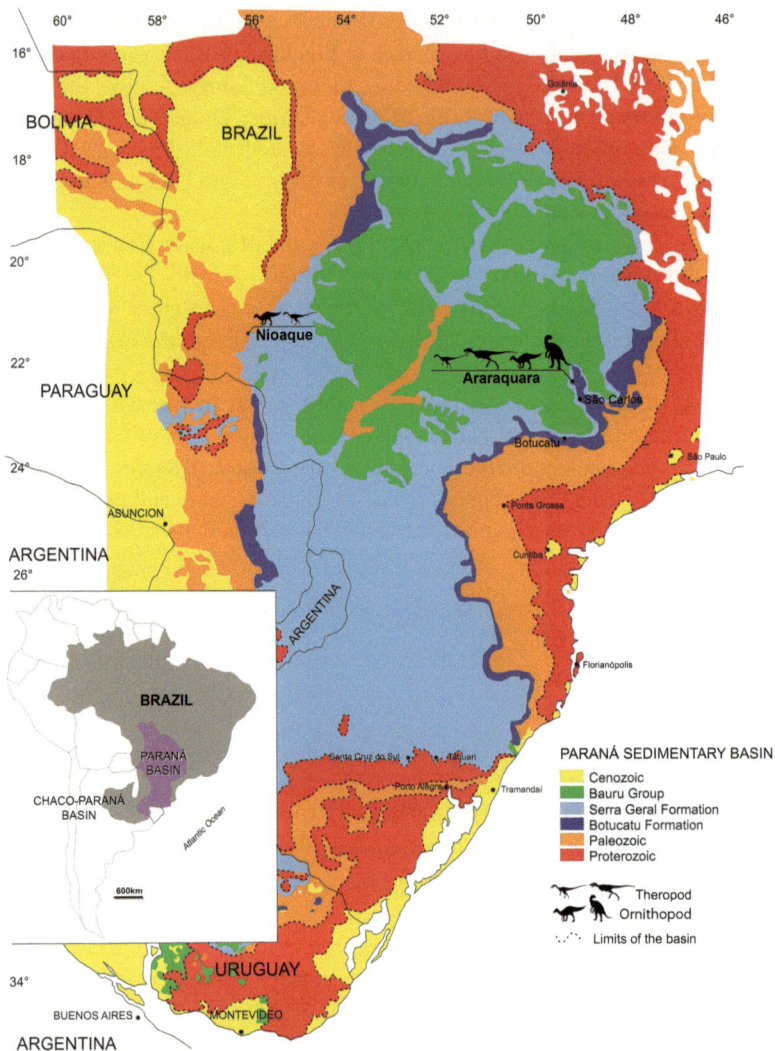

Fig. 4.1 Geological map of Paraná Basin and distribution area of Botucatu Formation, with the main localities where dinosaur footprints were found in situ [modified and adapted from "Mapa Geológico da Bacia do Paraná" (Almeida and Melo 1981) and Leonardi et al. (2024)

produced by torrential episodes, in a fluvial context, in which there are also footprints of bipedal dinosaurs (Manes et al. 2021). However this unit is generally composed of medium to fine sandstones, with rounded grains, reddish to brownish, with tangential cross-stratification (with a mean ~30° dip) and respectively topsets (with a mean ~5° dip) interpreted as a dune field. The cementation of the sandstones is due to micro-crystalline quartz, chalcedony, opal, celadonite, and calcite. There is the presence of infiltrated clays (Rios et al. 2023). Near the top of the formation, mudstones, silt-stones and fine sandstones with parallel stratification (Almeida and Melo 1981) are interpreted as interdune, dry or humid, environments with intermittent playa-lakes (Rios et al. 2023). Deserts are dynamic systems with active eolian sedimentation because of wind speed, low effective humidity, and limited vegetation cover (Rios et al. 2023; Lancaster 1988, 1995; Kocurek and Lancaster 1999).

The Botucatu deposits were extensive dune fields (Fig. 4.2a, b), with a possible alternation of wet and dry moments as a function of changes in the climatic conditions and eventual atmospheric precipitations, which may lead to areas of humid interdunes. Very possibly the tracks of mammals, dinosaurs and small invertebrates preserved in these sandstones originated in places of greater humidity of this aeolian system (Leonardi and Carvalho 2002; Francischini et al. 2015). The Botucatu Formation presents itself as a broad domain of the dune fields: an extraordinarily huge erg. A typical example of this environment is that represented by the transversal dune of São Bento quarry at Ouro-Araraquara, so incomparably rich in fossil tracks of vertebrates and invertebrates. This dune was studied by Leonardi (1980). Currently, the dune is almost destroyed by the progressive excavation for the production of street paving material and other building uses. The reduced part that remains of it would deserve to be studied in detail as well as what remains of other quarries in the same district of Ouro-Araraquara (Leonardi 1994).

The Botucatu Formation is crossed by frequent diabase subvolcanic dikes, which represent the exit routes of the eruptive material of the fissural volcanism of the Serra Geral Formation, which covered and often intruded sills and lenticular bodies, between the units of the Botucatu sandstone succession, also producing contact metamorphism. Lenticular bodies of the Botucatu sandstone often remained trapped among the igneous rocks of the Serra Geral Formation (intertrap deposits; Nogueira et al. 2021). Based on isotopic analyses, the most probable age for the great eruptive phenomenon of subcontinental character ranges from 127 to 137 million years old (Scherer 2000, 2002; Scherer et al. 2002). It was the largest in the history of the world considering the geographical extension and magma volume (excluding the oceanic basaltic bottom volcanism). The lava flows trapped in the first aeolian deposits of the Botucatu Formation indicate that the transformation of most of the Paraná Basin, from a semi-arid region to an area occupied by a huge erg, is around 134.5 ± 2.1 Ma (Brückmann et al. 2014; Renne et al. 1992).

4.3 Footprints: Diversity and Paleobiological Interpretation

Based on the studies of Leonardi (1989), Leonardi and Oliveira (1990), and Thulborn (1990), it is possible to effectively identify two different categories for the dinosaur ichnofauna with paleobiological affinities for the Botucatu Formation: Theropoda and Ornithopoda.

It is not easy to attribute a track to a trackmaker, and this is particularly true for the ichnofaunas of Botucatu Formation, and it is even harder when it comes to theropods.

Fig. 4.2 São Bento Quarry in Araraquara, São Paulo State, featuring the exposure of Botucatu Formation sandstones (paleodune), and its foresets with a dip of 29° and a height of 15–20 m

When theropod tracks began to be discovered in the Botucatu sandstones in the 1970s, they were identified as Coelurosaurs (the small theropods) and Carnosaurs (the large theropods). The taxonomic and cladistic situation today is much different and more complicated. Few skeletons of Late Jurassic and Early Cretaceous Brazilian theropods are known. The only dinosaur notable fauna (as known from bones) from the Early Cretaceous terrains of South America is that of the La Amarga Formation in the Neuquén Basin, Patagonia, Argentina. The common characteristics of South American dinosaurs are mostly very different from those of the northern continents, because of the probable geographic and biogeographic isolation of South America and, more generally, Gondwanan dinosaur faunas from those of northern continents

(Laurasia) during a large part of the Jurassic and almost all the Cretaceous. Besides, reasonably complete foot skeletons are rarely found, and they are usually conservative in structure (Leonardi and Carvalho 2021).

Theropods include several distinct clades and subclades, more than twenty at the level of family and superfamily, according to Hendrickx et al. (2015). However, because they are found in the Lower Cretaceous terrains of South America, many tracks are attributed to trackmakers of the family Abelisauridae Bonaparte and Novas 1985, probably part of the Ceratosauria. Small Spinosauridae, the sister group of Megalosauridae, are other possible theropod trackmakers in Botucatu Formation, as specimens similar to *Irritator* Martill et al. 1996.

Since the theropod footprints of the Botucatu Formation are always small, they may belong to Ceratosauria. Likely producers, as seen for *Farlowichnus* (Leonardi et al. 2024), of such southern small- and medium-size theropod tracks are the abelisauroid Noasauridae Bonaparte and Powell 1980 and Bonaparte 1991, for example, *Vespersaurus* Langer et al. 2019. Also, *Ligabueino* Bonaparte 1996 (Abelisauridae or, more probably, Noasauridae) from La Amarga Formation, Province of Neuquén, Argentina) may be a good candidate, even if one does not know well its feet. There are also theropods similar to the later *Santanaraptor* and *Mirischia* Naish et al. 2004, which is probably a Compsognathidae (Bonaparte 1991).

In accordance with Weishampel (1990), Ornithopoda represents groups of herbivorous dinosaurs whose hind-feet present three robust and functional digits (II, II and IV). As stated by Thulborn (1990), digit III is larger and more robust, while digits II and IV are shorter and slightly similar in length. The extremities of the digits always present a rounded shape, a trait of short nails or hooves, no claws. Among the Ornithopoda, the iguanodontids have a 0–3–4–5–0 phalangeal formula, with dorsoventrally flattened nails which are similar to hooves. Generally, the width of the Ornithopoda footprints is around 90–115% the size of their length, frequently being rounded. The total divergence within digits II and IV is commonly 60° but can reach 80°. The interdigital angles are normally similar. The posterior part of the Ornithopoda footprints, much like the hypex, are "U" shaped.

According to Leonardi and Carvalho (2002), the tetrapod trackways of the Ouro region's quarries (and in the sidewalks of the cities of Araraquara and São Carlos) are almost always (90–95%) of low quality, with the appearance of rounded or elliptical cavities without morphological details. These cavities are frequently accompanied by a sandstone crest in the shape of a half-moon or crescent, frequently in the direction of the strata's dip, which represents a dislodging of sand by the animal's feet, when in progression through the dunes. Despite the low quality of preservation of the footprints, the parameters of the trackways can frequently still allow their classification.

Concerning the shape of the footprints, many of them may differ within the same group, or even within the same individual animal, depending on how and in what direction it walked, if it was running or trotting, if it was going uphill, downhill, diagonally or laterally on the dune. The preservational variations of the fossilized footprints in the Botucatu Formation's sandstones match with aspects related to the consistency of the substrate in function of humidity and the different speeds and directions, adopted by the animals during their course through the paleodunes.

At the São Bento quarry, the tracks are disposed in preferential directions. Most of the animals of the Botucatu paleodesert crossed the large transverse dunes diagonally. In 80% of the occurrences, the preferential direction of their movement was between 300 and 330°, with some trackways in the opposite direction of 120–150°. Surely the direction of least effort on a slope is diagonal, however, it has still not been possible to establish the reason as to why there was this directional preference; maybe this was a direction that led directly to a water point, analogous to a small oasis.

According to the dispositions of the digits on the footprints, a group presents toes with sharp extremities, suggesting the presence of claws and a "V" shaped (acute) hypex, as in most theropods. The other group presents digits with very rounded, hoof-like extreme phalanges, much like their "U" shaped (obtuse) hypex, characteristic of ornithopods.

Considering the interdigital angles, we can group these footprints of Theropoda dinosaurs from the Botucatu Formation in two subgroups or rather, classify them into morphotypes: one with angles around 20 and 30° (Morphotype I) and the other with angles around 30 and 40°, (Morphotype II). In this second case there are two recurring morphological variations of the footprints, possibly because of the direction of movement on a slope and the speed developed during walking or running on the foreset, or maybe because of the foot anatomy itself.

According to the same interdigital angle traits, the Ornithopod dinosaur footprints may also be organized in two morphotypes, one with interdigital angles around 20 and 30° (Morphotype I) for larger forms, and around 30 and 45° (Morphotype II) for the smaller forms.

A third type of footprint with extremely elevated interdigital angles and divergence, to the order of 50°, has also been found, however since it is one single sample of an isolated footprint, it was not possible to establish with exactness a classifying pattern.

4.3.1 *Theropods: Morphotype I* - **Farlowichnus rapidus**

Diagnostic traits: footprints are longer than they are wide, digitigrade and mesaxonic. Footprint length/footprint width ratio is about 1.6. Digit III has a slight curvature in relation to the footprint's axis. Digits II and IV short in comparison to digit III, however digit II is always longer than IV. Apparently digits II and IV present a small curvature internal to the footprint. In most footprints the claw of digit III is very evident and curved. In small forms the three digits present evident claws. Hypex is "V" shaped. There is a widening of the proximal portion of digit III, however with a pronounced thinning at the extremities, where the claws are inserted. Interdigital angles around 20 and 30°. It presents total divergence within digits of 40–50° (Leonardi et al. 2024; Fig. 4.3).

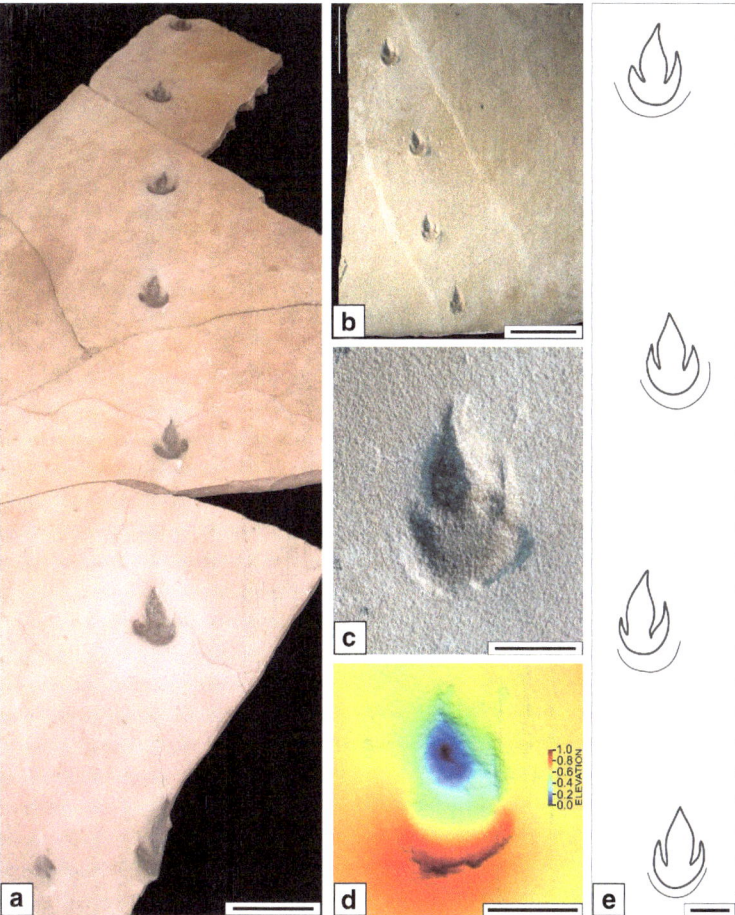

Fig. 4.3 Theropoda, Morphotype I, *Farlowichnus rapidus*, characterized by waterdrop-shaped footprints. **a** trackway with seven footprints; **b** four footprints of an individual ontogenetically younger than the one in Figure **a**; **c** detail of the fourth footprint of the trackway in Figure **b**, with two smaller outer digits II and IV, and the longest and larger digit III; **d** detail in photogrammetry technique, of the same footprint of the Figure **c**; **e** interpretative drawing of the footprints in Figure **b**. Scale bars: **a**: 20 cm, **b**: 10 cm, **c–e**: 2 cm

According to Thulborn (1998), the footprints of small theropods of the Triassic of Gondwana were attributed to the ichnogenus *Grallator*. Thulborn (1990) suggested, as a general rule, that theropod footprints with length smaller than 25 cm were attributed to coelurosaurs, and those with sizes greater than this were to carnosaurs. However, the morphometric traits of these two categories (Theropoda–morphotype I and morphotype II, variation I and variation II), are too discrepant, for that reason they are still grouped as morphotypes. This is not about Triassic footprints; but still, in front of these observations, none of the samples of Morphotype I of Theropoda corresponds to the morphological aspects of Grallatoridae, since digit II is always longer than digit IV and the II-IV interdigital angles are in the order of 25°, in addition to these digits present slightly convergent terminal portions. This allowed Leonardi et al. (2024) to insert this Morphotype I of Botucatu Theropoda into a new ichnogenus and ichnospecies, typical of the Botucatu Formation.

There is a possibility that digit II had a much larger claw (such as in the case of dromaeosaurids, cf. Ostrom 1990) that, in the moment of the step, sank much deeper in the sand, for instance in the case of the trackway in Fig. 4.3b, c. In all of the observed occurrences, the tapered groove in the sand caused by digit II was larger than that of digit IV.

For Morphotype I of Theropoda, by the similarities of morphometric patterns, differing only in footprint size, there can be suggested a possible ontogenical series, that is, different stages of development (Fig. 4.3a, b), within the same taxon. Therefore, forms considered juvenile and adults for *Farlowichnus rapidus* coexisted in this paleoenvironmental context. The theropod footprints referring to Morphotype I was named *Farlowichnus rapidus* by Leonardi et al. (2024). The producers of such footprints and trackways would have been small dinosaurs adapted to life in the desert, ancestors to noasaurs and velocisaurs which, according to the footstep angle, progressed rapidly through the dunes of Botucatu's paleodesert.

4.3.2   Theropods: Morphotype II

Diagnostic traits: tridactyl footprints slightly longer than they are wide, digitigrade and mesaxonic. Footprint length/footprint width ratio is around 1.1. Digit III is practically straight and on the same axis as the footprint. Digits II and IV short in comparison to digit III, on most of the footprints the claw of digit III is evident. The hypex is "V" shaped. There is widening of the proximal portion of digit III, making it wider than the other digits. Interdigital angles are between 30 and 40°. This track presents total divergence within digits II and IV of 50 to 70° (Figs. 4.4 and 4.5). It is possible to distinguish two variations.

4.3.2.1   Theropods: Morphotype II–Variation I

Diagnostic traits: tridactyl footprints slightly longer than they are wide, digitigrade and mesaxonic. The footprint length/footprint width ratio is around 1.0. Digit III has a larger curvature in relation to the footprint's axis. Digit II also presents curvature towards the same direction as digit III. Digits II and IV are shorter in comparison to digit III. On most of the footprints, the claw of digit III is more evident than on the others digits. Hypexes are "V" shaped. There is a widening of the proximal portion of digit III, with a pronounced tapering at the extremities where claws are inserted. Interdigital angles between 35 and 50°. The total divergence between digits II and IV is around 60 to 80° (Fig. 4.6).

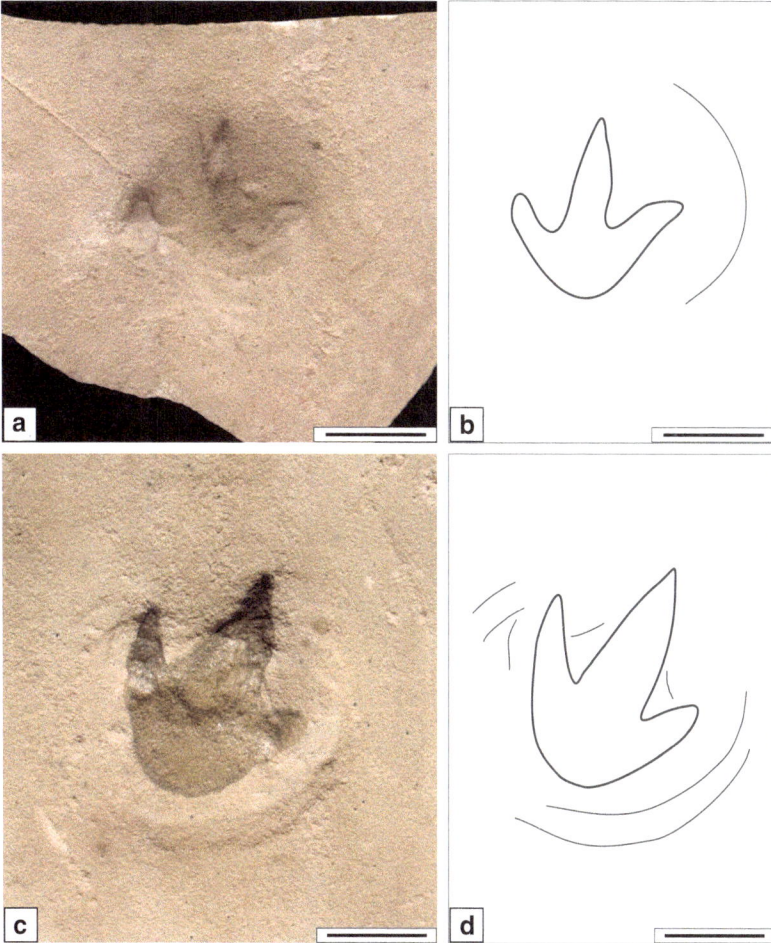

Fig. 4.4 Theropoda, Morphotype II. **a** isolated footprint, with elevated interdigital angles, when compared to Morphotype I of Theropoda; **b** isolated footprint; **c, d** interpretative drawings of the footprints; **c** footprint from Figure **a**; **d** footprint from Figure **b**. Scale bars: 5 cm

4.3.2.2   Theropods: Morphotype II–Variation II

Diagnostic traits: tridactyl footprints generally longer than they are wide, digitigrade and mesaxonic. Footprint length/footprint width ratio is around 1.2. Digit III has a larger curvature in relation to the footprint's axis. Digit II also presents curvature towards the same direction as digit III, in this case, digit IV also presents slight curvature towards the same direction. Digits II and IV are shorter in comparison to digit III, however digit II is larger than digit IV. On most footprints, the claw of digit III is more evident than on the other digits. Hypex is "V" shaped. There is a widening of the proximal portion of digit III, with a pronounced tapering at the

Fig. 4.5 Theropoda, Morphotype II, from the territory around Nioaque, Mato Grosso do Sul, western flank of the Paraná basin (Manes et al. 2021). **a** isolated footprint from locality MSNI01; **b** interpretative drawing of the footprint. Facies associated to fluvial systems, corresponding to sinuous and asymmetric current ripples and current climbing ripples structures, from the oldest part of the Botucatu Formation, locality MSNI26 near Nioaque, Mato Grosso do Sul State. Scale bars: **a, b**: 5 cm. Courtesy: Maria Izabel Lima de Manes

extremities where claws are inserted. Interdigital angles between 30 and 40°. The total divergence between digits II and IV is around 50 and 70° (Fig. 4.7).

Since Morphotype II of Theropoda doesn't have trackways, only isolated footprints, and considering that this group was not adapted to the same kind of environment as *Farlowichnus rapidus*, a similar interpretation was applied for the identification of left and right autopodia, with digit II in relation to digit IV.

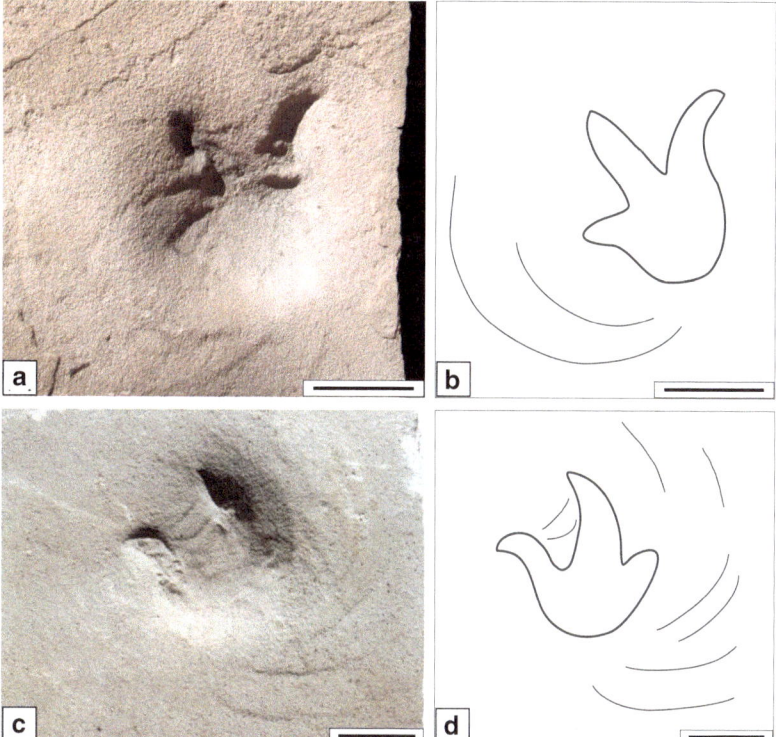

Fig. 4.6 Theropoda, Morphotype II–Variation I. **a, c** isolated footprints, with curved digits. **b, d** interpretative drawings of the footprints, with the position of the digits; **b** footprint from Figure **a**, showing the greater curvature of digit; **d** footprint from Figure **c**, showing the greater curvature of digits. Scale bars: 5 cm

As stated by Carvalho and Kattah (1998), the triangular shape of the posterior half of the footprint (backfoot triangle) and the symmetrical mirroring of the digits can be associated with dinosaurs of small dimensions, as is in the case of Morphotype II's footprints and variations I and II of Theropoda. In *F. rapidus*, the angles between digits II and III are 10 to 15% larger than that between digits III and IV. Within Morphotype II there doesn't seem to be any interdigital dominance, that is, larger angles between digits II and III, as seen in *F. rapidus*.

The footprints of Morphotype II–variation I may only be variations of Morphotype II itself, however they were placed in a separate group due to accentuated curvature of digit II and bigger footprint width, compared to Morphotype II–variation II. Interdigital angles on Morphotype II–variation I–are elevated, being larger or equal to the digits of Morphotype II's footprints. The Theropoda digits of Morphotype II–variation I seem to have a preferential curvature inwards in relation to the trackway, considering digit II, longer than IV and also curved.

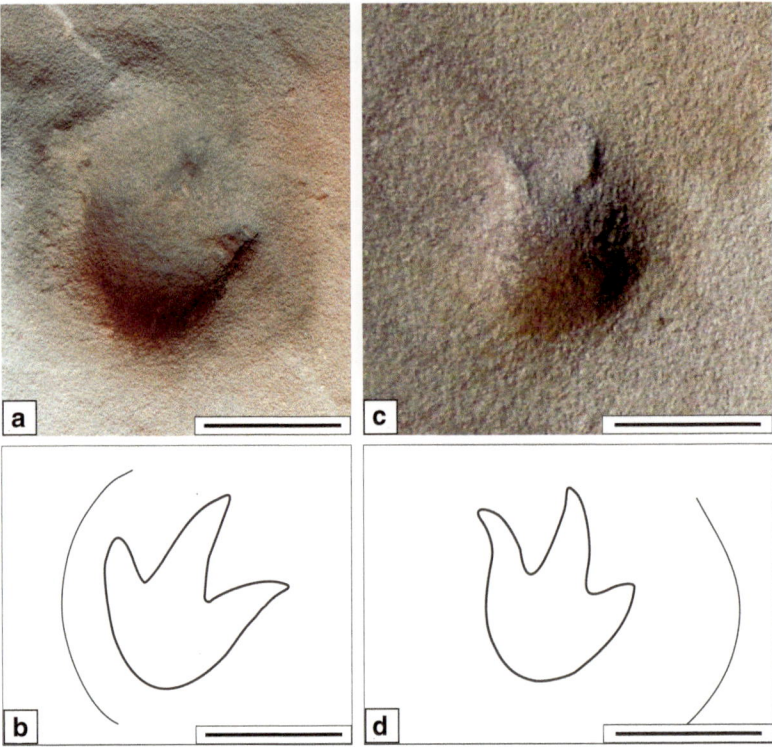

Fig. 4.7 Theropoda, Morphotype II–Variation II. **a** isolated footprint, with slightly curved digits; **b** interpretative drawing of footprint from Figure **a**, showing the curvature of digits; **c** another isolated footprint, with slightly curved digits; **d** interpretative drawing of footprint from Figure **c**, showing the small curvature of digits. Scale bars: **a, b**: 5 cm, **c, d**: 2.5 cm

4.3.3   Ornithopods: Morphotype I

Diagnostic traits: tridactyl footprints, generally as long as they are wide, larger than 30 cm in length. Footprint length/footprint width ratio is around 0.9. All are digitigrade and mesaxonic. Digits are short, in the shape of a hoof. Digit III is practically straight. Digits II and IV are shorter when compared to digit III. There are no claws at the digits' extremities, which are rounded, hoof-like. Hypex is "U" shaped. Presents the pattern of footprints attributable to the clade of ornithopods. Interdigital angles between 20 and 30°. Presents total divergence between digits II and IV around 60–80° (Fig. 4.8a, d, e).

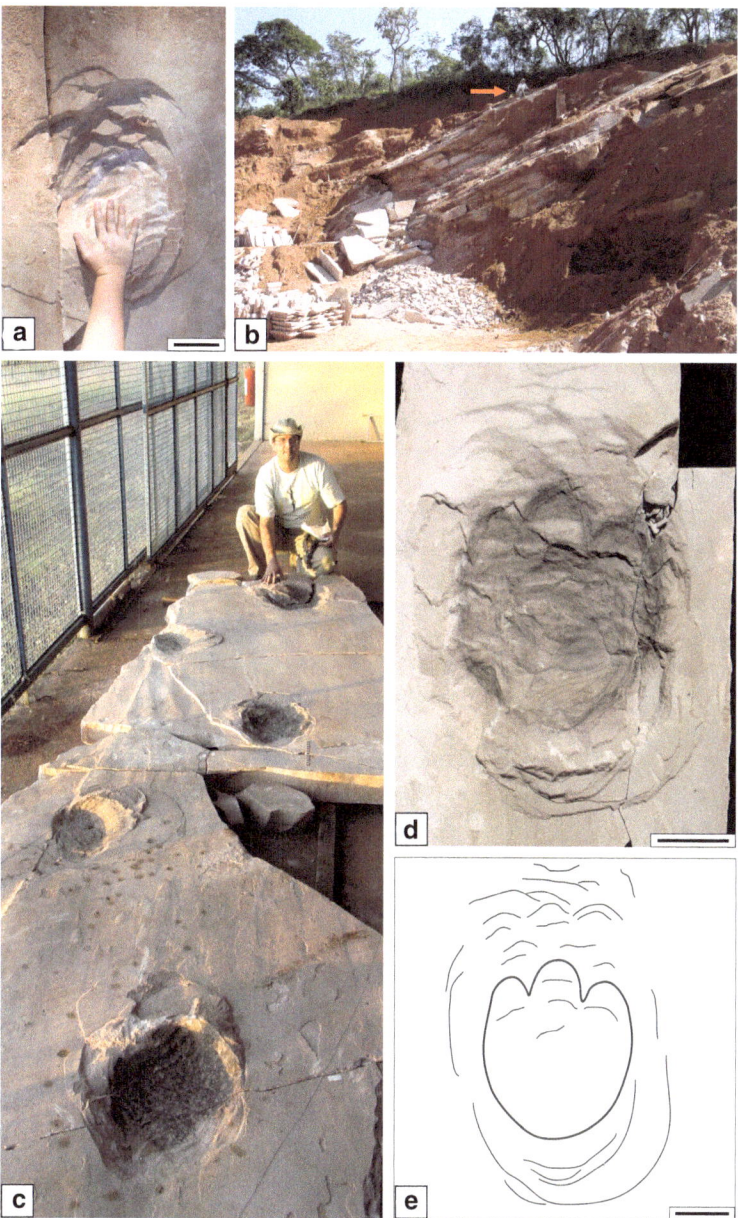

Fig. 4.8 Ornithopoda, Morphotype I. **a** isolated in situ footprint, showing crenulations (microtec-tonics with small faults) in front of the impression of three digits; **b** São Bento Quarry–the orange arrow indicates the location of the layer bearing the trackway with the footprint of figure **a**; **c** the same trackway with five footprints collected at São Bento Quarry and housed in the Science Museum of São Carlos; **d** isolated footprint with three "short" digits and crescent-shaped deformations in the posterior portion; **e** interpretative drawing for footprint in Figure **d**, showing crenulations in front, and deformations in the posterior part. Scale bars: 10 cm

4.3.4   Ornithopods: Morphotype II

Diagnostic traits: tridactyl footprints, generally slightly longer than wide. Footprint length/footprint width ratio is around 1.1. All footprints are digitigrade and mesaxonic. Digit III is practically straight, much larger than the others. Digits II and IV are shorter when compared to digit III. There are no claws at the digits' extremities, which are rounded and hoof-shaped. Hypex is "U" shaped. The footprints present a pattern attributable to the clade of ornithopods. The interdigital angles between digits II-III and III-IV are 30 and 45°. The total divergence between digits II and IV is around 50–80° (Figs. 4.9c, d and 4.10).

For the dinosaur whose digits present rounded extremities and hypexes, therefore without the presence of claws, such as the ornithopods, two morphotypes have been identified being very similar among themselves, except for the much larger size of digit III in some cases (Morphotype II) and very short digits for Morphotype I. Footprint length/footprint width was also considered as a diagnostic trait, whereas in morphotype I the footprints are wider or partially wider than they are long, especially in the larger forms (Fig. 4.8a, c and d). In this category are the largest specimens found till this day in the Botucatu Formation and one of the largest of Southeastern Brazil, with footprints up to 35 cm in length, a "giant" ornithopod of Botucatu's paleodesert, estimated to be around 5 m in length. Thus, the proposal by Leonardi (1989) about a dwarf desert fauna is not corroborated in this case.

Due to the excessive weight on a sandy substrate, the trackmaker of these Morphotype I Ornithopod footprints provoked a deformation of the lower layers of sediment, transmitting a subsurface impression thus generating an undertrack with many crenulations. However, this isn't the case of a true undertrack, but a contact of the foot with the subsurface, where there was sinking through the layer of dry sand. The real length of the largest footprint's axis might have changed when lifting the foot to take a step. This type of formation for footprints happens when there is dislodging on an apparently dry foreset, however when the foot sinks into subsurface there is contact and real deformation of lower substrate, marking and crenulations the sand around it. Sometimes, this crenulation corresponds to a true micro-tectonic, with small faults, all around the footprint, often like a flower system of micro-faults.

Among the occurrences of this kind of ornithopod footprints, only one trackway, with five footprints, was found, specifically of Morphotype I (Fig. 4.8c). At the quarry's workstation, the slabs would break apart during extraction or were split by the workers into smaller slabs (Fig. 4.8b) and then reassembled in the laboratory and gallery (Fig. 4.8c).

As for Morphotype II of Ornithopoda, there were found predominantly isolated footprints. In a singular occasion, there were registered three pairs of footprints of three distinct animals which ran or leaped in the same direction, side to side (Fig. 4.9a), obtained in situ. One pair was collected (Fig. 4.9b) and it was found that a thin layer of sand covered those footprints simultaneously, which leads to interpret the animals walked through the locality almost at the same time. The preferential

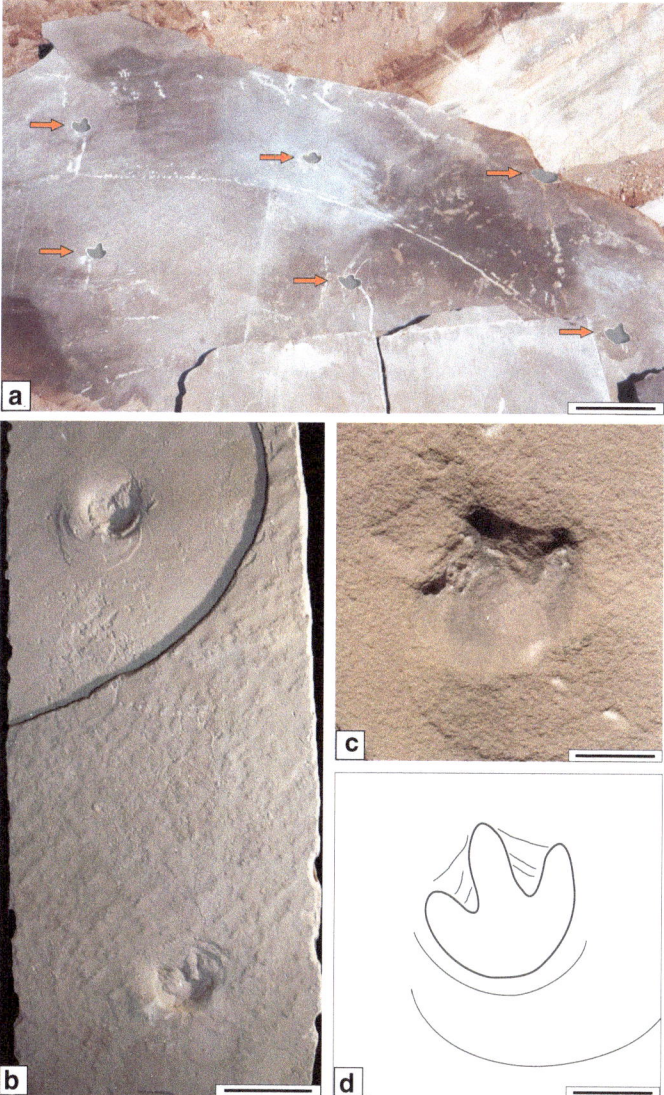

Fig. 4.9 Ornithopoda, Morphotype II. **a** footprints in sets of two, three partial trackways in situ, corresponding to three distinct animals walking side by side, indicative of gregarious behavior (details for the orange arrows), at São Bento Quarry, in Araraquara; **b** two footprints collected from that set; **c** an isolated footprint showing the three digits with rounded terminal portions; **d** interpretative drawing of the footprint in Figure **c**, with the three rounded digits. Scale bars: **a**: 50 cm, **b**: 20 cm, **c, d**: 5 cm

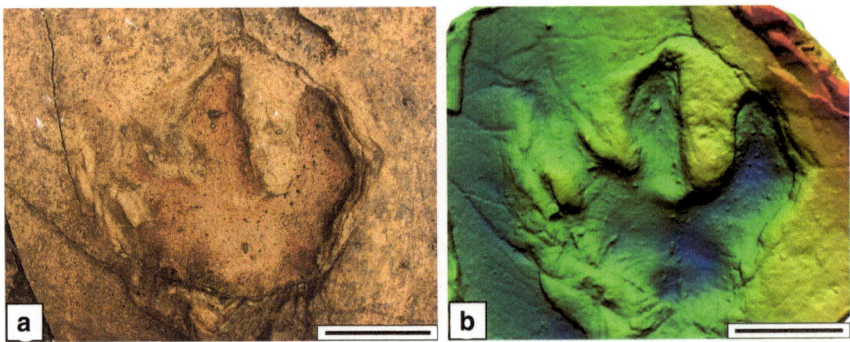

Fig. 4.10 Ornithopoda, Morphotype II, from the territory around Nioaque, Mato Grosso do Sul, loc. MSNI01-2a (Manes et al. 2021). **a** photograph of a dinosaur footprint obtained in the field. **b** 3D false color depth detail in photogrammetry technique, of the same footprint. Scale bars: 5 cm. Courtesy: Maria Izabel Lima de Manes

orientation of the dinosaur trackways may be used as a favorable argument towards gregarious habits, corroborating Carvalho (1995).

4.3.5   Undetermined Footprint, Possibly from a Dinosaur

Diagnostic traits: tridactyl footprint, wider than it is long. Footprint length/footprint ratio is around 0.89. It is a digitigrade and mesaxonic footprint. The three digits are comparatively thinner and practically of the same size, except for digit III which is slightly bigger than the other digits. Digit III is practically straight and on the same axis as the footprint. Digits II and IV don't present any internal or external rotation in relation to the footprint. There are no claws at the digit's extremities, and they are rounded. Hypex is "U" shaped. Interdigital angles are larger than 50°, a fact that makes it harder any comparison with the morphotypes identified thus far. The total divergence between digits II and IV is higher than 100° (Fig. 4.11).

This unique and isolated footprint presents a longer digit III and it is not possible to establish which digits correspond to digits II and IV. The length corresponding to digit III is 11.1 cm and width is 12.5 cm. The other digits have a length of 8.7 and 8.9 cm. Interdigital angles are 52 and 55° and total divergence between digits II and IV is 107°. Footprint length/footprint width ratio is 0.88. There is no hallux impression. According to Currie (1981), who described trackways of Canada's British Columbia birds of the Lower Cretaceous, the divergence between digits II and IV in all forms of small-sized theropods never surpass 100°. In comparison, total divergence of modern birds does surpass 100°; their footprint length/footprint width ratio is around 0.84. The footprint in question may represent the impression of a bird autopodium, since the diagnostic traits corroborate the description by Currie (1981).

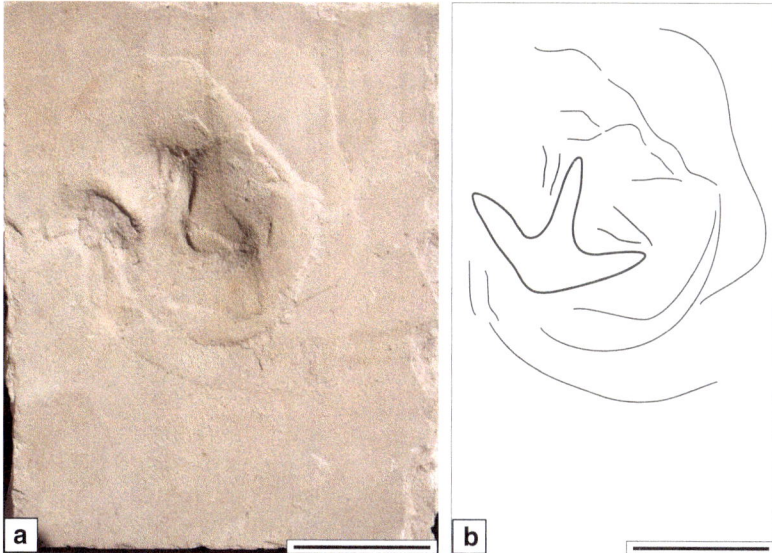

Fig. 4.11 Undetermined footprint, possibly from a dinosaur; **a** footprint with a longer digit III; **b** interpretative drawing of the footprint, with three slender digits and relatively elevated interdigital angles. Scale bars: 10 cm

Preservation of biogenic structures in aeolian sandstones is related to the condition of cohesion of sandy sediments when the organism's activity occurred. In relative humidity conditions and rapid burial, there is a larger probability of animal footprints to be preserved (Ahlbrandt et al. 1978). As stated by Winker et al. (1991), in a desert, these environmental conditions are more probable to happen within leeward and interdune areas. According to Leonardi (1980), there are two possibilities to explain the preservation of the footprints found in Botucatu Formation's sandstones. One of them would be the cloudless humidity (dew) of nighttime being responsible for the preservation of morphological details of footprints. The other possibility, arising of comparison with modern analogous environments, would be the existence of subsurface humidity from water table under the dune, which would enable the preservation of footprints on leeward layers (foresets). Due to the humidity at the interdune regions (Winkler et al. 1991), there is a possibility of the existence of ephemeral ponds with a temporary biota, a fact which would explain the large occurrence of footprints preferentially directed along the paleodune's foreset.

When a heavy animal steps on very dry sand, there cannot be seen any morpho-structural traits preserved on the surface, but only more rounded forms, with a half-moon on the area of most effort, with no evidence of digits. Preservation becomes more evident in the subsurface, with probable digit impressions because of the presence of humidity. The crenulations, which are a reflection of the autopodia impact on the substrate, are much more evident in medium to large-sized animals. The weight would provide a larger load on the substrate and therefore break the "surface tension" of unconsolidated layers of sand, with partially cohesive grains due to subsurface moisture.

Undertracks occur in distinct preservational degrees. In virtue of the animal's weight and substrate consistency facing the humidity conditions, evidences of undertracks are: partial or complete absence of crenulations, shallow proportional depth in relation to the footprints size, half-moons of locomotion effort are not very evident, and rounded shapes without footprint morphological details. Subsurface impressions and undertracks prove more significant on the topic of preservation of autopodial impression morphology, in relation to actual footprints. This fact suggests a more humid state on a layer of sand under the surface which eventually would be dry. This can be corroborated by observing associated arthropod tracks, whose weight would not be sufficient to break the tension of moist sand, being certainly formed on dry sandy surfaces.

For a lightweight animal or of small size, such as some juvenile forms of *Brasilichnium elusivum* or arthropods, the impressions in dry sand would be sufficient to transfer morphological traits of the autopodia and partially preserve digit forms. Subsequent contact with humidity by subsurface percolation or by deposition of air humidity, would aid on preservation. For a heavier animal, whose feet were relatively smaller (smaller autopodial area), considering locomotion effort, the autopodia could "sink" deeper in the sand and reach the moist subsurface, marking the shapes of digits. The impressions under the drier layer would become more representative of foot morphology, not being characterized directly as an undertrack, but a subsurface footprint. When there is movement at the foreset, the more preserved footprints are the ones at the lower levels of the dune, next to the humid interdune, lowering on a gradient in direction to its top.

Leonardi (1989), when studying south-American ichnofauna, observed a frequency of theropods of up to 87% for desert paleoenvironments in the whole continent, suggesting that this group would be better adapted to arid and desert regions than other groups of dinosaurs. However, this could be a reflection of conditional preservation of the environment, which favored preserving of footprints, or undertracks, with smaller proportions produced by dinosaurs with low body mass and also on higher or lower subsurface moisture.

The footprints found at the paleodesert that corresponds to the Areado Group of the Sanfranciscana Basin (Upper Jurassic-Lower Cretaceous) occurred on a moist interdune context, with erosion by wind with smaller and less dense particles (Carvalho and Kattah 1998; see also chap. 5, this book). Preservation of these footprints was interpreted as a result of fluctuations of the phreatic level during its formation, since those preserved in moments of higher humidity present more defined morphology. The same interpretation may be inferred for the context of the Botucatu Formation: the relatively well-preserved footprints would be associated to higher contact with humidity, be it from the influence of a shallow phreatic level, or by the action of precipitation of humidity with nighttime humid air.

In Brazil, there is a modern aeolian environment analogous to the Botucatu paleodesert, the Parque Nacional dos Lençóis Maranhenses (PNLM). By observing the formation of modern footprints at the PNLM, it was possible to corroborate the different preservation states of various animal footprints in relation to higher or lower humidity levels. At these dunes, even at the top, moist sediments are found at 6 to 8 cm of depth. Unconsolidated and dry surface sand is easily deformable because of the extreme plasticity it is encountered in. Animals of small size such as arthropods and small lacertoid reptiles hardly ever sink their feet through superficial sand, not reaching the subsurface. Larger animals on the other hand, which are heavier, impress their footprints directly under the drier layer of sand that suffers deformation, without preservation of foot morphology on those surface layers.

In the aeolian environment of PNLM there is partial preservation of organism's footprints in the function of subsurface humidity, which migrates by capillary to the upper portions of the dunes during the night. During daytime, the sun's heat "dries" the surface sand on a deepness gradient. At the surfaces near the interdunes, the sand depth is around 3 cm, and deeper there is moist sand.

The dunes of Badain Jaran, the second largest desert of inner Mongolia, contain giant subterraneous water reservoirs that could supply the chronic lack of water in northern China. In this desert, there are signs of humidity around 20 cm under the dry surface sand. This fact explains why the most elevated dunes in the world, of up to 500 m in height, resist aeolian erosion. This subterraneous water acts as a binding agent, giving dunes resistance against erosion and sand slides (Chen et al. 2004).

Just like in modern analogous environments, the thickness of the dry sand layer in the Botucatu desert may have been variable, in function of the temperature during the day, migration of interstitial water, intensity of winds, and season of the year.

4.4 Paleogeographical Distribution of the Footprints

The Botucatu sandstone covers most of the Paraná Basin, Eastern Paraguay, northern Uruguay and Northeastern Argentina, consisting of an area of more than 1,300,000 km^2, constituting what was one of the largest continuous arid deposits in the ancient world. In the paleontological context of the Botucatu Formation, aside from ichnofossils, there were minor body fossils of conchostracans and some silicified Conypherophyta logs at the northwestern part of the Formation, in the Triângulo Mineiro, in Uberlândia county (Almeida and Melo 1981).

According to Leonardi (1980) and Leonardi and Oliveira (1990), the ichnofauna attributed to the Botucatu Formation is endemic. This fact causes problems for the classification and interpretation of tracks but adds a lot to its interest. Ichnofossils of predominant tetrapods are attributed to primitive mammals, especially *Brasilichnium elusivum*, accompanied by bipedal dinosaur tracks (Ornithopoda and Theropoda).

At the northern portion of the Paraná Basin, where currently are located the states of São Paulo and Paraná, there should exist an increase in humidity due to the elevation of the groundwater table in deposits of sediments preceding the Botucatu Formation. In the southern portion of Paraná Basin, where the state of Rio Grande do Sul is located, the topographical changes during the Early Cretaceous would have produced environmental conditions unfavorable for preservation and fossilization at the paleodesert. This area would be hotter and drier, in contrast, for example, with the region where today the São Bento quarry's outcrop is situated. This fact would justify the largest occurrence of footprints more to the north of the basin.

A very different situation, from a paleoclimatic and paleoenvironmental perspective, has been discovered and studied in the western flank of the Paraná Basin, in the Mato Grosso do Sul State, in the region around the town of Nioaque. It also represents the deepest and oldest portion of the Botucatu Formation, but also has remarkable footprints of bipedal dinosaurs (Manes et al. 2021) (Figs. 4.5a, b; 4.10a, b), undoubtedly older and with different environmental specialization, compared to those of the eastern flank of the crown of outcrops of the Botucatu Formation.

The period of volcanism, known as the Paraná-Etendeka event, seems to have been preceded by global warming, as shown by Price et al. (2018), with their analysis by means of isotopes. This study would explain the gradual increase in aridity, causing the transition from fluvial (like in the Mato Grosso do Sul situation, discovered by Martins 1990 and Manes et al. 2021; Fig. 4.5c) to aeolian deposits or great dune-field, which were considered the normal environmental setting for the Botucatu Formation (Fig. 4.12). This fact suggests a climate change event between the Late Jurassic and the Early Cretaceous, changing conditions from semi-arid to hyper-arid and warmer climate.

4.5 Paleoenvironmental and Paleoclimatic Contexts

The Botucatu Formation (with the significant exception for the Mato Grosso do Sul State) represents an extensive dune field, with many sub environments of a large climactic desert, of growing aridness beginning during the Jurassic Period, over the old Gondwanic continent. This paleodesert was buried during the Early Cretaceous by the most voluminous episode of basaltic intracontinental volcanism of the planet, registered by the Serra Geral Formation.

During transportation and deposition of the Botucatu sediments, the grains suffered strong aeolian abrasive action and the minerals of low hardness were eliminated, with only the more resistant minerals remaining (Caetano-Chang and Wu 1992). The climate was an important controller of aeolian sedimentation due to the paleogeographic conditions of Gondwana (Scherer et al. 2002). The presence of the pre-Andean mountain range must have impeded that the winds from the west (westerlies) dislocated humidity to the inner part of the continent, producing an ample strip of aridness and generating aeolian dune fields that covered a great part of the South-American Platform. In addition to this, at the edge of the chain of mountains, an ample system of alluvial fans was developed, which served as a source area for the Botucatu Formation's aeolian sands (Milani 1997). The winds that moved the dunes of the Botucatu paleodesert in the State of São Paulo blew primordially from N to NNE (Bigarella and Salamuni 1961; Leonardi 1980).

The silicification process of the Botucatu Formation's aeolian dunes occurred due to the temperature elevation of subterraneous water, associated with volcanic

Fig. 4.12
Paleoenvironmental
reconstruction of the
Botucatu Paleodesert in the
Araraquara region.
Theropods and ornithopods
gather around a freshwater
point in interdune
environment. Nearby, some
diverse theropods are
roaming, interested in water,
but also in some prey. In the
foreground, the trackmaker
of *Farlowichnus rapidus*. Art
by Guilherme Gehr

elements (ionically charged solutions). The hydrothermal solution migrated to the surface by capillarity contributing to the cementing of the sand grains, making the preservation of paleodunes possible and therefore, of the fossil footprints in the deposited sand layers. The devitrification of glassy clasts was enhanced by the heat flow and hydrothermal activity of magmatic and volcanic rocks, particularly intruded sills and dikes, releasing silica and precipitating low temperature authigenic mineral assemblages. Chalcedony plays an important role in this cementing of the inter-trap sandstones, retaining the initial open structure of the sediment while filling the primary porosity. Massive precipitation of early diagenetic-hydrothermal minerals was the most common way for filling pores; these minerals included smectite, chalcedony, zeolite, mega-quartz, and hematite, resulting in a partial obliteration of the primary porosity and permeability (Nogueira et al. 2021). This process contributed to making the slabs of sandstone a hard material for paving and building.

Aside from the ichnological occurrences, there are numerous natural non-biogenic facies found in the Botucatu Formation's sandstones, which contributed to the paleoenvironmental interpretations. Many aeolian ripple-marks were found on the sandstones, both in forest as in topset, and indicate the direction and predominance of the paleowinds of NE to SW directions. However, the direction of the local main paleowind for the São Bento Quarry, taken with the analysis by Leonardi (1980), of 58 successive foresets, would be from N4°E to S4°W. The adhesion ripples, indicative of high moisture levels, would form on the foreset and are frequently found on the Botucatu Formation's sandstones. The presence of mud cracks in association with adhesion ripples reinforces the prerogative of a moist interdune as stated by Leonardi (1980).

Rare imprints from raindrops were preserved in the paleodesert's sands, as small impact craters, and corroborate the hypothesis that the occasional humidity would occur in determined times of the year. After the rain, the wind would cover the moisturized dunes with dry sand, creating surfaces with preserved raindrop marks associated with ripple marks. This water could supply subterranean reservoirs before the surface's total evaporation, as it does currently at the Parque Nacional dos Lençóis Maranhenses (Maranhão State, Brazil).

At the aeolian sandstones of the São Bento quarry are found very frequently ferromanganesian concretions, which formed inside the sedimentary layers in postdepositional events, but before complete lithification. With the interference of percolating humidity, oxidation of those particles created a radial dispersion of the oxides, staining the sand around the grain, and making concretions. These are the so-called dendrites or "flowers" of MnO_2, or nests of MnO_2 granules.

In some of the sandstone's strata observed at the São Bento quarry, there are differences in the thickness of the layers. The sandstone layers can have a millimetric thickness that can go up to 20 cm at certain points. A larger number of laminations by layers can be related to the increase in humidity during deposition. The cross strata that occur in the unconsolidated sands of a recent analogous environment, such as those in Lençóis Maranhenses, are extremely similar to those found in the Botucatu Formation. Field observations indicate that the conservation of these sedimentary strata only exists due to humidity. Tonality and thickness differences reflect both the mineral composition, intensity of winds, and humidity levels. The stratification amidst the white sands of the Lençóis Maranhenses is evident from the particulate material containing clay, associated with organic matter particles transported by the wind, under oxidative and humidity actions. A very similar fact probably occurred during the deposition of the Botucatu Formation, whose strata also vary in tonality and thickness. In this case, the reddish color is due to ferromanganese particulates coming from the Serra Geral's fissural volcanism, transported by the wind, and constantly deposited over the paleodesert's sand. Thus, one should not think that the sands and dunes of the paleodesert Botucatu should be reddish, like the rocks of the corresponding present rock formation. Their original color had to be rather from gray to yellowish.

New environments in which it will be appropriate to search and possibly discover fossil footprints of tetrapods are, firstly, the intertraps sediments, in which it appears that they have not yet been found, but they can undoubtedly be discovered. Secondly, in the volcano-sedimentary features overlying and underlying the flows, formed by the interaction of consolidated or unconsolidated tuffaceous sediments, saturated or not in the water (Rios et al. 2023).

An important element that concerns the paleoclimate, is that of the presence or absence of water in the present and fossil dunes, whether it is rainwater, whether on the rise by means of porosity or other, of the water table. In fact, the processes and products from interactions sand/water are still not fully understood (Rios et al. 2023). In particular, one should study the morphological quality of the footprints, current and fossil, in the presence or absence of water. Especially, it would be very important to study, in the case of footprints imprinted on dunes normally arid, and therefore in dunes not coastal but located in the hinterland, the difference of the characteristic "crescents", which accompany the footprints, current or fossil, in dry dunes. In the fossil footprints of the Botucatu Formation, the object of this chapter, the displacement rims (DR) with its typical crescent form are sometimes narrow and high, and one gets the impression that the sand was wet; other times they present themselves as true sand-slides, sometimes even double or repeated several times (Figs. 4.6a–d and 4.9b–d). The DRs are in any case always produced in the direction of the dip of the foreset or topset of the dune.

And finally, the magmatic and volcanic events of the Serra Geral Formation were also associated with rapid climate changes and mass extinction events. The great activity of these magmatic provinces must have influenced markedly the various terrestrial and marine environments. Abrupt changes in the concentration of green-house gases in the atmosphere occurred episodically throughout the history of the Earth, and it seems that much of the climate changes and environmental ruptures were related to the Large Igneous Provinces (Carvalho 2022).

4.6 Conclusions

The tetrapod ichnofauna represented by tracks in the Botucatu paleodesert is predom-inantly endemic, which make it harder comparison with other Mesozoic occur-rences. In the Botucatu paleodesert small to mid-sized mammals, lizard-like reptiles, theropods (insectivores and small-sized carnivores) and ornithopods (small to mid-sized herbivores), and invertebrates such as beetles and arachnids all coexisted. Those occurrences suggest a very diverse paleoecological relationship, with detritivore and coprophagic organisms like beetles; scorpions that would feed on those beetles; mammals that would eat those beetles and scorpions; herbivore dinosaurs that would graze at the margins of small ponds formed in periods of higher humidity, or that lived in marginal areas of interdunes, with vegetation, and were crossing through the dunes; small sized carnivore dinosaurs that would feed on the small mammals, beetles, and scorpions.

In this environment there were registered at least four different dinosaur groups, two Theropoda, with morphological variations, and two Ornithopoda, which lived and walked through the dunes, leaving their preserved footprints on foreset areas or on moist interdunes, as verified by the minor vestiges of raindrops, adhesion ripples and mud-cracks associated to the animal's footprints. There have never been any registered footprints attributed to sauropods for the Botucatu Formation and the fact is certainly related to the aridness of the environment with the scarcity of food for the referred group of dinosaurs.

Alterations in speed during locomotion, direction of movement on the foreset and preservational conditions, are decisive factors for the formation and preservation of tracks with distinct patterns that can be produced by the same animal. Preservation of footprint morphology on the dunes at leeway occurs due to the presence of humidity that clumps the sand grains because of the water's surface tension. This water could come from the nighttime humid air, dew, or by percolation, in function of evaporation, with the migration of interstitial water, especially during the nighttime. During the day, with the heating of the surface and action from the wind, successive layers of drier sand removed from the dune's windward and deposited over the track, would help protect the morphology of the footprint and would therefore make its definitive preservation possible.

References

Ahlbrandt TS, Andrews S, Gwynne DT (1978) Bioturbation in eolian deposits. J Sediment Res 48(3):839–848. https://doi.org/10.1306/212F7586-2B24-11D7-8648000102C1865D

Almeida FFM, Melo MS (1981) A Bacia do Paraná e o vulcanismo mesozóico. In: Mapa Geológico do Estado de São Paulo. Escala l:500.000, vol 1. Instituto de Pesquisas Tecnológicas do Estado de São Paulo S/A, São Paulo, pp 46–81

Assine ML, Piranha JM, Carneiro CDR (2004) Os paleodesertos Pirambóia e Botucatu. In: Mantesso-Neto V, Bartorelli A, Carneiro CDR, Brito-Neves BB (eds) Geologia do Continente Sul Americano: evolucão da obra de Fernando Flàvio Marques de Almeida. Beca, São Paulo, pp 77–93

Bigarella JJ, Salamuni R (1961) Early Mesozoic wind patterns as suggested by dune bedding in the Botucatu Sandstone of Brazil and Uruguay. GSA Bull 72(7):1089–1105. https://doi.org/10.1130/0016-7606(1961)72[1089:EMWPAS]2.0.CO;2

Boaparte JF (1996) Cretaceous tetrapods of Argentina. Münchner Geowiss Abh 30:73–130

Bonaparte JF (1991) Los vertebrados fósiles de la Formación Río Colorado, de la ciudad de Neuquén y cercanías, Cretácico superior, Argentina. Rev Mus Argent Cienc Nat 4(3):15–123

Bonaparte JF, Powell JE (1980) A continental assemblage of tetrapods from the Upper Cretaceous beds of El Brete, Northwestern Argentina (Sauropoda, Coelurosauria, Carnosauria, Aves). Mémoires De La Société Géologique De France 139:19–28

Bonaparte JF, Novas FE (1985) Abelisaurus comahuensis, n. g., n. sp., carnosauria from the late Cretaceous of Patagonia. Ameghiniana 21(2–4):259–265

Brückmann M, Hartmann LA, Tassinari CCG, Sato K, Baggio SB (2014) The duration of magmatism in the Serra Geral Group, Paraná volcanic province. In: Hartmann LA, Baggio SB (eds) Metallogeny and crustal evolution of the Serra Geral Group, 1st edn. IGeo, UFRGS, Porto Alegre, pp 507–518

Buck PV, Ghilardi AM, Peixoto BCPEM, Fernandes LBR, Fernandes MA (2017a) A new tetrapod ichnotaxon from Botucatu Formation, Lower Cretaceous (Neocomian), Brazil, with comments on fossil track preservation on inclined planes and local paleoecology. Palaeogeogr Palaeoclimatol Palaeoecol 466:21–37. https://doi.org/10.1016/j.palaeo.2016.11.009

Buck PV, Ghilardi AM, Fernandes LBR, Fernandes MA (2017b) A new ichnotaxon classification of large mammaliform trackways from the Lower Cretaceous Botucatu Formation, Paraná Basin, Brazil. Palaeogeogr. Palaeoclimatol. Palaeoecol. 485:377–388. https://doi.org/10.1016/j.palaeo.2017.06.027

Buck PV, Ghilardi AM, Peixoto BCPEM, Aureliano T, Fernandes MA (2022) Lacertoid tracks from the Botucatu Formation (Lower Cretaceous) with different locomotor behaviors: a new trackmaker with novel paleoecological implications. J South Am Earth Sci 116:103825. https://doi.org/10.1016/j.jsames.2022.103825

Caetano-Chang MR, Wu FT (1992) Estudo mineralógico dos arenitos das formações Pirambóia e Botucatu no Centro-Leste do Estado de São Paulo. Rev Inst Geol 13(1):58–68. https://doi.org/10.5935/0100-929X.19920004

Carvalho IS (1995) As pistas de dinossauros da Ponta da Guia (Bacia de São Luis, Cretáceo Superior–Maranhão, Brasil). An Acad Bras Ciênc 67(4):413–431

Carvalho IS, Kattah SS (1998) As pegadas fósseis do Paleodeserto da Bacia Sanfranciscana (Jurássico Superior-Cretáceo Inferior, Minas Gerais). An Acad Bras Ciênc 70(1):53–67

Carvalho IS (2022) Paleogeografia. Cenários da Terra. Interciência, Rio de Janeiro

Chen JS, Li L, Wang JY, Barry DA, Sheng XF, Gu WZ, Zhao X, Chen L (2004) Water resources: groundwater maintains dune landscape. Nature 432:459–460. https://doi.org/10.1038/432459a

Currie PJ (1981) Bird footprints from the Gething Formation (Aptian, Lower Cretaceous) of northeastern British Columbia, Canada. J Vertebr Paleontol 1(3/4):257–264

D'Orazi-Porchetti S, Bertini RJ, Langer MC (2017) Proposal for ichnotaxonomic allocation of therapsid footprints from the Botucatu Formation (Brazil). Ichnos 25(2–3):192–207. https://doi.org/10.1080/10420940.2017.1308929

Fernandes MA, Carvalho IS (2008) Revisão diagnóstica para a icnoespécie de tetrápode Mesozóico *Brasilichnium elusivum* (Leonardi, 1981) (Mammalia) da Formação Botucatu, Bacia do Paraná, Brasil. Ameghiniana 45:167–173

Fernandes ACS, Netto RG, Carvalho IS (1988) O icnogênero *Taenidium* na Formação Botucatu. An Acad Bras Cienc 60(4):493

Fernandes MA, Fernandes LBR, Souto PRF (2004) Occurrence of urolites related to dinosaurs in the Lower Cretaceous of Botucatu Formation, Paraná Basin, São Paulo State, Brazil. Rev Bras Paleontol 7(2):263–268. https://doi.org/10.4072/rbp.2004.2.20

Fernandes MA, Carvalho IS (2007) Pegadas fósseis da Formação Botucatu (Jurássico Superior? Cretáceo Inferior): o registro de um grande dinossauro Ornithopoda na Bacia do Paraná. In: Carvalho IS, Cassab RCT, Schwanke C, Carvalho MA, Fernandes ACS, Rodrigues MAC, Carvalho MSS, Arai M, Oliveira MEQ (eds) Paleontologia: Cenários de Vida, vol 1, 1st edn. Interciência, Rio de Janeiro, pp 425–432

Fernandes MA (2020) Micturálitos. In: Sedorko D, Francischini H (eds) Icnologia: interações entre organismos e substratos, 1st edn. CRV, Curitiba, pp 277–282

Francischini H, Dentzien-Dias PC, Fernandes MA, Schultz CL (2015) Dinosaur ichnofauna of the Upper Jurassic/Lower Cretaceous of the Paraná Basin (Brazil and Uruguay). J South Am Earth Sci 63:180–190. https://doi.org/10.1016/j.jsames.2015.07.016

Hendrickx C, Hartman SA, Mateus O (2015) An overview of non-avian theropod discoveries and classification. PalArch's J Vertebr Palaeontol 12(1):1–73

Huene F (1931) Verschiedene mesozoische Wierbeltierreste aus Südamerika. Neues Jahrbuch für Mineralogie, Geologie, Paläontologie Monatshefte 66(B):181–198

Kocurek G, Lancaster N (1999) Aeolian system sediment state: theory and Mojave Desert Kelso dune field example. Sedimentology 46:505–515. https://doi.org/10.1046/j.1365-3091.1999.00227.x

Lancaster N (1988) Development of linear dunes in the southwestern Kalahari, Southern Africa. J Arid Environ 14:233–244. https://doi.org/10.1016/S0140-1963(18)31070-X

Lancaster N (1995) The geomorphology of desert dunes. Routledge, London

Langer M, Martins NO, Manzig PC et al (2019) (2019) A new desert-dwelling dinosaur (Theropoda, Noasaurinae) from the Cretaceous of south Brazil. Sci Rep 9:9379. https://doi.org/10.1038/s41 598-019-45306-9

Leonardi G (1977) On a new occurrence of tetrapod trackways in the Botucatu Formation in the State of São Paulo, Brazil. Dusenia 10:181–183

Leonardi G (1981) Novo Ichnogênero de Tetrápode Mesozóico da Formação Botucatu, Araraquara, SP. An Acad Bras Ciênc 53(4):793–805

Leonardi G (1989) Inventory and statistics of the South American dinosaurian ichnofauna and its paleobiological interpretation. In: Gillette DD, Lockley MG (eds) Dinosaur tracks and traces, 1st edn. Cambridge University Press, pp 165–178

Leonardi G, Oliveira FH (1990) A revision of the Triassic and Jurassic tetrapod footprints of Argentina and a new approach on the age and meaning of Botucatu Formation footprints (Brazil). Rev Bras Geociênc 2(K1–4):216–229

Leonardi G, Sarjeant WAS (1986) Footprints representing a new Mesozoic vertebrate fauna from Brazil. Mod Geol 10:73–84

Leonardi G, Carvalho IS (2002) Jazigo icnofossilífero do Ouro, Araraquara, SP: ricas pistas de tetrápodes do Jurássico. In: Schobbenhaus C, Campos DA, Queiroz ET, Winge M, Berbert-Born MLC (eds) Sítios geológicos e paleontológicos do Brasil. DNPM, Brasília, pp 39–48

Leonardi G, Carvalho IS (2021) Dinosaur Tracks from Brazil: A Lost World of Gondwana. Indiana University Press

Leonardi G, Carvalho IS, Fernandes MA (2007). The desert ichnofauna from Botucatu Formation (Upper Jurassic- Lower Cretaceous). In: Carvalho IS, Cassab RCT, Schwanke C, Carvalho MA, Fernandes ACS, Rodrigues MAC, Carvalho MSS, Arai M, Oliveira MEQ (eds) Paleontologia: Cenários de Vida, vol 1, 1st edn. Interciência, Rio de Janeiro, pp 379–392

Leonardi G, Fernandes MA, Carvalho IS, Schutzer JB, Silva RC (2024) *Farlowichnus rapidus* new ichnogen., new ichnosp.: a speedy and small theropod in the Early Cretaceous Botucatu paleodesert (Paraná Basin), Brazil. Cretaceous Res 153:105720. https://doi.org/10.1016/j.cre tres.2023.105720

Leonardi G (1980) On the discovery of an abundant ichno-fauna (vertebrates and invertebrates) in the Botucatu Formation *s.s.* in Araraquara, São Paulo, Brazil. An Acad Bras Ciênc 52(3):559–567

Leonardi G (1994) Annotated Atlas of South America Tetrapod Footprints (Devonian to Holocene) with an appendix on Mexico and Central America. Companhia de Pesquisa de Recursos Minerais, Brasília, Brazil, p 248

Manes MIL, Silva RC, Scheffler SM (2021) Dinosaurs and rivers on the edge of a desert: a first recognition of fluvial deposits associated to the Botucatu Formation (Jurassic/Cretaceous), Brazil. J S Am Earth Sci 110:103339. https://doi.org/10.1016/j.jsames.2021.103339

Martill D, Cruickshank ARI, Frey E, Small PG, Clarke M (1996) A new crested maniraptoran dinosaur from the Santana Formation (Lower Cretaceous) of Brazil. J Geol Soc 153(1):5–8. https://doi.org/10.1144/gsjgs.153.1.0005

Martins GR (1990) Relatório de registro do Sítio Paleontológico "MS-NI-01." Rev Científ 5(1):7–12

Milani EJ, Melo JHG, Souza PA, Fernandes LA, França AB (2007) Bacia do Paraná. Bol De Geocienc Petrobras 15(2):265–287

Milani EJ (1997) Evolução tectono-estratigráfica da Bacia do Paraná e seu relacionamento com a geodinâmica fanerozóica do Gondwana sul-ocidental. Dissertation, Universidade Federal do Rio Grande do Sul

Naish D, Martill DM, Frey E (2004) Ecology, systematics and biogeographical relationships of dinosaurs, including a new Theropod, from the Santana Formation (?Albian, Early Cretaceous) of Brazil. Hist Biol 16(2):57–70. https://doi.org/10.1080/08912960410001674200

Nogueira ACR, Rabelo CEN, Góes AM et al (2021) (2021) Evolution of Jurassic intertrap deposits in the Parnaíba Basin, northern Brazil: the last sediment-lava interaction linked to the CAMP in West Gondwana. Palaeogeogr Palaeoclimatol Palaeoecol 572:110370. https://doi.org/10.1016/j.palaeo.2021.110370

Ostrom JH (1990) Dromaeosauridae. In: Weishampel DB, Dodson P, Osmólska H (eds) The Dinosauria. University of California Press, pp 269–279

Pacheco JAA (1913) Notas sobre a geologia do Valle do Rio Grande a partir da Fóz do Rio Pardo até a sua confluência com o Rio Parahyba. In: Comissão Geográphica e Geológica do Estado de São Paulo. Exploração do Rio Grande e de seus afluentes. São José dos Dourados. Comissão Geográphica e Geológica do Estado de São Paulo, São Paulo, pp 33–38

Peixoto BCPEM, Mángano MG, Minter NJ, Fernandes LBR, Fernandes MA (2020) A new insect trackway from the Upper Jurassic-Lower Cretaceous eolian sandstones of São Paulo State, Brazil: implications for reconstructing desert paleoecology. PeerJ 8:e8880. https://doi.org/10.7717/peerj.8880

Price GD, Janssen NMM, Martinez M, Company M, Vandevelde JH, Grimes ST (2018) A high-resolution belemnite geochemical analysis of Early Cretaceous (Valanginian-Hauterivian) environmental and climatic perturbations. Geochem Geophys Geosyst. https://doi.org/10.1029/2018GC007676

Renne PR, Ernesto M, Pacca IG, Coe RS, Glen JM, Prevot M, Perrin M (1992) The age of Paraná flood vulcanism, rifting of Gondwanaland, and the Jurassic-Cretaceous boundary. Science 258(5084):975–979. https://doi.org/10.1126/science.258.5084.975

Rios FR, Mizusaki AMP, Michelin CRL (2023) Rodrigues IC (2023) Volcano-sedimentary interactions–a key to understand Cretaceous paleoenvironments in the Paraná basin (southern Brazil). J S Am Earth Sci 128:104483. https://doi.org/10.1016/j.jsames.2023.104483

Scherer CMS (2000) Eolian dunes of the Botucatu formation (Cretaceous) in Southernmost Brazil: morphology and origin. Sediment Geol 137:63–84. https://doi.org/10.1016/S0037-0738(00)00135-4

Scherer CMS (2002) Preservation of aeolian genetic units by lava flows in the Lower Cretaceous of the Paraná Basin, Southern Brazil. Sedimentology 49(1):97–116. https://doi.org/10.1046/j.1365-3091.2002.00434.x

Scherer CMS, Faccini UF, Lavina E (2002) Arcabouço estratigráfico do Mesozóico da Bacia do Paraná. In: Holz M, De Ros LF (eds). Geologia do Rio Grande do Sul. Instituto de Geociências/CIGO, Porto Alegre, pp 335–354

Schneider RL, Mühlmann H, Tommasi E, Medeiros RA, Daemon RF, Nogueira AA (1974) Revisão estratigráfica da Bacia do Paraná. In: 28 Congresso Brasileiro De Geologia, Porto Alegre 1974. Sociedade Brasileira de Geologia, São Paulo 1:41–65

Thulborn T (1990) Dinosaur tracks. British Library, Chapman and Hall, London

Thulborn T (1998) Australia's earliest theropods: footprint evidence in the Ipswich coal measures (Upper Triassic) of Queensland. Gaia 15:301–311

Turner S, Regelous M, Hawkesworth C, Montovani M (1994) Magmatism and continental break-up in the South Atlantic: high precision ^{40}Ar-^{49}Ar geochronology. Earth Planet Sci Lett 121:333–348

Weishampel DB (1990) Ornithopoda. In: Weishampel DB, Dodson P, Osmólska H (eds) The Dinosauria. University of California Press, pp 484–485

Winkler DA, Jacobs LL, Congleton JD, Downs WR (1991) Life in a sand sea: biota from Jurassic interdunes. Geology 19(9):889–892. https://doi.org/10.1130/0091-7613(1991)019%3c0889:LIASSB%3e2.3.CO;2

Chapter 5
The Dinosaur Footprints
in the Cretaceous Aeolian Deposits
of Sanfranciscana Basin

Ismar de Souza Carvalho and Senira Kattah

5.1   Introduction

The Sanfranciscana Basin is one of the Brazilian intracratonic basins, located in central-eastern Brazil, oriented in the N–S direction with approximately 1,100 km long and 200 km wide (Cabral and Mescolotti 2021), occurring in the states of Minas Gerais, Goiás, Bahia, Tocantins, Piauí and Maranhão (Fig. 5.1). It covers 220,000 km^2 of the São Francisco Craton with Paleozoic (Santa Fé Group, Permo-Carboniferous) and Mesozoic rocks, ranging from the Late Jurassic to the Late Cretaceous (Campos and Dardenne 1997a; Sgarbi et al. 2001) and encompasses the Abaeté (south) and Urucuia (north) sub-basins (Campos and Dardenne 1997b). The Cretaceous sedimentary record is constituted by the Areado (Lower Cretaceous), Mata da Corda (Upper Cretaceous), and Urucuia (Upper Cretaceous) groups, deposited from Barremian to Maastrichtian and is one of the extensive events of Gondwanan continental sedimentation (Carmo et al. 2004; Mescolotti et al. 2019; Nascimento et al. 2022; Sgarbi et al. 2001).

The dinosaur footprints from the Sanfranciscana Basin are found in the Lower Cretaceous aeolian deposits of the Areado Group that is constituted by the Abaeté, Quiricó and Três Barras formations. The Abaeté Formation is composed of matrix-supported and clast-supported conglomerates, resulted from gravity-controlled processes in alluvial fans and wadi in desert conditions (Campos and Dardenne 1997a; Sgarbi et al. 2001). The Quiricó Formation is composed of mudstone and

I. S. Carvalho (✉)
CCMN/IGEO, Departamento de Geologia, Universidade Federal do Rio de Janeiro, 21.910-200 Cidade Universitária, Ilha do Fundão, Rio de Janeiro, Estado do Rio de Janeiro, Brazil
e-mail: ismar@geologia.ufrj.br

Centro de Geociências, Universidade de Coimbra, Rua Sílvio Lima, 3030-790 Coimbra, Portugal

S. Kattah
3128 Edgewater Drive, Austin, TX 78733, USA

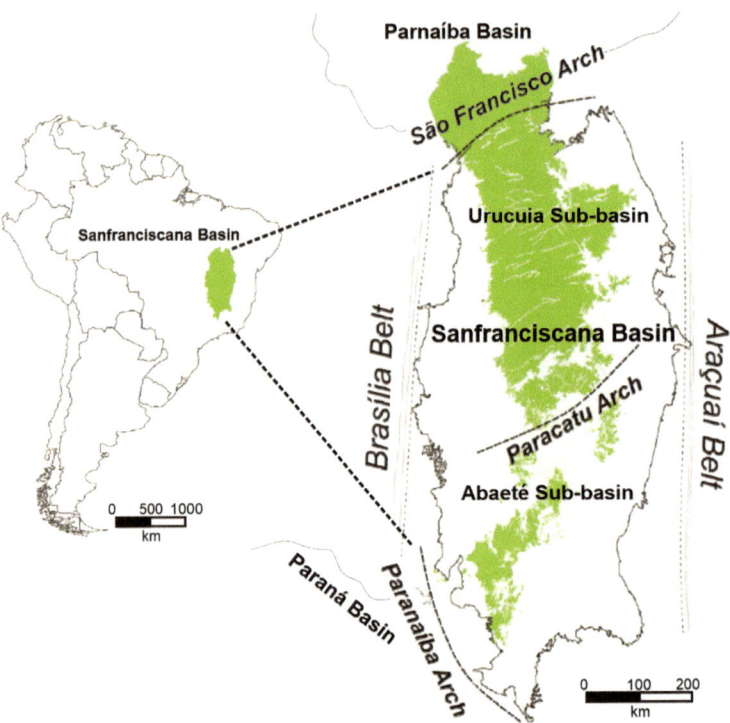

Fig. 5.1 Geographic context of Sanfranciscana Basin in Central-East Brazil. Structural archs and subdivision into the Urucuia and Abaeté sub-basins. 2022 Modified from Cardoso and Basilici (2022)

heterolithic facies (Mescolotti et al. 2019) interpreted as deposited in lacustrine, fluvial and aeolian environments (Campos and Dardenne 1997b). The Três Barras Formation is comprised mainly of sandstones, with associated conglomerate and fine-grained deposits resulted from fluvial-aeolian processes (Campos and Dardenne 1997a; Sgarbi et al. 2001). This unit was divided by Mescolotti et al. (2019) in two depositional units. The Lower Unit (Barremian/Aptian) encompasses a wet aeolian system composed of dunes, interdunes, and ephemeral alluvial deposits. In the upper part of the Lower Unit a continuous paleosol records dune stabilization and the end of aeolian accumulation. Follows a stratigraphic gap (Cenomanian to Coniacian) that coincides with the Cretaceous Thermal Maximum. The Upper Unit (Santonian?/ Campanian) comprises dune fields of a dry aeolian system capped by volcanic rocks (Mescolotti et al. 2019).

The footprints from the Sanfranciscana Basin occur in the aeolian deposits, interpreted as interdune deposits, an adequate area to flourish life, especially due the moister environment and the availability of water. The footprints are found in the Três Barras Formation and they are in the same environmental context of the Botucatu Formation and Caiuá Group (Paraná and Bauru basins), where also occur similar theropod footprints.

5.2 Geological Context

The Lower Cretaceous rock successions registered extremely arid conditions in the Gondwana. In the Paraná Basin there are the deposits of the most extensive paleodesert of Earth history—the Botucatu Paleodesert, where fossil footprints are found in the Botucatu Formation. The Cretaceous deposits of the Sanfranciscana Basin (mainly the Areado Group, Fig. 5.2) also record semi-arid to desert settings associated to the hot and dry climatic conditions (Grossi Sad et al. 1971; Suguio and Barcelos 1983).

The Três Barras Formation (Areado Group) is dominantly composed of sandstones that interfinger and conformably overlie the deposits of Quiricó Formation and it is overlain by the Upper Cretaceous volcanic alkaline rocks of Mata da Corda Group (~80 Ma, Sgarbi et al. 2004). Ten sedimentary facies were recognized in the Três Barras Formation by Mescolotti et al. (2019), including: Medium- to coarse-grained trough cross-bedded sandstone; meter-scale trough cross-bedded sandstone; planar cross-bedded sandstone; parallel-laminated sandstone; climbing

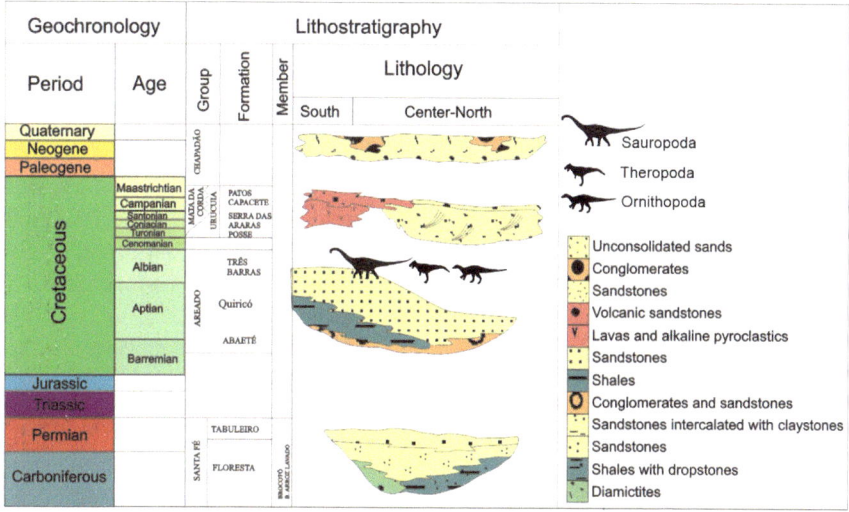

Fig. 5.2 Stratigraphic chart of the Sanfranciscana Basin (modified from Leite and Carmo 2021)

ripple cross-laminated sandstone; sandstone with deformational structures; paleosol; massive mudstone; laminated mudstone; and interlaminated fine-grained sandstones and mudstones with heterolithic lamination.

The fossils of the Areado Group are restricted to the Quiricó and Três Barras Formation (Carvalho et al. 1994). The mudstones, shales, siltstones and carbonates of Quiricó Formation (?Hauterivian-Aptian) are a mixed siliciclastic–evaporitic succession interpreted as fluvio-lacustrine and playa lake paleoenvironment (Cardoso et al. 2022, 2023). They contain palynomorphs, leaves of gymnosperms and angiosperms, annelids, insects, ostracods, spinicaudatans, elasmobranchs, actinopterygians, coelacanthiformes and dinosaur bones (Scorza and Santos 1955; Barbosa 1965; Duarte, 1968; Barbosa et al. 1970; Santos 1971; Lima 1979; Duarte 1985a, b; Santos 1985; Carvalho et al. 1994, 1995; Arai et al. 1995; Duarte 1997; Barbosa et al. 1997; Delício et al. 1998; Carmo et al. 2004; Gallego and Martins-Neto 2006; Carvalho and Maisey 2008; Zaher et al. 2011, 2020; Leite et al. 2018; Fragoso et al. 2019; Brito et al. 2020; Bittencourt and Brandão 2021; Bittencourt et al. 2015, 2018; Ribeiro et al. 2018; Coimbra 2020; Carvalho and Santucci 2021; Leite and Carmo 2021). The shales of Quiricó Formation can be organic-rich locally, and temporally record permanent dysoxic/anoxic environments. The succession is characterized by mudstones with scattered grains of fine sand and desiccation cracks, evaporites interpreted as ephemeral shallow lakes with events of subaerial exposure and rapid modifications of the water chemistry (Mescolotti et al. 2019; Cardoso et al. 2022). In the Três Barras Formation (Barremian-Albian), silexilite units occur in association with sandstones and mudstones. In these silica-rich rocks, radiolarians and other marine fossils are present (Kattah 1992; Kattah and Koutsoukos 1992). The Três Barras Formation (Fig. 5.3) comprises siliciclastic deposits interpreted as fluvial, deltaic, marine and aeolian sedimentation. The presence of dinoflagellates, radiolarians, spicules of sponges, carapaces of foraminifera and possible acritarchs in silexilite levels were interpreted as indicative of marine environments (Kattah 1991; Kattah and Koutsoukos 1992; Pessagno and Dias-Brito, 1996; Pessagno et al. 1997; Dias-Brito et al. 1999) such as restricted platform (Kattah 1992), marine transitional marginal (Castro 1996) and neritic or oceanic environments (Arai 1999; Dias-Brito et al. 1999; Arai 2009, 2014a, b). Carvalho and Kattah (1998) described eleven dinosaur footprints in the Três Barras Formation. They are in deposits of an aeolian facies of a humid interdune environment (Kattah 1993; Carvalho and Kattah 1998). Other footprints were identified as cross-section structures, concave-upward and asymmetric soft-sediment deformation structures, interpreted as produced in wet interdune facies, in the interdune-dune contact and in dune foresets (Mescolotti et al. 2019).

5.3   Footprints: Diversity and Paleobiological Interpretation

The fossil footprints are found in two distinct ichnosites: João Pinheiro and Presidente Olegário, both in the homonymous municipalities of the Northwest of the state of Minas Gerais, microregion Paracatu. There are eleven dinosaur footprints in the João

Fig. 5.3 Outcrop of the Três Barras Formation where occur the footprints of the João Pinheiro ichnosite

Pinheiro ichnosite preserved as epireliefs filled with a fine to coarse-grained reddish sandstone similar to the surrounding matrix. Three smaller of them are part of a short trackway (Carvalho and Kattah 1988). There are also cross-section footprints observed as concave upward deformation structures and narrow shaft straight borders from the Presidente Olegário ichnosite (Mescolotti et al. 2019).

5.3.1   João Pinheiro Ichnosite

The short track of three consecutive footprints (JPAR 01, JPAR 02 and JPAR 03) shows an oblique pace of 60 cm and step angle of 155°. The stride is 112.5 cm long. These footprints are tridactyl, mesaxonic and digitigrade with evidence of claws, with 15 cm of width and length. The hypexes angles between digits II-III and III-V are acute, with an average value of 30°. Footprint JPAR 01 is preserved as convex epirelief and footprints JPAR 02 and JPAR 03 as concave epirelief. Two footprints from this short track (JPAR 01 and JPAR 02) show a projection on the rear border that could correspond to digit I. The digits are tapered and the extremities of footprints JPAR 01 and JPAR 03 show claw imprints (Figs. 5.4 and 5.5).

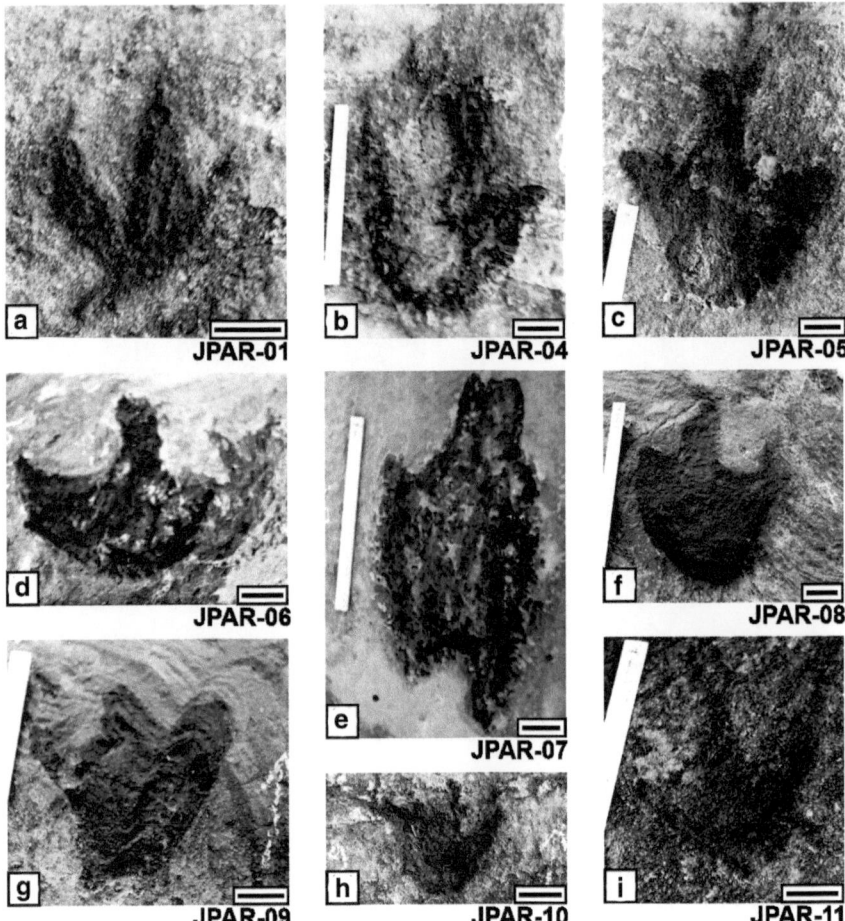

Fig. 5.4 Footprints from the João Pinheiro ichnosite that are interpreted as tracks of small and large-sized theropods and ornithopod trackmakers preserved in the Três Barras Formation, Areado Group. JPAR—João Pinheiro County, Areado Group. Footprints in dark color due the use of water to highlight them. Scale bar: 5 cm

The isolated footprints are bigger than these last ones. They are also tridactyl, mesaxonic and with digit III projecting more than digits II and IV (JPAR 04, JPAR 05, JPAR 07, JPAR 08, JPAR11) or presenting almost the same size (JPAR 06, JPAR 09, JPAR 10).

The footprint JPAR 04 is isolated, tridactyl and mesaxonic preserved in concave epirelief without clear morphological details. The footprint shows 25 cm in width and 30 cm in length. Digit III is the bigger, very large in its base and pointed in the anterior extremity. The two other digits show almost the same size, shorter and more pointed, suggesting the presence of claws. The hypexes between digits II-III and III-IV are acute (35°).

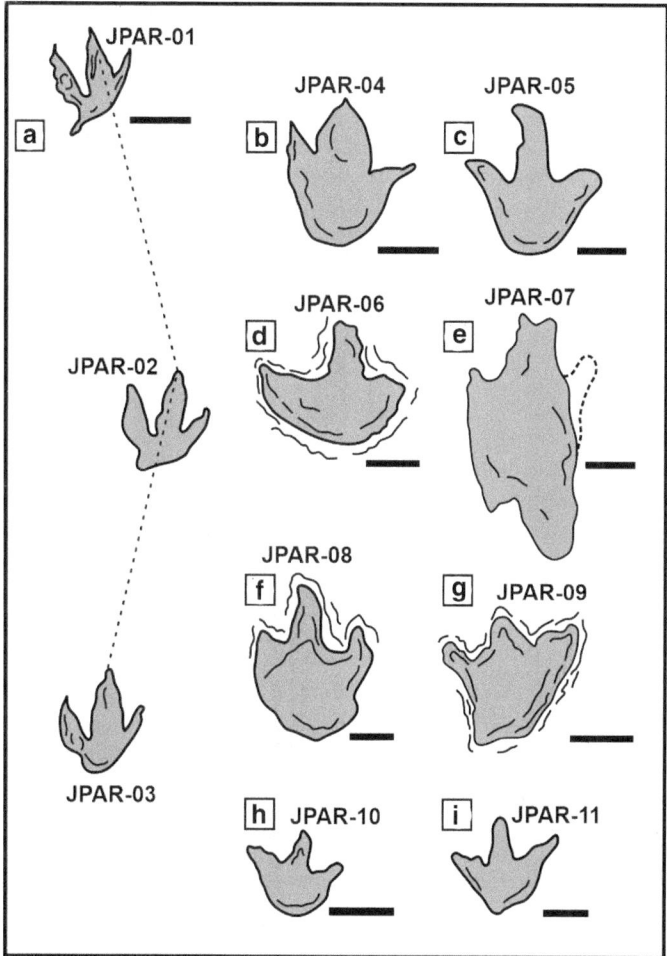

Fig. 5.5 Morphological aspects of the dinosaur footprints from the João Pinheiro ichnosite. **a** Small-sized theropod footprints in a short trackway (JPAR 01, JPAR 02, JPAR03); **b** (JPAR 04), **c** (JPAR 05) theropod footprints; **d** (JPAR 06), an isolated ornithopod footprint; **e** (JPAR 07), a dubious trackmaker (ornithopod or large-sized theropod) due the preservation conditions of the footprint; **f** (JPAR 08), **g** (JPAR 09) and **h** (JPAR 10) probable theropod footprints; **i** (JPAR 11), an isolated large-sized theropod footprints probably related to Abelisauroidea; Modified from Carvalho and Kattah (1998). JPAR – João Pinheiro County, Areado Group. Scale bar: 10 cm

The footprint JPAR 05 is preserved as concave epirelief. It is an isolated footprint, tridactyl and mesaxonic with 26 cm in width and 30 cm in length. Digit III is the longest in size, slightly curved in the anterior region. Digits II and IV show the same size, while digit III is the most pointed. The hypexes are acute, with 30° of angular value between digits II-III and III-IV. The rear border of the footprint is well defined and rounded.

JPAR 06 is also an isolated footprint, tridactyl and mesaxonic, preserved as concave epirelief, with 38 cm in width and 24 cm in length. The index l/w is here 0.63, a rare condition. It is, in fact, a very short and wide footprint, perhaps because of the collapse effect. Surrounding the digits and the rear border there is the deformation of the sediments with a crenulation aspect. It corresponds to a fluidization process related to the load of the trackmaker over the substrate. Digit IV is not preserved. Digit II is pointed, showing bigger dimensions than the other two digits. Only the lower portion of digit III is partially preserved. The hypex between digits II-III is wide and rounded and the III-IV angle is 40°. The rear border of the footprint is a wide continuous curve.

JPAR 07 is an isolated footprint, 20 cm in width and 48 cm in length, preserved as concave epirelief, without clear morphological details and contour. Digits II and III are partially preserved, with a wide elongation of the posterior margin of the footprint. The interdigital angle between digits II-III is 35° and the hypex is parabolic.

JPAR 08 is an isolated tridactyl and mesaxonic footprint preserved as a convex epirelief. There is also as in JPAR 06 the deformation of the sediments, with a crenulation aspect, surrounding the digits. It is an evidence of fluidization induced by the trackmaker load. The footprint is 25 cm width and 38 cm in length with a rounded rear outline. The digits show a larger lower portion and a pointing aspect in the distal region. They present almost the same size, although digit III is slightly longer (10 cm in length) than the others and shows a small curvature in the distal extremity. The angular value is 30° between digits II-III and 50° between digits III–IV.

JPAR 09 is an isolated tetradactyl and mesaxonic footprint preserved as a convex epirelief. It measures 24 cm in width and 24 cm in length. Surrounding the footprint there are crenulated deformations, especially on the borders of digit IV. The three digits present almost the same length (12 cm) with a larger lower portion. The distal ends are more pointed. The hypexes are wide, with 50° between digits II-III and 40° between digits III-IV. The rear margin presents a short spur, probably corresponding to digit I.

The footprint JPAR 10 is isolated, tridactyl and mesaxonic, 15 cm in width and in length, preserved as concave epirelief. The posterior margin is rounded and shows a continuous curve between digits II and IV. The digits are pointed, almost with the same length (7 cm), and a probable claw impression in digit III. The hypexes are acute, and the angle value between digits II-III and III-IV is 40°.

JPAR 11 is an isolated, tridactyl and mesaxonic footprint preserved as concave epirelief. The footprint measures 22 cm in width and 22 cm in length. Although the three digits are pointed they show rounded extremities, without the presence of claw marks. The hypices between II-III and III-IV are acutes and they show the same angle value (45°). The rear margin of the footprint is narrow and slightly pointed.

The trackmaker interpretation of the footprints from the João Pinheiro ichnosite is difficult. There is only one theropod based on osteological data described to the Sanfranciscana Basin, the Abelisauroidea *Spectrovenator ragei* (Zaher et al. 2020) from the Quiricó Formation. Other dinosaur remains (bone fragments and teeth) were preliminarily attributed to Abelisauridae, Dromaeosauridae (Carvalho et al. 2012; Santucci et al. 2014), Abelisauroidea (Bittencourt and Langer 2011; Zaher et al.

2011; Pires-Domingues 2009) and Noasauridade (Silva 2013). The short track with three small fooprints is classified as a small biped dinosaur. In these footprints it is possible to observe claws in the digits II, III and IV (JPAR 01, JPAR 02 and JPAR 03) and a digit I (hallux) in the rear border. The general morphology allows attributing these footprints to a small-sized theropod. Carvalho and Kattah (1998) suggested to this track an origin "coelurosauriform" (term that had another meaning, at that time), and the Aptian small theropods from the Araripe Basin, similar to *Mirischia asymmetrica* Naish et al. (2004), probably a coelurosaurid or *Santanaraptor placidus* Kellner (1999), classified as coelurosaurid or as a noasaurids (Naish et al. 2004; Kellner 1999), could be good examples of trackmakers.

The footprints from the Três Barras Formation present similarities with some footprints found in the Botucatu Formation, Paraná Basin (Leonardi and Godoy 1980; Leonardi and Lima 1990). There are also features in common with footprints from China (Zhen et al. 1991) that show a wide posterior margin (*Zhengichnus jinningensis* Zhen et al. 1986 and *Xiangxipus chenxiensis* Zeng 1982), in addition to those that present a well-defined metatarsal region, similarly to the Sanfranciscana Basin footprints JPAR 06 and JPAR 07. The tracks are generally of small, cursorial theropod, related to Gondwanan clades such as noasaurids or velocisaurids. Leonardi (1977) and Langer et al. (2019) interpreted the recurrence of a similar desert-dwelling fauna based on similar tracks from the Botucatu Formation that are also found in the Caiuá Group. These authors also observed that the functionally monodactyl theropod footprints from the Botucatu Formation, Caiuá Group (Paraná Basin) and La Matilde Formation (Argentina; Casamiquela 1964) are consistent with the foot morphology of the noasaurid *Vespersaurus paranaensis*. Noasauridae is a clade of theropod dinosaurs nested within Abelisauroidea mainly known by rather incomplete records from the Cretaceous of Gondwana. They are small to medium-sized dinosaurs with necks, arms, and skulls relatively longer than those of other abelisauroid clades (Carrano et al. 2011; Barbosa et al. 2023) and a species such as *Vespersaurus paranaensis* was not a top predator, having a possibly generalist diet and an opportunistic feeding strategy (Barbosa et al. 2023).

The footprints JPAR 10 and JPAR 11, that show an acute rear border and bigger dimensions than the short track (JPAR 01, JPAR 02 and JPAR 03), are related to large-sized theropod dinosaurs such as the Abelisauroidea. The same classification is assigned to the footprints JPAR 04, JPAR 05, JPAR 08 and JPAR 09, with pointed digits and digit III (JPAR 04, JPAR 05, JPAR 08) longer than the others. In the Sanfranciscana Basin the unique possible comparison could be with *Spectrovenator ragei* (Zaher et al. 2020). Although it is an Abelisauridae from the Quiricó Formation (Aptian) it is correlative in time to these footprints from the Três Barras Formation. There are few postcranial remains of *Spectrovenator* to estimate its length: an astragalus is fused to the calcaneum; astragalar ascending process is tall and laminar and lacks the fusion with the fíbula; the metatarsus is gracile, lacking the reduced width of metatarsal II present in noasaurids; the pedal unguals have proximally bifurcated grooves (Y-shaped) and a flexor tubercle with an associated ventral depression (Zaher et al. 2020). It is a medium-sized Abelisauridae (2.5 m in length), which general dimensions fit in the larger isolated footprints proportions (20–28 cm in width and 23–48 cm in length) of the Três Barras Formation.

The footprint JPAR 06 is the one with the most distinct morphological pattern among the Sanfranciscana Basin footprints. The rear border and the obtuse angled hypex between digits II-III suggest an autopodium with a digital membrane as identified by Reynolds (1991) and Zhen et al. (1991). Therefore Reynolds (1991) suggested that the large width of the footprint could be related to a locomotor response to the sand surface where the trackmaker walked, and not necessarily related to the presence of a digital membrane between digits. The absence of sharp claws indicates that the footprints can be assigned to the ornithopods described and illustrated by Leonardi (1994).

Concerning footprint JPAR 07, the stretching of the rear border could represent very distinct groups. The impression of a metatarsal, which is common in Lower Jurassic ornithopods (Thulborn 1990; Ellenberger 1970) such as *Anomoepus scambus* and *Moyenisauropus natator*, is also present in Triassic and Cretaceous theropods (Lockley and Gillette 1991). Kuban (1991) considered that the metatarsal impression in biped dinosaurs could be a behavioral response to the unstable condition of the substrate or a posture change of the trackmaker during the walk.

5.3.2 Presidente Olegário Ichnosite

The footprints from the Presidente Olegário ichnosite are preserved as cross-section structures concave-upward, symmetric and asymmetric deformation structures, and a narrow shaft structure in the facies meter-scale trough cross-bedded sandstone (St2) proposed by Mescolotti et al. (2019). The concave-upward footprints range 55–60 cm in width and 20 cm in penetration depth, despite the depth of the bedding deformation reach 75 cm. The narrow shaft shows straight borders, without the bending of the surrounding sandstone beds. It is slightly curved and pointed in the lower border. It has 36 cm in width and 80 cm in depth. The load of the trackmakers induced a successive deformation in the lower levels of reddish sandstones, fine- to medium grained. In the case of the large concave-upward footprints there isn't the rupture of the lower strata. Instead, the narrow shaft shows a clear rupture of the substrate lamina succession with the association of microfaults. The environment where these tracks were preserved was interpreted as the migration of large sinuous-crested aeolian dunes, sometimes with sand liquefaction at the base of foresets due to gravitational instability, in a context of wet interdune facies (Mescolotti et al. 2019) (Fig. 5.6).

There are also concave-upward symmetric and asymmetric cross-section tracks in the interdune-dune contact. They range from 25–45 cm in width and 32–37 cm penetration depth, despite the depth of the bedding deformation reaches 65 cm. They are found in fine- to medium-grained parallel laminated sandstones with two-fold grain size segregation, arranged in 0.5- to 2-m-thick horizontal to low angle (less than 5°) strata (SI facies, Mescolotti et al. 2019). The paleoenvironment is interpreted as

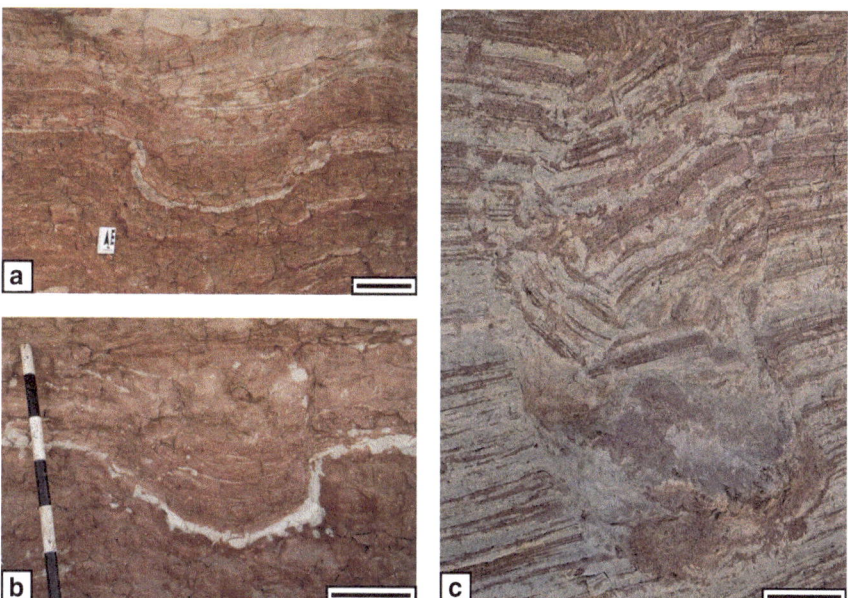

Fig. 5.6 Cross-section footprints from the wet aeolian system of Três Barras Formation, Presidente Olegário ichnosite. **a, b** Concave upward deformation structures with bedding deformation interpreted as a sauropod footprint in the wet interdune facies; **c** Narrow shaft structures, with a pointed lower extremity, this last being interpreted as the impression of a theropod digit III in the dune foresets. Facies St2—meter-scale trough cross-bedded sandstone. Photographs courtesy by Patricia Colombo Mescolotti. **a** Scale bar: 20 cm; **b** Scale bar: 2 cm; **c** Scale bar: 5 cm

protodunes migration across deflation sand flats under high ratio of wind velocity/ sedimentary supply (Mescolotti et al. 2019).

The cross-section concave-upward footprints (Fig. 5.7) from the Presidente Olegário ichnosite were interpreted as produced by sauropods, considering the morphology and dimensions of similar occurrences (Avanzini 1998; Gatesy 2003; Romano and Whyte 2003; García-Ramos et al. 2006; Falkingham et al. 2016; Sanz et al. 2016; Campos-Soto et al. 2017; Díaz-Martínez et al. 2018; Mescolotti et al. 2019; Carvalho et al. 2021a). The trackmaker could be a large sauropod similar to *Tapuiasaurus macedoi*, a Titanosauria of the family Nemegtosauridade found in the Aptian of Quiricó Formation (Zaher et al. 2011). The other cross-section footprint that is a narrow shaft, slightly curved with a pointed impression is related to a tridactyl trackmaker, in which digit III supported a greater load and consequently its marked impression in the substrate. The medium-sized Abelisauroidea *Spectrovenator ragei* from the Quiricó Formation (Zaher et al. 2020) could be a good example of trackmaker to this footprint morphology. In bipedal dinosaurs, the digit III exerts a higher pressure on the substrate, conducting to a greater deformation in the central area of the cast as suggested by Gatesy et al. (1999), Milàn and Bromley (2006), Falkingham

Fig. 5.7 Two cross-section concave upward footprints in the interdune-dune contact. The probable trackmaker is a sauropod. Facies Sl—parallel-laminated sandstone. Photograph courtesy by Patricia Colombo Mescolotti

and Gatesy (2014) and Milàn et al. (2004, 2006) that analyzed the sediment deformation induced by theropod foot movements during a stride. Besides, as proposed by Delcourt et al. (2024) the distribution of abelisaurids was driven by climatic conditions because they were well adapted to semi-arid conditions and diversified in size and number of species.

5.4 Paleogeographical Distribution of the Footprints

At the end of the Jurassic and beginning of the Cretaceous, a region of great aridity was established in the interior of the Gondwana continent. The eolic and fluvial sandstones found in Rio Grande do Sul State (Guará Formation, Late Jurassic), indicate a semi-arid climate (Scherer and Lavina 2005, 2006). At the beginning of the Cretaceous (Scherer 2000, 2002; Scherer et al. 2002), there was a change in climate, with hyperarid conditions, establishing a wide desert known as the Botucatu Paleodesert (Almeida 1953; Bertolini et al. 2021). The origin of the Botucatu Paleodesert (Botucatu Formation) resulted from the geographic configuration defined from the end of the Permian, with the origin of Pangea. The area occupied by this paleodesert reached about 1,300,000 km^2, covering central-southern Brazil, northern Uruguay, eastern Paraguay and eastern Argentina. The ichnofauna of the Botucatu Formation is an endemic fauna of an extremely arid environment (Leonardi and Sarjeant 1986) including "coelurosaurs", "carnosaurs", ornithopods, and mammals trackmakers (Leonardi 1980; Leonardi 1981a, b; Leonardi 1991; Leonardi and Godoy 1980). The presence of conchostracans indicates the existence of temporary lakes (Cardoso 1965).

The paleogeographical position of continental landmasses, the low eustatic sea level and atmospheric currents, defined a climate of great aridity during almost all of the Mesozoic (Almeida and Carneiro 1998). This climate changed after the breakup of Gondwana during the Cretaceous, which led to a modification in the atmospheric and oceanic currents, humidity and temperature and, therefore, to the disappearance of this great desert.

During the Late Cretaceous, regional arid conditions are still persistent as observed in the deposits of the Caiuá Group. It is a quartz and feldspathic sandstones succession, fine to-medium grained size, reddish in color, with large cross stratifications presenting an ichnofauna of theropods and mammals (Leonardi 1977; Leonardi 1981a, b; Leonardi 1991). The dinosaur tracks in this unit are few, with only one large theropod trackway, one trackway and an isolated footprint of a "coelurosauriform". None of these are similar to those found in the Sanfranciscana Basin. Their morphologies are masked by a wide deformation zone on the surrounding sandy matrix, not allowing clearly recognizing the contours of the footprints and their digits.

The stratigraphic levels of the Sanfranciscana Basin (Três Barras Formation) where the fossil footprints are found are indicative of humid interdunes areas, with subordinated deflationary pavements (Kattah 1991). The footprints with a clear morphology would indicate the moments of greater wetness of the substrate, while without clear morphology would have formed in conditions of dryness. During the living time (Barremian-Albian) of the trackmakers occurred great environmental changes related to the South Atlantic opening (Mescolotti et al. 2019), especially a more humid climate and the first marine ingressions from the Equatorial region.

5.5 Paleoenvironmental and Paleoclimatic Contexts

The more diverse biologic activity in deserts occurs in interdune areas. The parameters controlling the kind of activity in those settings are moisture, sedimentation rate, and the size of surrounding dunes. Especially in the interdune environments, such as the wadi, there is a high potential to the preservation of the fossil footprints. The maintenance of tracks in aeolian sandstones is directly related to the cohesion of the sand, related to a high humidity, when the trackmaker induces the deformation of the substrate (Ahlbrandt et al. 1978). Furthermore the fast burial also plays an important role in the preservation, especially in the leeward and interdune areas, which are the most conducive to preserving footprints.

The origin of the wetness in the sand dunes that allow the footprints preservation is controversy. Leonardi (1980) considered that the preservation of footprints in the Botucatu Paleodesert would be related to a higher humidity (because of dew) of the sandy surface during the night, when the footprints would be produced and at dawn there would be their subsequent covering by dry sand. Another possibility would be the existence of a high groundwater level, which would allow a greater wetness of the sediments and, consequently, the preservation of the footprints even in the dunes foreset. However, the predictive model of Winkler et al. (1991), of restricting occurrences of fossil footprints to interdune areas, is the most common. Interdune deposits represent a complex sedimentation of aeolian and non-aeolian environments ranging from deflation surfaces to ephemeral lakes. Due to the high local sedimentation rate and grain size of the surrounding active dunes, the potential for fossil footprints preservation in interdune areas is high. In addition, observation of current environments demonstrates that the wetness conditions in these areas allow the proliferation of a diverse biota (Carvalho and Kattah 1998).

In this way, the lack of fossils in deposits adjacent to the interdunes would reflect the low original biological productivity and the low diversity of large animals in an area of active dunes. Winkler et al. (1991) postulated that the presence of fossil tracks and fossils of large vertebrates in the interdune deposits of the Navajo Sandstone (Middle Jurassic, Arizona, USA) would imply the existence of a complex food chain, and may even represent temporal intervals not typical of the hyper aridity of deserts, a moment where the dunes are most active. Leonardi (1991) considered that the dinosaur tracks in the desert paleoenvironments of South America generally belong to bipedal and small-sized forms. The frequency of theropods is high (87%) with a predominance of "coelurosauriforms", absence of sauropods and rare ornithopods, which possibly would have been the largest dinosaurs in arid environments. A similar conclusion was also reached by Lockley and Conrad (1991) in the study of the Glen Canyon and San Rafael groups (Jurassic, USA), in which most of the fossil footprints in desert paleoenvironments are those produced by theropods. There would be a preferential distribution of these in the playa lake and interdune fluvial deposits, when compared with the paleoenvironments that represent the dune fields.

The footprints of the Três Barras Formation (João Pinheiro ichnosite) are in a context of sandstones with fine to medium grain size, poorly selected. There is inter-calation of centimetric levels of coarse-grained sandstones, claystones and siltstones. The color is pink to reddish. Under the microscope, crystals of calcite and dolomite are observed in the siltstones. The sedimentary structures are dissecation cracks, symmet-rical and climbing wavy ripples, and channel cross-bedding. The pavement on which the footprints are found is poor selected quartz sandstone, locally with ripple marks and tabular cross-stratifications. Above the level of footprints there is conglomeratic sandstone with faceted pebbles. Kattah (1991) interpreted this stratigraphic level as indicative of an area of humid interdunes, with subordinated deflationary pavements. Most footprints preserved on this surface do not have clear morphological details, which must be related to a very loose sandy substrate. However in some cases (JPAR 01 and JPAR 03) the anatomical details such as claws and phalangeal pads are quite evident. Reynolds (1991) and Lockley (1991) consider that this fact results from fluctuations in the hydrostatic level during the formation of footprints. The foot-prints with a clear morphology would indicate the moments of greater moisture of the substrate. The others would have formed in conditions of greater aridity, when the groundwater is lower. Drier sand would not have been cohesive enough to allow for better footprint preservation.

The fossil footprints from the João Pinheiro ichnosite were recognized by morpho-logical data from the autopodia on the bedding plane. Otherwise the footprints from the Presidente Olegário ichnosite reach many sedimentary levels beyond the surface and they are preserved as deformation structures which can be observed in cross-sections. They are directly related with the trampling by terrestrial vertebrates and the pressure generated during the contact between a tetrapod autopodium and the substrate, leading to the origin of load structures with successive laminae deformation (Carvalho et al. 2021a, b, 2022). Deformation of the print-bearing surface, by the dinosaur trampling, favors the preservation of underprints and transmitted prints on bedding planes beneath the primary footprint-bearing surface as "undertracks" or "ghost prints" (Sarjeant and Leonardi 1987). The stratigraphic analysis of Mescolotti et al. (2019) considered a wet aeolian system including a wet interdune facies, interdune-dune contact and dune foresets as the environments where these footprints were preserved (Fig. 5.8).

5.6 Conclusions

The desert environment is a challenging geological context to the preservation of footprints. The dry sands, the low humidity, the inconstancy and fast changes in the substrate, the deflation, generally destroy the tracks or just allow a deformation structure in the contact between the trackmaker's feet and the superficial layers. Despite these restrictions, some environmental settings especially in the wet interdune areas, where there is a more diverse biologic activity in deserts, present a high potential to the preservation of the footprints. Interdune deposits represent a complex sedimentation of aeolian and non-aeolian environments ranging from deflation surfaces to ephemeral lakes. Due to the high local sedimentation rate and grain size of the surrounding active dunes, the potential for fossil footprints preservation in the interdune areas is high. In addition, observation of current environments demonstrates that the moisture conditions in these areas allow the proliferation of a diverse biota. Then, the maintenance of tracks in aeolian sandstones is directly related to a high humidity that allows the cohesion of the sand. The fast burial also plays an important role in the preservation, especially in the leeward and interdune areas, which are the most conducive to preserving footprints.

The footprints from the Sanfranciscana Basin are found in two localities where the Três Barras Formation outcrops: João Pinheiro and Presidente Olegário ichnosites. There are eleven dinosaur footprints in the João Pinheiro ichnosite preserved as concave epilelief with infilling of fine to coarse-grained reddish sandstone similar to the surrounding matrix. In the Presidente Olegário ichnosite the footprints are observed as cross-section concave upward deformation structures and narrow shaft straight borders. These two sets of footprints indicate small and medium-sized theropods, ornithopods and sauropods. Aptian age deposits of the Sanfranciscana Basin (Quiricó Formation) present osteological remains of Abelisauroidea (Noasauridae and Abelisauridae), Dromaeosauridae and Titanosauria (Nemegtosauridae) that could be related to some of the trackmakers of João Pinheiro and Presidente Olegário ichnosites.

Fig. 5.8 Environmental reconstruction of the wet aeolian system of Três Barras Formation, Sanfranciscana Basin, and the dinosaur trackmakers (Art by Guilherme Gehr)

References

Ahlbrandt TS, Andrews S, Gwyne DT (1978) J Sediment Petrol 48(3):839–848
Almeida FFM (1953) Botucatu, a Triassic desert of South America. In: 19th International Geological Congress, Argel, 1953. Comptes Rendus, XIX Sessão, fasc. VII. p 9–24

Almeida FFM, Carneiro CDR (1998) Botucatu: o grande deserto brasileiro. Ciência Hoje 24(143):36–43

Arai M (1999) A transgressão marinha Mesocretácea: sua implicação no paradigma da reconstitu-ição paleogeográfica do Cretáceo no Brasil In: 5° Simpósio sobre o Cretáceo do Brasil, Boletim, Serra Negra, Sociedade Brasileira de Geologia, p 577–582

Arai M (2009) Paleogeografia do Atlântico Sul no Aptiano: um novo modelo a partir de dados micropaleontológicos recentes. Boletim de Geociências da Petrobrás 17:331–351

Arai M (2014a) Aptian/Albian (Early Cretaceous) paleogeography of the South Atlantic: a pale-ontological perspective. Braz J Geol 44(2):339–350. https://doi.org/10.5327/Z2317-488920140 0020012

Arai, M (2014b) Reconstituições paleo-oceanográfica e paleoclimática do Oceano Atlântico no Cretáceo, baseadas em dinoflagelados. In: Carvalho IS, Garcia MJ, Lana CC, Strohschoen Jr O (eds) Paleontologia: Cenários de Vida - Paleoclimas. 1ed., vol 5. Rio de Janeiro: Editora Interciência. pp 45–62

Arai M, Dino R, Milhomem PS, Sgarbi GNC (1995) Micropaleontologia da Formação Areado, Cretáceo da Bacia Sanfranciscana: estudo dos ostracodes e palinologia. In: 14° Congresso Brasileiro de Paleontologia, Atas, Uberaba, Sociedade Brasileira de Paleontologia, pp 1–2

Avanzini M (1998) Anatomy of a footprint: Bioturbation as a key to understanding dinosaur walk dynamics. Ichnos 6:129–139

Barbosa EM, Delicio MP, Coimbra JC (1997) Conchostracans and ostracodes indicators of pale-oenvironment in the Alto Sanfranciscana Basin, Olhos d'Agua Area, Northwest of Minas Gerais state Brazil. In: The first international conference application of micropaleontology in environmental sciences. Tel Aviv, pp 37–38

Barbosa GG, Langer MC, Martins NO (2023) Montefeltro FC (2023) Assessing the palaeobiology of *Vespersaurus paranaensis* (Theropoda, Noasauridae), Cretaceous, Bauru Basin - Brazil, using Finite Element Analysis. Cretac Res 150:105594. https://doi.org/10.1016/j.cretres.2023.105594

Barbosa O (1965) Geologia econômica e aplicada a uma parte do Planalto Central brasileiro. In: 19° Congresso Brasileiro de Geologia, Rio de Janeiro, Sociedade Brasileira de Geologia, Anais pp 1–11

Barbosa O, Braun OPG, Dyer RC, da Cunha CABR (1970) Geologia da região do Triângulo Mineiro. Boletim da Divisão de Fomento da Produção Mineral 136:1–140

Bertolini G, Marques JC, Hartley AJ, Basei MAS, Frantz JC (2021) Santos PR (2021) Determing sediment provenance history in a Gondwanan erg: Botucatu Formation, Northern Paraná Basin, Brazil. Sediment Geol. https://doi.org/10.1016/j.sedgeo.2021.105883

Bittencourt JS, Brandão NCA (2021) A segmented wormlike specimen from the Lower Cretaceous lacustrine shales of the Sanfranciscana Basin in southeast Brazil. J S Amn Earth Sci 111 (2021) 103488. https://doi.org/10.1016/j.jsames.2021.103488

Bittencourt JS, Langer MC (2011) Mesozoic dinosaurs from Brazil and their biogeographic implications. Anais da Academia Brasileira de Ciências 83 (1):23–60. ISSN 0001-3765

Bittencourt JS, Kuchenbecker M, Vasconcelos AG, Meyer KEB (2015) O registro fóssil das coberturas sedimentares do cráton do São Francisco em Minas Gerais. Geonomos 23(2):39–62

Bittencourt JS, Rohn R, Gallego OF, Monferra MD, Uhlein A (2018) The morphology and system-atics of the clam shrimp *Platyestheria* gen. nov. *abaetensis* (Cardoso) (Crustacea, Spinicau-data) from the Lower Cretaceous of the Sanfranciscana Basin, southeast Brazil. Cretac Res 91:274–286

Brito PM, Figueiredo FJ, Leal MEC (2020) A revision of *Laeliichthys ancestralis* Santos, 1985 (Teleostei: Osteoglossomorpha) from the Lower Cretaceous of Brazil: Phylogenetic relation-ships and biogeographical implications. PLoS ONE 15(10):e0241009. https://doi.org/10.1371/journal.pone.0241009

Cabral VC, Mescolotti PC (2021) Varejão FG (2021) Sedimentary facies and depositional model of the Lower Cretaceous Quiricó Formation (Sanfranciscana Basin, Brazil) and their implication for the occurrence of vertebrate fauna at the Coração de Jesus region. J S Am Earth Sci 112:103632. https://doi.org/10.1016/j.jsames.2021.103632

Campos JEG, Dardenne MA (1997a) Estratigrafia e sedimentação da Bacia Sanfranciscana: uma revisão. Revista Brasileira de Geociências 27(3):269–282

Campos JEG, Dardenne MA (1997b) Origem e evolução tectônica da Bacia Sanfranciscana. Revista Brasileira de Geociências 27(3):283–294

Campos-Soto S, Cobos A, Caus E, Benito MI, Fernández-Labrador L, Suarez-Gonzalez P, Quijada IE, Mas R, Royo-Torres R, Alcalá L (2017) Jurassic coastal park: a great diversity of palaeoenvironments for the dinosaurs of the Villar del Arzobispo Formation (Teruel, eastern Spain). Palaeogeogr Palaeoclimatol Palaeoecol. https://doi.org/10.1016/j.palaeo.2017.06.010

Cardoso RN (1965) Sobre a ocorrência no Brasil de Monoleiolophinae e Afrograptidae, conchostráceos carenados. Divisão de Geologia e Mineralogia, Departamento Nacional da Produção Mineral, Boletim 221, 35p

Cardoso AR, Basilici G (2022) Silva, PAS (2022) Early diagenetic calcite replacement of evaporites in playa lakes of the Quiricó Formation (Lower Cretaceous, SE Brazil). Sed Geol 438:106212. https://doi.org/10.1016/j.sedgeo.2022.106212

Cardoso AR, Basilici G, Mesquita ÁF (2023) The Cretaceous palaeodesert of the Sanfranciscana Basin (SE Brazil): a key record to track dissolved evaporites in the West Gondwana. Cretac Res. https://doi.org/10.1016/j.cretres.2023.105788

Carmo DA, Tomassi HZ, Oliveira SBSG (2004) Taxonomia e distribuição estratigráfica dos ostracodes da Formação Quiricó, Grupo Areado (Cretáceo Inferior), Bacia Sanfranciscana, Brasil. Revista Brasileira de Paleontologia 7(2):139–149

Carrano MT, Loewen MA, Sertich JJW (2011) New materials of *Masiakasaurus knopfleri* Sampson, Carrano and Forster, 2001, and implications for the morphology of the Noasauridae. Smithson Contrib Paleobiol 95:1–53

Carvalho AB, Zaher H, da Silva RR, Nascimento PM (2012) Análise morfológica dos dentes de terópodes (Dinosauria: Saurischia) da Formação Quiricó, Eocretáceo da Bacia Sanfranciscana, estado de Minas Gerais. In: 8° Simpósio Brasileiro de Paleontologia de Vertebrados, Recife, Boletim de Resumos, p 109

Carvalho IS, Kattah SS (1998) As pegadas fósseis do Paleodeserto da Bacia Sanfranciscana (Jurássico Superior-Cretáceo Inferior, Minas Gerais). An Acad Bras Ciênc 70(1):53–67

Carvalho IS, Bertolino LC, Borghi LF, Duarte L, Carvalho MSS, Cassab RCT (1994) Range charts of the fossils of the Cretaceous interior basins—the São Francisco Basin. In: Beurlen G, Campos DA, Viviers MC (eds) Stratigraphic range of Cretaceous mega- and microfossils of Brazil. UFRJ, Rio de Janeiro, pp 333–352

Carvalho IS, Leonardi G, Rios-Netto AM, Borghi L, Paula Freitas A, Andrade JA, Freitas, FI (2021a) Dinosaur trampling from the Aptian of Araripe Basin, NE Brazil, as tools for paleoenvironmental interpretation. Cretac Res 117:104626. https://doi.org/10.1016/j.cretres.2020.104626

Carvalho IS, Leonardi G, Andrade JAFG, Freitas FI, Borghi L, Rios-Netto AM, Figueiredo SMD, Cunha PP (2021b) Dinoturbation in the Exu Formation (Cenomanian, Upper Cretaceous) from the Araripe Basin, Brazil. In: 3rd Virtual Paleontological congress, 2021. Book of Abstracts 3rd paleontological virtual congress, vol 1. pp 193–193

Carvalho IS, Neto, JX, Leonardi G, Andrade JAFG, Freitas FI, Borghi L, Rios-Netto AM, Figueiredo SMD, Cunha PP (2022) Estruturas de Dinoturbação do Cretáceo Superior da Bacia do Araripe (Formação Exu), Brasil. In: 27° Congresso Brasileiro de Paleontologia, Cuiabá, 2022. Paleontologia em Destaque, 37 (Ed. Especial), p 133. https://doi.org/10.4072/paleodest.2022.37.ed.especial

Carvalho JC, Santucci RM (2021) New fish remains from the Quiricó Formation (Lower Cretaceous, Sanfranciscana Basin), Minas Gerais, Brazil. Journal of South American Earth Sciences 111:103430. https://doi.org/10.1016/j.jsames.2021.103430

Carvalho MSS, Campos DA, Dardenne MA, Sgarbi GNC, Campos JEG, Cartelle C (1995) Ocorrência de celacantídeo *Mawsonia* nos sedimentos lacustres da Bacia Sanfranciscana, noroeste do Estado de Minas Gerais. In: XIV Congresso Brasileiro de Paleontologia, Uberaba, p 35

Carvalho MSS, Maisey JG (2008) New occurrence of *Mawsonia* (Sarcopterygii: Actinistia) from the Early Cretaceous of the Sanfranciscana Basin, Minas Gerais, southeastern Brazil. Geol Soc Spec Pub 295:109–144

Casamiquela RM (1964) Estudios Icnológicos. Problemas y métodos de la Icnología con aplicación al estudio de pisadas mesozóicas (Reptilia, Mammalia) de la Patagónia. Buenos Aires, p 229, 26 plates

Castro JC (1996) O Cretáceo da porção sul da Bacia Sanfranciscana: uma breve revisão. In: 4º Simpósio sobre o Cretáceo do Brasil, Águas de São Pedro, Sociedade Brasileira de Geologia, Boletim pp 209–211

Coimbra JC (2020) The Genus *Cypridea* (Crustacea, Ostracoda) and the age of the Quiricó Formation, SE Brazil: a critical review. Revista Brasileira de Paleontologia 23(2):90–96. https://doi.org/10.4072/rbp.2020.2.02

Delcourt R, Brilhante N, Pires-Domingues RA, Hendrickx C, Grillo ON, Augusta BG, Maciel BS, Ghilardi AM, Ricardi-Branco F. (2024) Biogeography of theropod dinosaurs during the Late Cretaceous: evidence from central South America. Zool J Linn Soc XX:1–40. https://doi.org/10.1093/zoolinnean/zlad184

Delicio MP, Barbosa EM, Coimbra JC, Vilella RA (1998) Ocorrência de conchostraceos e ostracodes em sedimentos Pós-Paleozóicos da Bacia do Alto Sanfranciscana, Olhos d´Água, noroeste de Minas Gerais, Brasil. Acta Geologia Leopoldensia 46(47):13–20

Dias-Brito D, Pessagno Jr. EA, Castro JC (1999) Novas considerações cronoestratigráficas sobre silexito a radiolários a sul da Bacia Sanfranciscana, Brasil, e a ocorrência de foraminíferos planctônicos nestes depósitos. In: 5º Simpósio sobre o Cretáceo do Brasil, Boletim, Rio Claro, Sociedade Brasileira de Geologia, pp 567–575

Díaz-Martínez I, Cónsole-Gonella C, de Valais S, Salgado L (2018) Vertebrate tracks from the Paso Córdoba fossiliferous site (Anacleto and Allen formations, Upper Cretaceous), Northern Patagonia, Argentina: preservational, environmental and palaeobiological implications. Cretac Res 83:207–220. https://doi.org/10.1016/j.cretres.2017.07.008

Duarte L (1968) Restos vegetais fósseis da Formação Areado. In: 22º Congresso Brasileiro de Geologia, Belo Horizonte, Sociedade Brasileira de Geologia, Anais, p 68

Duarte L (1985a) Vegetais fósseis da Chapada do Araripe. In: Campos DA, Ferreira CS, Brito IM, Viana CF (eds) Coletânea de trabalhos paleontológicos. DNPM, Rio de Janeiro, pp 557–563

Duarte L (1985b) Vegetais fósseis da Formação Areado, município de Presidente Olegário, Minas Gerais. In: 9º Congresso Brasileiro de Paleontologia, Fortaleza, Sociedade Brasileira de Paleontologia, Resumos p 59

Duarte L (1997) Vegetais do Cretáceo Inferior (Aptiano) da Formação Areado, município de Presidente Olegário, Estado de Minas Gerais. An Acad Bras Ciênc 69:495–503

Ellemberger P (1970) Le niveaux paléontologiques de première apparition des mammifères primordiaux en Afrique du Sud e leur ichnologie. Establissement de zones stratigraphiques detaillées dans Le Stormberg du Lesotho (Afrique du Sud) (Trias Supérieur à Jurassique). In: Haughton SH (ed) Proceedings and papers of the Second Gondwana Symposium, South Africa, 1970. South African Council for Scientific and Industrial Research, Pretoria, pp 343–370

Falkingham PL, Gatesy SM (2014) The birth of a dinosaur footprint: subsurface 3D motion reconstruction and discrete element simulation reveal track ontogeny. PNAS 111(51):18279–18284

Falkingham PL, Marty D, Richter A (eds) (2016) Dinosaur tracks: the next steps. Indiana University Press, p 428. ISBN-10: 9780253021021

Fragoso LGC, Bittencourt JS, Mateus ALD, Cozzuol MA, Richter M (2019) Shark (Chondrichthyes) microremains from the Lower Cretaceous Quiricó Formation, Sanfranciscana Basin, Southeast Brazil Shark (Chondrichthyes) microremains from the Lower Cretaceous Quiricó Formation, Sanfranciscana Basin, Southeast Brazil. Historical Biology https://doi.org/10.1080/08912963.2019.1692830

Gallego OF, Martins-Neto RG (2006) The Brazilian Mesozoic conchostracan faunas: its geological history as an alternative tool for stratigraphic correlations. Geociências 25:231–239

García-Ramos JC, Piñuela L, Lires JL (2006) Atlas del Jurásico de Asturias. Oviedo. Ediciones Nobel. p 228. ISBN: 9788484596578

Gatesy SM (2003) Direct and Indirect tracks features: what sediment did a dinosaur touch? Ichnos 10:91–98

Gatesy SM, Middleton KM, Jenkins FA Jr, Shubin NH (1999) Three-dimensional preservation of foot movements in Triassic theropod dinosaurs. Nature 399(6732):141–144

Grossi Sad JH, Cardoso RN, da Costa MT (1971) Formações cretácicas em Minas Gerais: uma revisão. Revista Brasileira de Geociências 1:2–13

Kattah SS (1991) Análise faciológica e estratigráfica do Jurássico Superior/Cretáceo Inferior na porção meridional da bacia do São Francisco, oeste do estado de Minas Gerais. Universidade Federal de Outro Preto, Escola de Minas, Departamento de Geologia (Dissertação de Mestrado), p 227

Kattah SS (1992) Novas considerações sobre o Aptiano-Albiano da Bacia Sanfranciscana. In: 2° Simpósio sobre as Bacias Cretácicas Brasileiras, Rio Claro, UNESP, pp 118–119

Kattah SS (1993) A ocorrência de pegadas de dinossauros no Grupo Areado, porção meridional da bacia Sanfranciscana, oeste de Minas Gerais. An Acad Bras Ciênc 65(2):217–218

Kattah SS, Koutsoukos EAM (1992) Ocorrências de radiolários em fácies de origem marinha no Mesozóico da Bacia Sanfranciscana. Revista da Escola de Minas 45:214

Kellner AWA (1999) Short note on a new dinosaur (Theropoda, Coelurosauria) from the Santana Formation (Romualdo Member, Albian), Northeastern Brazil. Boletim Do Museu Nacional 49:1–8

Kuban GJ (1991) Elongate dinosaur tracks. In: Gillette DD, Lockley MG (eds) Dinosaur tracks and traces. Cambridge University Press, Cambridge, pp 57–72

Langer MC, Martins NO, Manzig PC, Ferreira GS, Marsola JCA, Fortes E, Lima R, Sant'ana LCF, Vidal LS, Lorencato RHS, Ezcurra MD (2019) A new desert-dwelling dinosaur (Theropoda, Noasaurinae) from the Cretaceous of south Brazil. Sci Rep 9:9379. https://doi.org/10.1038/s41 598-019-45306-9

Leite AM, Carmo DA (2021) Description of the stratotype section and proposal of hypostratotype section of the Lower Cretaceous Quiricó formation, São Francisco Basin, Brazil. Anais da Academia Brasileira de Ciências 93(Suppl. 2):e20201296. https://doi.org/10.1590/0001-376 5202120201296

Leite AM, Carmo DA, Ress CB, Pessoa M, Caixeta GM, Denezine M, Adorno RR, Antonietto LS (2018) Taxonomy of limnic Ostracoda (Crustacea) from the Quiricó Formation, Lower Cretaceous, São Francisco basin, Minas Gerais State, Southeast Brazil. J Paleontol 1–20. https:// doi.org/10.1017/jpa.2018.1

Leonardi G (1977) Two new ichnofaunas (vertebrates and invertebrates) in the eolian Cretaceous sandstones of the Caiuá Formation in northwest Paraná. In: 1° Simpósio Regional de Geologia, São Paulo, 1977, Sociedade Brasileira de Geologia, Actas, pp 112–128

Leonardi G (1980) On the discovery of an abundant ichnofauna (vertebrates and invertebrates) in the Botucatu Formation s.s. in Araraquara, São Paulo, Brazil. Anais da Academia Brasileira de Ciências 52(3):559–567

Leonardi G (1981a) As localidades com rastros fósseis de tetrápodes na América Latina. In: 2° Congresso Latinoamericano de Paleontologia, Porto Alegre, 1981, Porto Alegre, Anais 2:929–940

Leonardi G (1981b) Novo icnogênero de tetrápode mesozoico da Formação Botucatu, Araraquara, SP. An Acad Bras Ciênc 53(4):793–805

Leonardi G (1991) Inventory and statistics of the South American dinosaurian ichnofauna and its paleobiological interpretation. In: Gillette DD, Lockley MG (eds) Dinosaur tracks and traces. Cambridge University Press, Cambridge, pp 165–178

Leonardi G (1994) Annotated Atlas of South American Tetrapod Footprints (Devonian to Holocene). Ministério das Minas e Energia, Rio de Janeiro, Companhia de Pesquisa e Recursos Minerais, p 247

Leonardi G, Godoy LC (1980) Novas pistas de tetrápodes da Formação Botucatu no estado de São Paulo. In: 31º Congresso Brasileiro de Geologia, Camboriú, 1980, Camboriú, Santa Catarina, Sociedade Brasileira de Geologia, Anais 5:3080–3089

Leonardi G, Lima FHO (1990) A revision of the Triassic and Jurassic tetrapod footprints of Argentina and a new approach on the age and meaning of the Botucatu Formation fooptrints (Brazil). Revista Brasileira de Geociências 20(1–4):216–229

Leonardi G, Sarjeant WAS (1986) Footprints representing a new Mesozoic vertebrate fauna from Brazil. Mod Geol 10:73–84

Lima MR (1979) Palinologia dos calcários laminados da Formação Areado, Cretáceo de Minas Gerais. In: 2º Simpósio Regional de Geologia, Rio Claro, Sociedade Brasileira de Geologia, Atas pp 203–216

Lockley M (1991) Tracking dinosaurs. Cambridge University Press, Cambridge, p 238

Lockley M, Conrad K (1991) The paleoenvironmental context, preservation and paleoecological significance of dinosaur tracksites in the Western USA. In: Gillette DD, Lockley MG (eds) Dinosaur tracks and traces. Cambridge University Press, Cambridge, pp 121–134

Lockley MG, Gillette DD (1991) Dinosaur tracks and traces: an overview. In: Gillette DD, Lockley MG (eds) Dinosaur tracks and traces. Cambridge University Press, Cambridge, pp 3–10

Mescolotti PC, Varejão FG, Warren LV, Ladeira FSB, Giannini PCF, Assine ML (2019) The sedimentary record of wet and dry eolian systems in the Cretaceous of Southeast Brazil. Braz J Geol 49(3) https://doi.org/10.1590/2317-4889201920190057.

Milàn J, Bromley RG (2006) True tracks, undertracks and eroded tracks, experimental work with tetrapod tracks in laboratory and field. Palaeogeogr Palaeoclimatol Palaeoecol 231:253–264

Milàn J, Clemmensen LB, Bonde N (2004) Vertical sections through dinosaur tracks (Late Triassic lake deposits, East Greenland)—undertracks and other subsurface deformation structures revealed. Lethaia 37:285–296

Milàn J, Avanzini M, Clemmensen LB, García-Ramos JC, Piñuela L (2006) Theropod foot movement recorded by Late Triassic, Early Jurassic and Late Jurassic fossil footprints. N M Mus Nat Hist Sci Bull 37:352–364

Naish D, Martill DM, Frey E (2004) Ecology, systematics and biogeographical relationships of dinosaurs, including a new theropod, from the Santana Formation (?Albian, Early Cretaceous) of Brazil. Hist Biol 18:1–14

Nascimento DL, Martinez P, Batezelli A, Ladeira F (2022) Corrêa L (2022) From the micromorphology of paleoweathering fronts to paleoenvironmental analysis: a case study of the Cretaceous dune fields of Sanfranciscana Basin, Brazil. Catena 211:106008

Pessagno Jr. EA, Dias-Brito D (1996) O silexito a radiolários do sul da Bacia Sanfranciscana, Brasil: idade, origem e significado. In: 4º Simpósio sobre o Cretáceo do Brasil, Águas de São Pedro, Sociedade Brasileira de Geologia, Boletim pp 213–221

Pessagno Jr. EA, Dias-Brito D, Castro JC (1997) Tectonostratigraphic significance of radiolarian chert in Lower Cretaceous continental sequence, Minas Gerais, Brazil. In: 29th GSA Annual Meeting, Utah, p 374

Pires-Domingues RA (2009) Paleogeografia do Alto de Paracatu: o registro geológico dos *bonebeds* de dinossauros da Bacia Sanfranciscana. Universidade de São Paulo, Tese de Mestrado, p 110

Reynolds RE (1991) Dinosaur trackways in the Lower Jurassic Aztec Sandstone of California. In: Gillette DD, Lockley MG (eds) Dinosaur tracks and traces. Cambridge University Press, Cambridge, pp 285–292

Ribeiro AC, Poyato-Ariza FJ, Bockmann FA, Carvalho MR (2018) Phylogenetic relationships of Chanidae (Teleostei: Gonorynchiformes) as impacted by *Dastilbe moraesi*, from the Sanfranciscana basin Early Cretaceous of Brazil. Neotrop Ichthyol 16(3):e180059. https://doi.org/10.1590/1982-0224-20180059

Romano M, Whyte MA (2003) Jurassic dinosaur tracks and trackways of the Cleveland Basin, Yorkshire: preservation, diversity and distribution. Proc Yorks Geol Soc 54:185–215

Santos MECM (1971) Um nôvo artrópodo da Formação Areado, Estado de Minas Gerais. An Acad Bras Ciênc 43:415–420

Santos RS (1985) *Laeliichthys ancestralis*, novo gênero e espécies de Osteoglossiformes do Aptiano da Formação Areado, estado de Minas Gerais, Brasil, In: Campos DA, Ferreira CS, Brito IM, Viana CF (eds) Paleontologia e Estratigrafia, Coletânea de Trabalhos Paleontológicos. DNPM, Brasília, pp 161–167

Santucci RM, Pinto RL, Almeida MF, Souza LM, Mineiro AS, Santos DM (2014) Um dente de terópode da Formação Quiricó, Bacia Sanfranciscana (Aptiano) do norte de Minas Gerais. In: 9° Simpósio Brasileiro de Paleontologia de Vertebrados, Vitória, pp 124

Sanz E, Arcos A, Pascual C, Pidal IM (2016) Three-dimensional elasto-plastic soil modelling and analysis of sauropod tracks. Acta Palaeontol Pol 61(2):387–402

Sarjeant WAS, Leonardi G (1987) Substrate and Footprints. In: Leonardi G (ed) Glossary and Manual of Tetrapod Footprint Palaeoichnology. Brasília, DNPM (Geological Survey of Brazil, Brasília)

Scherer CMS (2000) Eolian dunes of the Botucatu formation (Cretaceous) in Southernmost Brazil: morphology and origin. Sed Geol 137:63–84

Scherer CMS (2002) Preservation of aeolian genetic units by lava flows in the Lower Cretaceous of the Paraná Basin Southern Brazil. Sedimentology 49(1):97–116

Scherer CMS, Lavina EL (2005) Sedimentary cycles and facies architecture of aeolian-fluvial strata of the Upper Jurassic Guará Formation, Southern Brazil. Sedimentology 52:1323–1341

Scherer CMS, Lavina EL (2006) Stratigraphic evolution of a fluvial-eolian succession: the example of the Upper Jurassic-Lower Cretaceous Guará and Botucatu formations, Paraná Basin, Southernmost Brazil. Gondwana Res 9:475–484

Scherer CMS, Faccini UF, Lavina EL (2002) Arcabouço estratigráfico do Mesozóico da Bacia do Paraná. In: Holz M, De Ros LF (eds) Geologia do Rio Grande do Sul. Instituto de Geociências/CIGO, pp 335–354

Scorza FP, Santos RS (1955) Ocorrência de folhelho fossilífero no município de Presidente Olegário, Minas Gerais. Boletim do Departamento Nacional de Produção Mineral—Divisão De Geologia e Mineralogia 155:1–27

Sgarbi GNC, Sgarbi PBA, Campos JEG, Dardenne MA, Penha UC (2001) Bacia Sanfranciscana: o registro Fanerozóico da Bacia do São Francisco In: Pinto CP, Martins-Neto MA (eds) Bacia do São Francisco: Geologia e Recursos Naturais. SBG, Belo Horizonte, pp 93–138

Sgarbi PCB, Heaman LM, Gaspar JC (2004) U-Pb perovskite ages for Brazilian kamafugitic rocks: further support for a temporal link to a mantle plume hotspot track. J S Am Earth Sci 16(8):715–724. https://doi.org/10.1016/j.jsames.2003.12.005

Silva RR (2013) Descrição osteológica e posicionamento filogenético de um terópode (Dinosauria, Saurischia) do Cretáceo Inferior da Bacia Sanfranciscana, município de Coração de Jesus, Minas Gerais. Tese de Mestrado, Universidade de São Paulo, Brasil, p 121

Suguio K, Barcelos JH (1983) Paleoclimatic evidence from the Areado Formation, Cretaceous of the Sanfranciscana Basin, State of Minas Gerais, Brazil. Revista Brasileira de Geociências 13:229–231

Thulborn T (1990) Dinosaur tracks. London, Chapman and Hall, 1st edn. 410

Winkler DA, Jacobs LL, Congleton JD, Downs WR (1991) Life in a sand sea: biota from Jurassic interdunes. Geology 19:889–892

Zaher H, Pol D, Carvalho AB, Nascimento PM, Riccomini C, Larson P, Juarez-Valieri R, Pires-Domingues R, Silva Jr. NJd, Campos DA (2011) A complete skull of an Early Cretaceous sauropod and the evolution of advanced titanosaurians. PlosOne 6:1–10

Zaher H, Pol D, Navarro BA, Delcourt R, Carvalho AB (2020) An Early Cretaceous theropod dinosaur from Brazil sheds light on the cranial evolution of the Abelisauridae. CR Palevol 19(6):101–115. https://doi.org/10.5852/cr-palevol2020v19a6

Zeng X (1982) Fossil handbook of hunan province (Geology Bureau of Hunan Province)

Zhen S, Zhen B, Rao C (1986) Dinosaur footprints of Jinning. Yunnan Mem Beijing Nat His Museum 33:1–18

Zhen S, Jianjun L, Chenggang R, Mateer NJ, Lockley MG (1991) A review of dinosaur footprints in China. In: Gillette DD, Lockley MG (eds) Dinosaur tracks and traces. Cambridge University Press, Cambridge, pp 187–197

Chapter 6
The Cretaceous Araripe Basin Dinosaur Tracks and Their Paleoenvironmental Meaning

Ismar de Souza Carvalho, Giuseppe Leonardi, and Jaime Joaquim Dias

6.1 Introduction

During the Mesozoic in South America, the terrestrial ecosystems were remodeled due to changes in the configuration of continents and oceans, particularly the opening of the South Atlantic Ocean by the Gondwana supercontinent rifting process (Matos 1992; Assine 1992, 2007; Marques et al. 2014). Within this context, continental depressions were formed analogous to pull-apart basins, with their genesis by transcurrent tectonics along faults during the opening of the Atlantic Ocean (Matos 1992). The Araripe Basin is the largest among these interior sedimentary basins in northeastern Brazil (Fig. 6.1), covering an approximate area of 12,200 km^2 in the southern part of the Ceará State, and portions of the Pernambuco and Piauí states (Carvalho et al. 2012; Fambrini et al. 2020; Dias et al. 2022). The Araripe Basin is not only important for understanding the environment and climate of the Brazilian Mesozoic but also stands out for the high quantity and quality of its fossils, including dinosaur footprints found in four lithostratigraphic units: Mauriti, Rio da Batateira, Crato, and Exu formations (Carvalho et al. 1995, 2018, 2019a, b, 2021a, b, 2022, 2023).

I. S. Carvalho (✉) · G. Leonardi · J. J. Dias
CCMN/IGEO, Departamento de Geologia, Universidade Federal do Rio de Janeiro, 21.910-200 Cidade Universitária, Ilha do Fundão, Rio de Janeiro, Estado do Rio de Janeiro, Brazil
e-mail: ismar@geologia.ufrj.br

I. S. Carvalho
Centro de Geociências, Universidade de Coimbra, Rua Sílvio Lima, 3030-790 Coimbra, Portugal

G. Leonardi
Istituto Cavanis, Dorsoduro 898, 30123 Venezia, Italy

J. J. Dias
CCMN/IGEO, Departamento de Geologia, Universidade Federal do Rio de Janeiro, 21.910-200 Cidade Universitária, Ilha do Fundão, Rio de Janeiro, Estado do Rio de Janeiro, Brazil

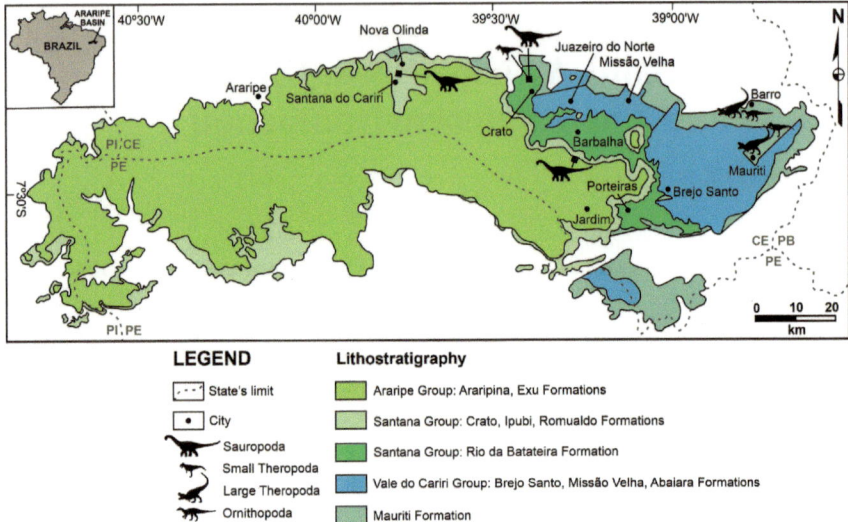

Fig. 6.1 Geological map of the Araripe Basin and the location of the dinosaur footprints ichnosites. The geochronological data, lithostratigraphic unit's limits and nomenclature were based in Ponte and Appi (1990), Fambrini et al. (2011), Rios-Netto et al. (2012), Assine (2007) and Arai and Assine (2020)

The footprints in the Araripe Basin are imprints in the upper bedding surface or even as structural deformations only visible in cross section. They allow the evaluation of substrate consistency besides the potential trackmaker identification. The environmental contexts of the dinosaur footprints from the Araripe Basin include the dinosaur trampling in fluvial sand bars, floodplains, deltas, and saline-alkaline lake borders. Then these footprints permit us to evaluate the diversity of the Cretaceous biota in this region and also discuss the environmental changes throughout the early and beginning of the Late Cretaceous in this region.

6.2 Geological Context

The sedimentary filling of the Araripe Basin begins with a controversial Paleozoic sequence (Carvalho et al. 2024), followed by three Mesozoic super sequences known as Pre-Rift, Rift, and Post-Rift (Ponte and Appi 1990; Assine 2007; Assine et al. 2014). The full sedimentary succession of the basin (Fig. 6.2) rests unconformably on the igneous and metamorphic rocks of the Precambrian Piancó-Alto Brígida Terrain, part of the Transversal Zone of the Borborema Province (Brito-Neves et al. 2000).

The Mauriti Formation footprints occur in two ichnosites: Milagres (Milagres County) and Mauriti (Mauriti County), both in the Ceará State, in a succession of coarse and fine-grained sandstones. The unit is constituted of conglomerate and

Fig. 6.2 Araripe Basin stratigraphical chart and the lithostratigraphic units with dinosaur footprints. Abbreviations: JURAS: Jurassic; TIT: Tithonian; BER: Berriasian; VAL: Valanginian; HAU: Hauterivian; BAR: Barremian; CEN: Cenomanian; LK: Lake; CST: Coastal environment; SAB: Sabhka; RIO DA BAT: Rio da Batateira Formation. The geochronological data and nomenclature of the lithostratigraphic units were based in Ponte and Appi (1990), Fambrini et al. (2011), Rios-Netto et al. (2012), Assine (2007) and Arai and Assine (2020)

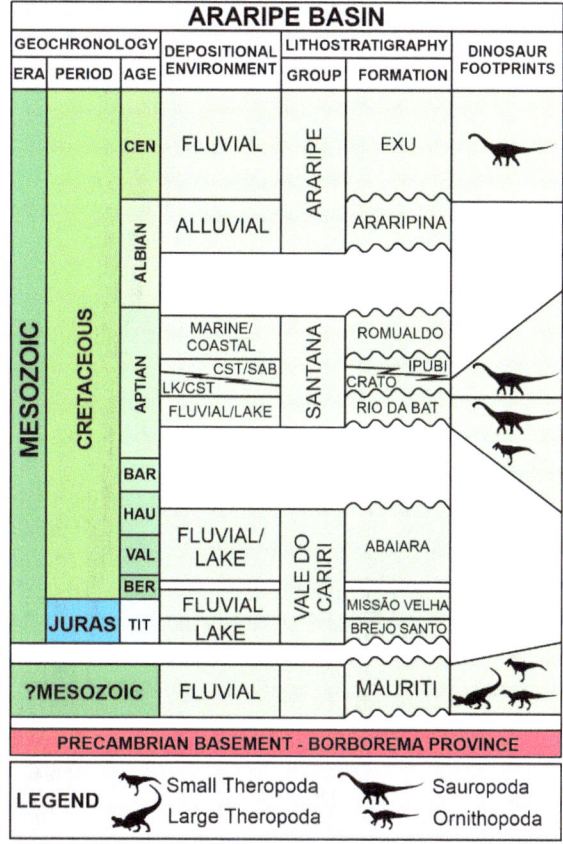

pebbly sandstone that grades into medium-to coarse-grained sandstone towards the top in a sedimentary succession interpreted as a braided fluvial system during a hot and dry climatic context. The fossil record is marked only by the presence of invertebrates and vertebrate ichnofossils, such as dinosaur footprints (Ponte and Appi 1990; Assine 1992; Carvalho et al. 1994, 1995, 2023, 2024; Batista et al. 2012; Cerri et al. 2022).

Initially designated as Cariri Formation, of Neocomian age (Beurlen, 1962), it was renamed as Mauriti Formation (Gaspary and Anjos, 1964) and has been interpreted as indicative of an initial depositional event in the Lower Paleozoic, Upper Ordovician to Lower Devonian, based on lithostratigraphic correlation with the Serra Grande Group (Parnaíba Basin) and the Tacaratu Formation of the Jatobá Basin (Ponte and Appi 1990; Assine 1992). Sedimentologic, stratigraphic, detrital zircon U–Pb dating and provenance approaches based on trace elements in detrital rutile established that the sedimentation of the Mauriti Formation started after the Late Cambrian, probably extending through the Ordovician (Cerri et al. 2022). However, the identification of theropod and possibly ornithopod tracks in the Milagres and Mauriti ichnosites may indicate a Neojurassic to Early Cretaceous age for the Mauriti Formation, suggesting that the beginning of deposition in the Araripe Basin was restricted to the last part of the Mesozoic (Carvalho et al. 1995, 2023). This Mesozoic age is also supported by tectonic and sedimentary analyses (Berthou 1990; Mabesoone 1990).

The Upper Jurassic Pre-Rift super sequence of the Araripe Basin (Assine 2007; Assine et al. 2014), consists of the Brejo Santo and Missão Velha formations, which include shales, mudstones, and locally conglomeratic sandstones interpreted as alluvial and lacustrine sedimentary systems (Ponte and Appi 1990; Assine 1992, 2007; Fambrini et al. 2011, 2013). Furthermore, the Rift super sequence is characterized by the facies of the Abaiara Formation, which was formed in shallow lakes and braided channel fluvial plains associated with the rifting of Gondwana in the Early Cretaceous (Assine 1992, 2007).

The Santana Group indicates the beginning of the Aptian Post-Rift I super sequence (Assine 2007; Assine et al. 2014), and consists of the Rio da Batateira, Crato, Ipubi, and Romualdo formations that encompass the Brazilian Alagoas local stage. The Rio da Batateira Formation (also called Barbalha Formation by Assine et al., 2014) presents sandstones, micro conglomerates, siltstones, carbonates, and bituminous shales of fluvial-lacustrine and deltaic origin (Chagas et al., 2007; Paula-Freitas et al. 2007; Paula Freitas and Borghi 2011). The multiproxy approach elaborated by Varejão et al. (2021a) includes sedimentological, paleontological, ichnological, and chemo-stratigraphic analyses through the upper portion of the Rio da Batateira Formation, the entire Crato Formation, and the lower portion of the Ipubi Formation. These authors recorded the first marine influence in the Araripe Basin through the uppermost portion of the Rio da Batateira Formation (also referred to as Barbalha Formation), in which syn-rift fluvial channels, and overbank deposits with sedimentary transport from north-west to south-east, were bounded upward by bayhead deltas, commonly developed in the innermost part of bays and estuaries in transgressive coastlines. The Fundão Member (Rios-Netto et al. 2012), comprises a very fossiliferous horizon in the Rio da Batateira Formation, in which algal laminations, coprolites (possibly from fish), ostracods, conchostracans, fish, amber, plant fragments, and dinosaur tracks (sauropods and theropods) are recorded (Hashimoto et al. 1987; Rios-Netto et al. 2012; Carvalho et al. 2021a).

During the deposition of the Crato Formation, there is a wide variety of continental and transitional environments, also with records of marine incursions (Neumann and Cabrera 2002a, b; Varejão et al. 2021a; Ribeiro et al. 2021). Within the lacustrine hypersaline succession (Varejão et al. 2021a), there are abundant and diverse exquisite preserved fossils that give the Lagerstätte status for the unit (Martill 2007; Selden and Nudds 2012). The Crato Biota, as referred by Dias et al. (2022), is characterized by fungi, plants, arthropods, fish, frogs, lizards, turtles, pterosaurs, non-avian dinosaurs, and birds living in a wetland-type ecosystem influenced by climatic oscillations between wetter and drier periods (Ribeiro et al. 2021). The marine influence in the Crato Formation is attested by tide-dominated bay facies, and confined bay with typical facies deposited in foreshore to upper shoreface conditions with storm deposits (hummocky cross-stratified sandstones). The maximum flooding surface is a dark shale below these foreshore-to-shoreface facies, marking the beginning of the Highstand Systems Tract that culminates with the deposition of the evaporites from the Ipubi Formation (Varejão et al. 2021a). The presence of micro foraminiferal linings in the transition between the Crato and Ipubi formations suggests that these evaporites from the Ipubi Formation may have been putatively precipitated by the evaporation of marine waters (Goldberg et al. 2019).

Although there is the influence of marine environments in other facies association (Varejão et al. 2021a, b) the dinosaur footprints of the Aptian in the Araripe Basin occur in floodplain areas of meandering rivers and low-energy lake deposits, without evidence of a marine influence. In the Rio da Batateira Formation, the tracks occur in fluvial siliciclastic successions (Carvalho et al. 2019a, b, 2021a). In the Crato Formation, however, the footprints are restricted to the locality of Três Irmãos mine in the carbonate deposits formed in hypersaline alkaline lakes and microbial-induced(Carvalho et al. 2021a).

The Aptian in the Araripe Basin also includes the Ipubi and Romualdo formations (Assine et al. 2014). In diastemic contact with the Crato Formation, the Ipubi Formation includes evaporite intercalations (gypsum and anhydrite) and green and/or pyrobituminous shales. They were deposited in a shallow and saline coastal environment, under a warmer climate with precipitation from brines (Assine et al. 2014; Bobco et al. 2017). The Alagoas local stage sedimentation of the Araripe Basin ends with the lagoonal and marine deposits of the Romualdo Formation, consisting of conglomerates, sandstones, marls, shales, and limestones. This unit also preserves an abundant and diverse biota, including foraminifera, palynomorphs, corals, mollusks, arthropods, echinoids, fishes, turtles, crocodyliforms, pterosaurs, dinosaurs, and plants (Abreu et al. 2020; Araripe et al. 2021; Lopes and Barreto 2021; Dias et al. 2022; Santana et al. 2022).

The Cretaceous sedimentation of the Araripe Basin finishes during the late Albian to early Cenomanian. This is recorded by the deposits of the Araripe Group, constituted by the Araripina and Exu formations (Assine 1992, 2007). The Araripina Formation are cyclic distal plain deposits of alluvial fan systems, while the Exu Formation is essentially fine-grained quartz sandstones with siltstones and occasional mudstones deposited in a fluvial environment (Assine 2007; Carvalho et al. 2021b, 2022). The dinosaur footprints are recorded in the fluvial floodplains and sand sheets of the Exu Formation, during episodes of hot climate (Carvalho et al. 2021b, 2022).

6.3 Footprints: Diversity and Paleobiological Interpretation

Although dinosaur tracks are commonly found in the surrounding basins of Sousa, Uiraúna-Brejo das Freiras, Malhada Vermelha, and Lima Campos, they are still rare in the Araripe Basin (Leonardi 1994; Leonardi and Spezzamonte 1994; Carvalho 2000; Carvalho et al. 2021a, b, 2022, 2023). The footprints from the Araripe Basin are found in four stratigraphic units: Mauriti, Rio da Batateira, Crato, and Exu formations, which were deposited in very distinct temporal, paleogeographical, and environmental contexts.

In the Mauriti Formation, the Milagres ichnosite (Fig. 6.3) presents theropod and ornithopod tracks. The theropod tracks are three isolated footprints (ARMI 01, ARMI 02, ARMI 04) and a short trackway with three footprints (ARMI 05). All of them are tridactyl, mesaxonic, with pointed digits, some of them with claw impressions. The rear borders of the footprints are V-shaped or angular (Fig. 6.4). The sandstone filling of the footprints is similar to the surrounding matrix. The footprints are large, ranging from 28–40 cm in length and 20–30 cm in width (Carvalho et al. 1995). The probable trackmakers are large theropods related to the groups that are already known in the Cretaceous deposits of the basin, such as the Spinosauridae *Angaturama limai* or *Irritator challengeri* (Kellner and Campos 1996; Martill et al. 1996). However, the footprints from Milagres ichnosite are certainly older than the Aptian-Albian age of these fossils. There is also an isolated footprint related to an ornithopod (ARMI 03). It is a tridactyl and mesaxonic footprint with rounded extremities of the three digits and wide concave hypexes (Fig. 6.3b). The digits II and IV are 5 cm in length and digit III is longer showing 10 cm in length. The footprint is 20 cm in width and its length is also 20 cm. The rear portion of the footprint shows an extrusion rim, with a rounded and wide crescent shape. Its color is more reddish than the surrounding matrix. The absence of claws, the wide concave hypexes, and the wide width allowed its interpretation as an ornithopod footprint (Carvalho et al. 1995). Osteological elements of this group are unknown in the Araripe Basin, although tracks are found in the surrounding Rio do Peixe basins.

The Mauriti ichnosite (Fig. 6.5), Maurity county, presents at least seven isolated footprints. There are four tridactyl, mesaxonic footprints with pointed (?theropod) and rounded digits (?ornithopod). The other imprints are rounded depressions with no clear digit impressions. The partial sandstone filling of the footprints is similar to the surrounding matrix. They range from 30–48 cm in length and 25–48 cm in width. The trackmakers of the theropod footprints could be the large theropods related to those already known in the Araripe Basin's Cretaceous formations (Carvalho et al. 2023).

The Rio da Batateira Formation tracks (Aptian) are observed as cross-section casts (Figs. 6.6 and 6.7). They are three-dimensional casts in cross-section, as pillar-like morphologies, small- and large-sized concave-up and sub-cylindrical structures. They allow examination of the deformation of the underlying layers and also how the footprints were filled by the sediments deposited afterward. The casts may also be presented as amorphous bulges or sedimentary layers deformed and downfolded, reaching one meter below the depositional surface.

The dinosaur tracks of the Rio da Batateira ichnosite (Fig. 6.6a) can seem to be simple load casts; however, they are interpreted as dinosaur trampling, and more in detail an association of distinct groups of dinosaurs. These load structures, interpreted as dinosaur footprints, measure 15–120 cm in length and 20–100 cm in depth, in fine-grained siliciclastic beds, such as shales, siltstones, and fine sandstones. The depth penetration, which can reach 100 cm, probably is due to the higher plasticity of the substrate, similar to some sauropod tracks from the Upper Jurassic of Spain

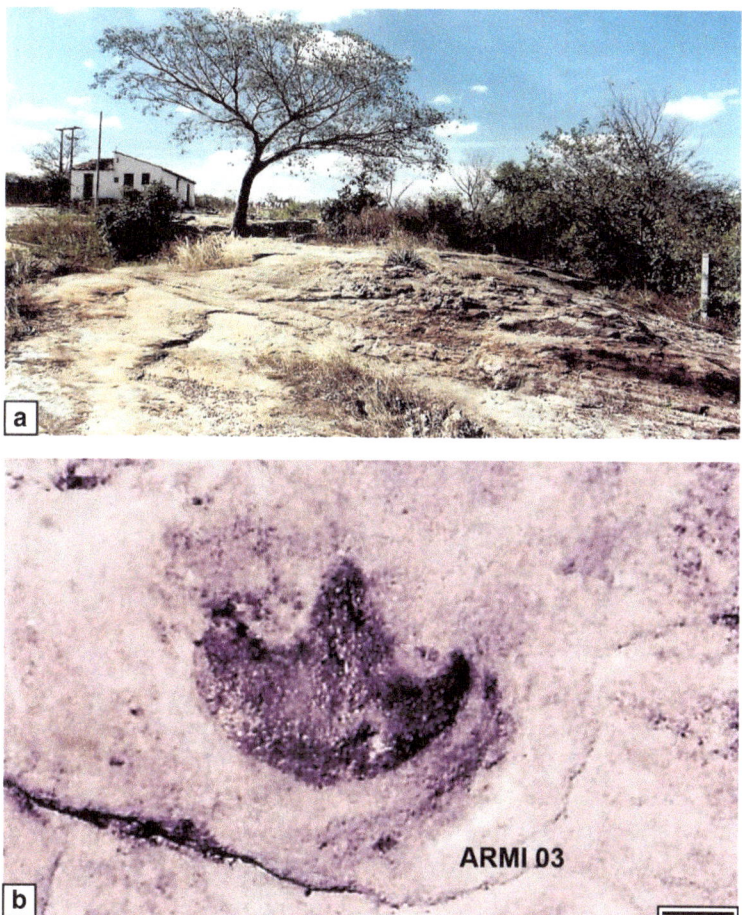

Fig. 6.3 Milagres ichnosite, in the Milagres ranch. **a** Outcrop of the Mauriti Formation, where the dinosaur tracks are found. **b** An isolated ornithopod footprint (ARMI 03) showing a contrasting color with the surrounding matrix. Scale bar: 5 cm

(Valenzuela et al. 1988; García-Ramos et al. 2006). In Rio da Batateira ichnosite it is also possible to observe the digit impressions in some of the casts (Fig. 6.6b), enabling their interpretation as belonging to bipedal or quadruped dinosaurs (Fig. 6.6b–e). The largest footprints are produced in an exposed waterlogged substrate or in a flooded area, where was possible the liquefaction of the sediments of sauropod trackmakers (Fig. 6.7). The smaller ones present a "V-shaped" cross-section with the evidence of a more prominent digit that exerts a higher pressure on the substrate, conducting a greater deformation in the central area of the cast. It probably corresponds to digit III of small theropods, like *Mirischia asymmetrica* or *Santanaraptor placidus* (Naish et al. 2004; Kellner 1999) or some small ornithopod. The interpretation of these tracks indicates the presence of quadrupedal (probably sauropod) and bipedal

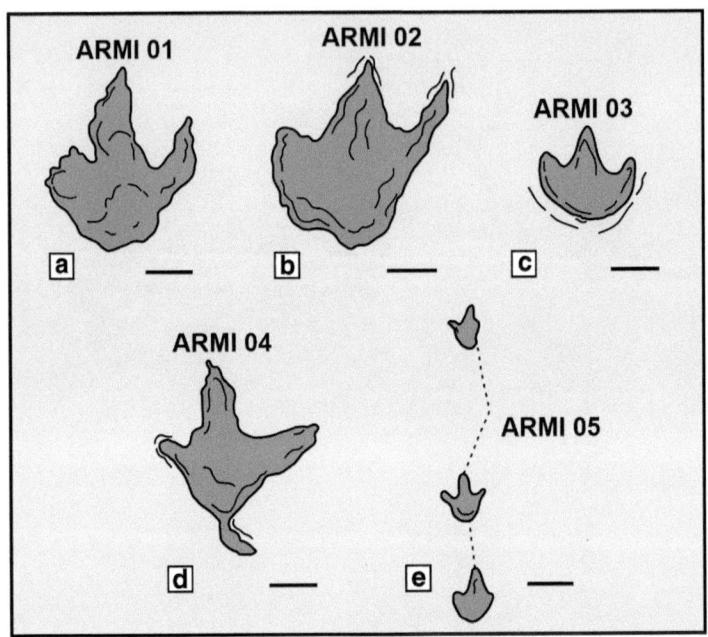

Fig. 6.4 Footprints from the Mauriti Formation, Milagres ichnosite. **a, b, d** Isolated theropod footprints (ARMI 01, ARMI 02, ARMI 04) with sharp pointed digits; **c** Ornithopod isolated footprint (ARMI 03); **e** Short track (ARMI 05) with three sequential tridactyl footprints interpreted as a theropod trackway. ARMI—Araripe Basin, Milagres ichnosite. Scale bars: **a–d** 10 cm; **e** 30 cm

(theropod and ornithopod) dinosaurs (Carvalho et al. 2019a, b, 2021a). These tracks are an important tool for the reconstruction of the terrestrial Cretaceous ecosystem in the context of the Araripe Basin. It is noteworthy that no sauropod body fossils were so far found either in this or other lithostratigraphic units of the Araripe Basin (Carvalho et al. 2019b).

In the Crato Formation (Nova Olinda County, Ceará State), in the Três Irmãos ichnosite, the dinosaur tracks are found in fine-grained sandstones, intercalated with shales and laminated carbonates (Fig. 6.8). They range from 35 to 100 cm in length and 30–50 cm in depth. The pressure that occurred during the contact of dinosaur feet and the substrate led to the deformation of the upper surface of the sediments, with the origin of load structures accompanying a concave aspect with successive lamina deformation. Tracks may occur as isolated or superimposed casts in cross-section, as pillar-like or concave-up morphologies, but casts are more commonly irregularly cylindrical to "U" shaped (with a larger basal diameter than at the top, as usually occurs with undertracks). Undulating forms that grade into load casts may be recognized as tracks when they occur along the same bedding plane adjacent to recognizable tracks, and when they have relief and dimensions similar to those of associated distinct tracks. The substrate should be soft and moist, with a relatively high cohesiveness (Carvalho et al. 2018) allowing for the deformation of successive

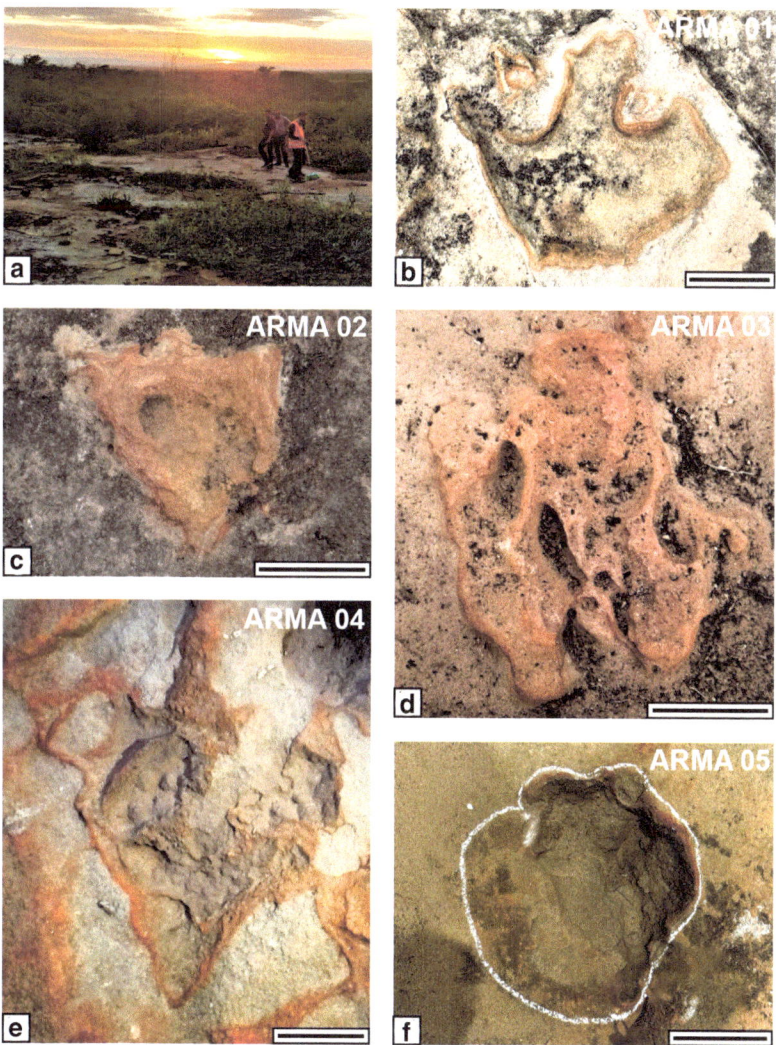

Fig. 6.5 Footprints from the Mauriti Formation, Mauriti ichnosite. **a** Outcrop of the Mauriti Formation with the dinosaur footprints; **b–e** Isolated footprints of small and large theropods (ARMA 01, ARMA 02, ARMA 03, ARMA 04). ARMA—Araripe Basin, Mauriti ichnosite. Scale bars: **b–e** 10 cm; **f** 20 cm

layers and developing undertracks. The dinoturbation index was defined as the degree of dinosaur trampling (Lockley and Conrad 1989) and its intensity over a surface (light: 0–33%, moderate: 34–66%, and heavy: 67–100%). In the Rio da Batateira Formation, the dinoturbation index can be considered heavy, while in the Crato Formation, it is light (Carvalho et al. 2021a).

Fig. 6.6 Footprints from the Rio da Batateira Formation, Rio da Batateira ichnosite. **a** Riacho da Batateira outcrop, Cascatinha locality where are found the cross-section footprints; **b** Section of a track cast with the digit imprints in the lower portion, indicating a probable sauropod track. Dashed line indicates the limits of the foot contact with the sediment, the original surface stepped on; **c** Digit III (indicated by an arrow) exerts a higher pressure on the substrate, conducting to a greater deformation in the central area of the cast, feature common in theropod footprints; **d** A large flattened depression, bordered by displacement rims (high declivity borders) interpreted as a footprint of a quadrupedal dinosaur, probably a sauropod; E. Cross-section of a small footprint with a rounded outline, showing the distinct patterns of deformation of the substrate by the dinosaur trampling. Scale bars: **b** 15 cm; **c** 10 cm; **d** 20 cm; **e** 10 cm

Fig. 6.7 Footprints from the Rio da Batateira Formation, Rio da Batateira ichnosite. **a** Outcrop with dinoturbation structures in the margins of Riacho da Batateira. **b** Disturbed layers resulted from the vertebrate trampling; **c** The cross-section through dinosaur tracks displays large structural and dimension variations indicated by arrows; **d** The high-water content induces the deformation by the foot impact up to one meter below the surface; **e** High deformation of the substrate including fluidization (indicated by an arrow) induced by the dinosaur trampling. Scale bars: 30 cm

Fig. 6.8 Footprints from the Crato Formation, Três Irmãos ichnosite. **a** Outcrop of Crato Formation, Três Irmãos Quarry, showing the succession of laminated carbonates and the level with cross-section tracks; **b** A cast of the true track with surrounding strongly bent and downfolded layers indicate a deformation due to a foot impact of a sauropod or a large quadrupedal dinosaur; **c** Concave deformations induced by the foot load pression. The cross-section footprints are eroded in its upper surface before the following deposition of fine sandstones. Arrows indicate the position of the footprints. Scale bars: **a** 1 m; **b–c** 30 cm

Marty (2011) and Marty et al. (2006) indicated that, after the foot impact, some structures are apparent on the surface (true track, overall track, underprint) and others are hidden within the substrate (undertrack, deep track). The types of deformation indicate that theropods adopted many walking strategies at different times, resulting in the formation of a stacked succession of undertracks that gradually becomes wider, shallower, and less detailed downward (Milàn et al. 2006). The deformation structures in a vertical section allow obtaining additional details about the walking kinematics that rarely could be available from the true track at the surface (Milàn and Bromley 2006), usefulness for the correct interpretation of the trackmaker and the substrate consistency (Milàn et al. 2004, 2006). Laboratory track simulations presented by Manning (2008) enabled the analysis of the magnitude and distribution of load acting on surface sediments, transmitting through and deforming subsequent layers. This aspect is clear in the tracks at the Rio da Batateira and Crato formations, due to the deformation produced in the lower sedimentary levels after the footprint impact (Carvalho et al. 2021a).

The Exu Formation footprints (Barbalha County, Ceará State), Barbalha ichnosite, are about 20 cm in height and 30 cm wide (Fig. 6.9). They are evident on a vertical cross-section of a sandstone bed as concave-up deformations of the lamina-set. Digit impressions or other morphological features of the footprints are not preserved (Carvalho et al. 2021b, 2022). Therefore, the geometry and dimensions of these structures allow us to interpret them as similar to dinoturbation structures produced by sauropods. These dinosaur footprints enhance the understanding of the genetic interpretation of deformational structures and paleoenvironmental scenarios of the Late Cretaceous from Northeastern Brazil.

6.4   Paleogeographical Distribution of the Footprints

The dinosaur footprints of the Araripe Basin are recorded in three unquestionably Cretaceous lithostratigraphic units (Rio da Batateira, Crato, and Exu formations), and one with controversial age but adopted as Jurassic-Cretaceous based on its dinosaur tracks (Mauriti Formation). Thus, the paleogeographic context during the Late Jurassic and Early Cretaceous of the Araripe Basin is linked to the rifting process of the Gondwana supercontinent, with distinct tectonic evolution during the pre-rift, rift, and post-rift phases, which influenced the pattern of biota dispersal and speciation processes.

The Mauriti Formation footprints of the Milagres and Mauriti ichnosites (Carvalho et al. 1994, 1995, 2023, 2024) present a temporal inconsistency, as the Mauriti Formation is frequently considered part of an Early Paleozoic depositional cycle (Ponte and Appi 1990; Assine 1992, 2007; Cerri et al. 2022). The NW paleoflow indicates that the main source areas for the Mauriti fluvial system are located in the Transversal and Southern zones of the Borborema Province (Cerri et al. 2022). Meanwhile, the presence of dinosaur footprints in the Mauriti Formation suggests that it is certainly a Mesozoic unit (Carvalho et al. 1995, 2023). Due to the geographical proximity with

Fig. 6.9 The Cenomanian Exu Formation and its footprints, Barbalha ichnosite. The Exu Formation is the last Cretaceous sedimentation cycle in the Araripe Basin, composed of coarse to fine-grained reddish sandstones. **a** Outcrop of the Exu Formation with a succession of channel cross-stratification and laminated bimodal sandstones; **b** Concave-up deformation of the lamina-set, 20 cm height and 30 cm wide, observed on a vertical cross-section. Digit impressions or other morphological features of the footprints are not preserved; **c** Deformations as concave-up features interpreted as dinoturbation structures observed in cross-section. Arrows indicate the position of the footprints. Scale bars: **c** 20 cm; d 30 cm

Rio do Peixe basins, that present similar dinosaur tracks, probably Early Cretaceous age is suggested (Carvalho et al. 2023, 2024). The importance of these two ichnosites confirms the need to revise the age of the Mauriti Formation and the paleogeographical context of these footprints, establishing a new stratigraphic framework for the lower successions of the Araripe Basin (Carvalho et al. 2023, 2024). If the Jurassic-Cretaceous age was confirmed, the paleogeography of the Mauriti Formation would not be related to the Western Paleozoic Gondwana, as indicated by Cerri et al. (2022), but with the Mesozoic Gondwana supercontinent evolution during the Jurassic and Cretaceous, as the others lithostratigraphic units of the Araripe Basin.

The break up between the South American and African continents occurred predominantly due to east-west divergent tectonic, during the Jurassic. The rupture began in the southern part of Gondwana and progressively extended northward throughout the Early Cretaceous, shaping itself along pre-existing weakness zones (Françolin and Szatmari 1987). Valuable paleobiogeographic data from Maisey (2011) show that in the pre-rift stage, there were no individual depocenters in the Brazilian sedimentary basins, resulting in low diversity of the Gondwanan biota at the genus and species levels. In Western Gondwana, Brazil and West Africa was a single continuous landmass, sometimes referred to as the Afro-Brazilian Depression, allowing for extensive taxa dispersion. This distinct pre-rift biota is Gondwanan, with a Pangaean origin, and various groups of vertebrates such as mawsoniid coelacanths, notosuchian crocodiles, and dinosaurs, diversified within Gondwana before the breakup (Maisey 2011). Within this context, the dinosaur footprints of the Mauriti Formation are included, and the large theropods and an ornithopod as probable trackmakers, would be in this terrestrial scenario of a single continuous continental mass before the rifting process of the Gondwana.

The Brazilian Alagoas local stage is associated with the breakup of the Gondwana supercontinent in the Mesozoic. This interval has special importance to the correlation of sub-aerial exposition surfaces throughout the basin. The Aptian dinosaur tracks from the Araripe Basin occur in the Rio da Batateira Formation (Fundão Member) and Crato Formation, in the Ceará State. They can be observed only in cross-section, as three-dimensional natural structures in siliciclastic and carbonate successions.

During the Gondwana breakup, rift valleys were formed and subsequently flooded by epicontinental seas during the Aptian, which separated Northeast Brazil from the rest of South America, but remained contiguous with Africa (Maisey 2011). The Aptian epicontinental seas were formed before the complete detachment process of South America and Africa. As a result, these marine incursions do not necessarily correspond to tectonic boundaries of rifting but rather represent distinct biogeographic provinces created through intracontinental vicariance processes due to land separations caused by the marine incursions (Maisey 2011). Within this context, the biotas of the Aptian formations in the Araripe Basin are included, such as Rio da Batateira and Crato, which feature small theropods and sauropods as possible trackmakers.

Taxonomic studies of ostracod fossils from the Rio da Batateira Formation were a subsidy for the understanding of the beginning of the opening of the South Atlantic Ocean during the Aptian in the Araripe Basin (Santos Filho et al. 2023). The ostracod assemblage from the Santo Antônio section presents nine typically brackish-marine species, associated with the first marine ingression in the interior of the continent during the beginning of the formation of the Atlantic Ocean (Tomé et al. 2022). Fauth et al. (2023) present three marine incursion events in the Batateiras unit (two of them in the Fundão Member), defined by benthonic and planktonic foraminifera, calcareous nannofossils, dinocysts, serpulid tubes, and a mass mortality event of mixohaline ostracods. The integration of paleontological, sedimentological, and ichnological data by Varejão et al. (2021a) also indicates the deposition under marine influence in the upper portion of the Rio da Batateira Formation. The stratigraphic architecture and paleocurrents data suggest that marine waters reached the basin from the south, with marine ingressions of an incipient South Atlantic Ocean over the interior basins of northeastern Brazil (Varejão et al. 2021a). Other sedimentary and stratigraphic studies of the Aptian in the Araripe Basin point to a more complex scenario, with distinct marine pulses from different directions (Custódio et al. 2017; Bom et al. 2021). However, the Tethyan origin of a wide variety of vertebrates, invertebrates, and microfossils groups recorded in the Araripe Basin and other chrono-related basins of Northeast Brazil suggest marine ingressions from North to South (Arai 2014; Pereira et al. 2017; Araripe et al. 2021; Lindoso and Carvalho 2021; Kroth et al. 2021).

Although the dinosaur footprints are still not recorded in the Aptian Romualdo Formation of the Araripe Basin, the unit contains five species of theropod dinosaurs (see Kellner and Campos 1996; Martill et al. 1996, 2000; Kellner 1999; Aureliano et al. 2018; Sayão et al. 2020) and one possible Ornithischia (Leonardi and Borgomanero 1981), later considered as a theropod bone (Batista and Kellner 2007). The Romualdo Formation records the last marine ingression within the Cretaceous interior basins of Northeastern Brazil, with, at least, two distinct pulses of marine incursions associated with the formation of a proto-Atlantic Ocean (Assine et al. 2014; Custódio et al. 2017; Teixeira et al. 2017; Fürsich et al. 2019; Bom et al. 2021; Kroth et al. 2021). The marine ingressions that reached the Araripe Basin during the late Aptian formed a vast epicontinental sea in Northeast Brazil, with an area much larger than the current Araripe Basin (Arai, 2014). These interior seas are characterized by water masses that rest directly on the continental crust and are commonly formed by short-term variations in sea level, resulting in abrupt fluctuations in water salinity, temperature, and oxygenation (Kroth et al. 2021) recorded in the Rio da Batateira, Crato, Ipubi, and Romualdo formations.

While the sedimentary succession of the Santana Group represents a transgressive–regressive cycle associated with sea-level variations during the Aptian, the sedimentary deposits of the Araripe Group (which includes the Araripina and Exu Formations) indicate a differentiated tectonic uplift during the late Albian and early Cenomanian (Assine 2007). During the Cenomanian, there was the establishment of oceanic crust with the separation of the South American and African continents (Maisey 2011). The last record of dinosaur footprints in the Araripe Basin occurs in the Exu Formation, a Cenomanian fluvial succession (Carvalho et al. 2021b, 2022) that represents the return to the strictly continental conditions in the Araripe Basin, not being directly related to the marine influence of the opening of the South Atlantic Ocean.

6.5 Paleoenvironmental and Paleoclimatic Contexts

6.5.1 Mauriti Formation

The paleoenvironmental interpretation of the deposits where the Mauriti Formation footprints are found is coalescent alluvial fans and a braided fluvial system with high energy (Fig. 6.10), formed in a hot and more arid climatic context (Ponte and Appi 1990; Carvalho et al. 1995; Batista et al. 2012; Carvalho et al 2024). The few isolated footprints and trackways in the Milagres and Mauriti ichnosites of the Mauriti Formation may reflect the time between periods of sediment accumulation and the nature of the substrate. It is possible that the high energy of the environment contexts, grain size, low water content, and lack of sediment plasticity did not allow the preservation of a large number of footprints, indicating a potential preservation bias (Carvalho et al. 1995, 2023, 2024).

6.5.2 The Aptian Rio da Batateira and Crato formations

The Aptian records of dinosaur tracks in the Araripe Basin occurred in a moment of environmental changes of transitional siliciclastic to carbonate environments related to the deposition in an endorheic lake, also during a hot and arid climate. As demonstrated by Moratalla et al. (1995) Avanzini et al. (2000), Leonardi and Mietto (2000), Marty (2008), Santos et al. (2013), and Campos-Soto et al. (2017), carbonate environments are important for the preservation of fossil tracks.

The Rio da Batateira sequence is interpreted as fluvial and clastic lake shore environments, including floodplain areas of meandering rivers and low-energy lacustrine environments. The stratigraphic data interpretation shows that the interval was subject to tectonic control (Paula Freitas 2010; Rios-Netto and Regali 2007; Rios-Netto et al. 2012). The low-energy lacustrine paleoenvironment was subjected to water level fluctuations and anoxic events (Assine et al. 2014). Although Varejão et al. (2021a) recorded a bayhead delta facies association deposited under a marine influence in the upper portion of the Rio da Batateira Formation, the dinosaur footprints recorded in this unit occur in fine sandstones and shale successions associated with floodplain areas of meandering rivers and low-energy lakes, without evidence of a marine influence (Fig. 6.11).

The footprints of the Rio da Batateira Formation were produced in an exposed waterlogged substrate or in a flooded area, where the sediments' liquefaction was possible. The evaluation of these tracks and their relationship with the substrate, allow the understanding of the deformation due to a foot impact, and the construction of a model for the cross-section track formation. They also show behavioral insights into the trackmaker biology, substrate properties, interaction among the producer, and environmental factors (Carvalho et al. 2019b; 2021a, b). The preservation of these dinosaur footprints is enhanced by specific environmental and more arid climatic

Fig. 6.10 Reconstruction of the environmental scenery of the Mauriti Formation and the dinosaur trackmakers. Exposed channel bars of braided rivers in a context of arid climate allowed the preservation of theropod and ornithopod footprints (Art by Guilherme Gehr)

conditions, which may have a biogenic component associated. Rapid and significant sedimentation with the track coating favors the preservation, thus, footprints are most commonly preserved in environments with cyclic sedimentation (Carvalho et al. 2021a), including water-level fluctuations, as described for the Rio da Batateira Formation depositional environments.

A key point for the final preservation of fossil vertebrate tracks in laminated sediments has been explained by the biostabilization process of the sediment surface by microbial mats (Carvalho et al. 2013). In the Lower Cretaceous Rio do Peixe Group in the Sousa Basin, also in the northeast of Brazil, microbial mats developed in the temporary and shallow lacustrine environments, during warm climate conditions. According to Carvalho et al. (2013), the footprint consolidation and its early lithification probably occurred due to the presence of microbial mats, which provided a more cohesive substrate, preventing the footprints from being eroded. The sediments were initially biostabilized by early cementation and the covering of the

Fig. 6.11 Reconstruction of the environmental scenery of the Rio da Batateira Formation and the dinosaur trackmakers. Trampling by dinosaurs in a floodplain area of meandering rivers and lacustrine environments induced the deformation of the substrate. The climate was hot and more humid as observed through the palynological assemblages (Art by Guilherme Gehr)

microbial mats over the footprints. Subsequent successive floods and the influx of sediments allowed the preservation of a large number of layers with dinosaur footprints. This same mechanism was discussed for the invertebrate trace fossil preservation in the Sousa Basin by Carvalho et al. (2017), giving light to the biostabilization process by microbial mats as one of the main responsible factors for the ichnofossils preservation. In this case, petrographic analysis showed microbially induced sedimentary structures (MISS), such as small pits, bumps, and crinkles, associated with microbiolaminations and dispersed microbial filaments (Carvalho et al. 2017).

Drawing a parallel with the similar proposed paleoenvironment for the Early Cretaceous Rio da Batateira Formation, it is plausible that these microbes also could have influenced the dinosaur footprints' preservation. So far, there are records of algal laminations in the bituminous shales of the Fundão Member, established by Rios-Netto et al. (2012). In the 3-D casts of the Rio da Batateira dinosaur tracks, there is a section that shows evidence of digits, indicating the high plasticity of the substrate where the track was produced (Carvalho et al. 2021a). This ductility of the substrate may have been provided by the extracellular polymeric substance (EPS) secreted by the microbial mats, which gives a plastic nature to the sedimentary substrate through mucilage production. The presence of these microbial communities could be enhanced by the drier climatic context and the oscillatory character of the environment. A possible future perspective of paleoichnological studies in Brazil is the investigation of the microbial role in the fossilization of the Rio da Batateira dinosaur tracks, in the same way as described for the dinosaur tracks and invertebrate ichnofossils of the Sousa Basin.

During the Aptian, alkaline lakes were one of the main depositional environments in the Araripe Basin, representing the Lagerstätte succession of the Crato Formation. Although there are records of marine facies association (Varejão et al. 2021a, b) the dinosaur footprints of the Crato Formation are recognized in deposits from the margins of alkaline and hypersaline lakes, where other exceptionally preserved fossils (Fig. 6.12) of vertebrates, invertebrates, plants, and even fungi are also found (Martill et al. 2007; Carvalho et al. 2021a; Dias et al. 2022, 2023). The tracks of the Crato Formation occur in the Três Irmãos Quarry, Nova Olinda County, which is mainly composed of micritic limestone, with levels of marls and fine-grained siliciclastic beds (Neumann and Cabrera 2002a, b). This Lagerstätte succession of the Crato Formation represents a hypersaline lacustrine environment based on the presence of evaporitic features (halite hoppers, gypsum beds, and isolated gypsum crystals), predominant terrestrial fossil fauna and flora content, absence of bioturbation, and presence of several structures that points to the microbial nature of the carbonates, such as peloids, amorphous organic matter, coccoid and filamentous cells embedded in EPS, and horizons of microbialites (Heimhofer et al. 2010; Catto et al. 2016; Warren et al. 2017; Varejão et al., 2019). The vertical passage from the underlying ephemeral lake and river-dominated delta facies association to the overlying hypersaline lacustrine deposits indicates an increase in the dry condition (Varejão et al. 2021a).

In shallow, perennial, and closed lacustrine water bodies, as proposed for the Lagerstätte succession of the Crato Formation, the register of climate oscillations is expected. These climatic variations have been suggested by Neumann et al. (2003), Osés et al. (2017), Gomes et al. (2021), Guerra-Sommer et al. (2021), and Dias and Carvalho (2022). During drier periods, there is a greater proliferation of microbial mats and carbonate precipitation (Varejão et al. 2019, 2021a; Dias and Carvalho 2020, 2022). In wetter periods, increased productivity in the water column can also generate thick carbonate layers, although without a significant influence of microbial mats on the genesis of these rocks (Heimhofer et al. 2010). Consequently, Dias and Carvalho (2022) and Dias et al. (2023) suggested a possible climate control influencing fossil preservation, which probably affected the record of the dinosaur footprints, both in the Aptian Rio da Batateira and Crato formations.

Distinctly of the Rio da Batateira Formation, the microbial role in the carbonate genesis and fossilization process in the Crato Formation is already been well-discussed (Catto et al. 2016; Osés et al. 2016, 2017; Warren et al. 2017; Varejão et al. 2019; Dias and Carvalho 2020, 2022; Iniesto et al. 2021; Prado et al. 2021;

Fig. 6.12 The fossils variety from Crato Formation. **a** Outcrop of fine laminated carbonates where occur the exquisite Aptian fossils, Três Irmãos Quarry; **b** Insect of the Hemiptera order (UFRJ-DG 1163 Ins); **c** Insect larvae of the Ephemeroptera order (UFRJ-DG 441 Ins); **d** *Brachyphyllum obesum* (UFRJ-DG 2424 Pb), a Coniferophyta; **e** Arachnid of the Araneae order (UFRJ-DG 42 Ac); **f** *Cratoavis cearensis* (UFRJ-DG 31 Av), an Enantiornithes; **g** The fish *Dastilbe crandalli* (UFRJ-DG 1898 P). Scale bars: **a** 50 cm; **b** 1 cm; **c** 2 mm; **d** 3 cm; **e** 2 mm; **f** 2 cm; **g** 4 cm

Dias et al. 2023). The high degree of morphological fidelity of fossils has been mainly attributed to the influence of microbial mats in the fossilization process, particularly in the covering and mineralization of organic remains when they reach the lacustrine substrate (Varejão et al. 2019; Dias and Carvalho 2020, 2022; Iniesto et al. 2021; Dias et al. 2023).

The dinosaur footprints of the Crato Formation probably could be included in this fossilization scenario (Fig. 6.13), since this same mechanism of track preservation mediated by the microbial mats has already been described for the Sousa Basin (Carvalho et al. 2013). Just as a microbial mat enhances the chance of preservation by covering the carcass of an organism and creating physicochemical conditions conducive to the mineralization of organic remains (Dias et al. 2023), the coverage of a dinosaur track by the mats could favor exquisite preservation.

In Vermelha Lagoon (Rio de Janeiro State), a hypersaline lagoon from the Quaternary of Brazil, there is extensive development of microbial mats, MISS (microbially induced sedimentary structures), and microbialites. On the margins of this environment, desiccation tracks and wrinkle marks are associated with the preservation of wave marks and already lithified footprints of humans wearing sneakers. on the substrate (Guedes et al. 2022; Dias et al. 2023). During the Early Cretaceous in the Araripe Basin, the preservation of these dinosaur footprints occurs especially in the lacustrine margins, where there is more development of microbial mats, the same way observable for the Vermelha Lagoon. The sealing effect after covering the track, and the microbial sarcophagus created by the microorganisms, are key factors for the

Fig. 6.13 Reconstruction of the environmental scenery of the Crato Formation and the dinosaur trackmakers. The footprints are found in a context of carbonate environments related to the deposition in an endorheic lake, during a hot and arid climate. Episodes of more humid events allowed the increase of life diversity (Art by Guilherme Gehr)

preservation. This biogenic influence is directly correlated with environmental and climatic controls, with fossil preservation associated with shallow and hypersaline lacustrine environments, during more hot and arid contexts.

According to Pérez-Lorente (2015), the bearing capacity in which a foot can sink into sediment occurs when the resistance to penetration of the foot is equal to the pressure applied, directly related to the substrate plasticity and consistency. Falkingham et al. (2011, 2014) discuss different substrate models on track formation potential. As the walls of some of the Crato Formation tracks are vertical, it is interpreted that the original substrate was soft, yet cohesive and competent. The foot could enter deeply, but the sediment stayed together leaving sharp walls. Otherwise, in the Rio da Bateira Formation footprints, the deformation can reach 100 cm depth as the result of the foot impact in a less firm substrate (Carvalho et al. 2021a).

Besides the microbial component in the fossilization, other factors are also important for the dinosaur track preservation in the fossil record. For Carvalho et al. (2021a), the abundance of vertebrate bioturbation depends upon rates of trampling, texture, and plasticity of the substrate, and also the subsequent permanent burial with a low reworking rate. The small grain size, consistency, plasticity, and water content of the sediments are determinants for the preservation of anatomical details.

6.5.3 The Cenomanian Exu Formation

The succession of the Exu Formation is interpreted as channel bars of ephemeral streams and floodplains under a more arid to semiarid climate during the Cenomanian. The sediments in the sand bars of the dry channel streams could be reworked by winds resulting in bimodal sandstone deposits (Carvalho et al. 2021b, 2022). The opening of the South Atlantic Ocean during the Cenomanian and the return to continental conditions in the Araripe Basin at this age were likely accompanied by an increase in aridity and average temperature of terrestrial ecosystems due to the continentalization process.

In the Exu Formation (Barbalha ichnosite), the digit impressions or other morphological features of the footprints are not recognized. Probably, the substrate where the tracks were produced was not as plastic as the substrate with footprints in the Aptian Rio da Bateira and Crato formations. The absence of more delicate features of the Exu footprints could be a preservation bias due to the environmental conditions (Fig. 6.14).

6.6 Conclusions

The dinosaur footprints found in the Araripe Basin are temporal markers of subaerial exposition surfaces throughout the basin, recording cyclical changes in the environments and climate. These footprints vary across four distinct lithostratigraphic units:

Fig. 6.14 Reconstruction of the environmental scenery of the Exu Formation (Cenomanian) and the dinosaur trackmakers. Channel bars of ephemeral streams and sand sheets in floodplains were reworked by the feet load of dinosaurs. The climate during this moment was hot and dry with more humid events (Art by Guilherme Gehr)

Mauriti, Rio da Batateira, Crato, and Exu formations, each associated with different environmental settings. The tracks in the Mauriti Formation are located in alluvial fans and braided river deposits, formed during a hot and arid climate. This environmental context limited the preservation of a larger number of footprints. In contrast, the Aptian dinosaur footprints of the Rio da Batateira and Crato Formations, which developed in fluvial-lacustrine settings under arid conditions, have better-preserved records. This preservation may have been influenced by the presence of microbial mats during the fossilization process. Other studies conducted with arthropod fossils from the Crato Formation highlight the significant role of these microbes in the coverage, sealing, and mineralization of organic remains. Extrapolating this mechanism for preserving dinosaur tracks offers valuable insights for future paleoichnological studies in the Araripe Basin. Finally, the dinoturbations observed in the Cenomanian Exu Formation do not exhibit the features seen in the Aptian record of the same basin. This discrepancy is likely due to preservation biases caused by the more unstable environment, similar to the Mauriti Formation.

References

Abreu D, Viana MSS, Oliveira PV, Viana GF, Borges-Nojosa DM (2020) First record of an amniotic egg from the Romualdo Formation (Lower Cretaceous, Araripe Basin, Brazil). Revista Brasileira de Paleontologia 23(3):185–193

Arai M (2014) Aptian-Albian (Early Cretaceous) paleogeography of the South Atlantic: a paleontological perspective. Braz J Geol 44(2):339–350

Arai M, Assine ML (2020) Chronostratigraphic constraints and paleoenvironmental interpretation of the Romualdo Formation (Santana Group, Araripe Basin, Northeastern Brazil) based on palynology. Cretac Res 116:104610

Araripe RC, Oliveira DHD, Tomé METR, Moura de Mello R, Barreto AMF (2021) Foraminifera and ostracoda from the Lower Cretaceous (Aptian-lower Albian) Romualdo Formation, Araripe Basin Northeast Brazil: Paleoenvironmental Inferences. Cretac Res 122:104766

Assine ML (1992) Análise estratigráfica da Bacia do Araripe, Nordeste do Brasil. Revista Brasileira de Geociências 22(3):289–300

Assine ML (2007) Bacia do Araripe. Boletim de Geociências da Petrobrás 15(2):371–389

Assine ML, Perinotto JAJ, Custódio MA, Neumann VH, Varejão FG, Mescolotti PC (2014) Sequências deposicionais do Andar Alagoas da Bacia do Araripe, Nordeste do Brasil. Boletim de Geociências da Petrobrás 22(1):3–28

Aureliano T, Ghilardi AM, Buck PV, Fabbri M, Samathi A, Delcourt R, Fernandes MA, Sander M (2018) Semi-aquatic adaptations in a spinosaur from the Lower Cretaceous of Brazil. Cretac Res 90:283–295

Avanzini M, Frisia S, Rinaldo M (2000) I Lavini di Marco nel Giurassico inferiore: la riconstruzione di um antico ambiente di vita. In: Leonardi G, Mietto P (eds) Dinosauri in Italia, Accademia Editoriale. Pisa-Roma, p 245–272

Batista EB, Kellner AWA (2007) Reavaliação da suposta ocorrência de Ornithischia na Formação Santana (Cretáceo Inferior), Bacia do Araripe, Nordeste Brasil. Anuário do Instituto de Geociências 30(1):224

Batista ZV, Valença LMM, Fambrini GL, Silva SM, Neumann VHML, Santos CA, Barros CL (2012) Análise de fácies da Formação Cariri, Bacia do Araripe Nordeste do Brasil. Estudos Geológicos 22(1):3–20

Berthou PY (1990) Le bassin d'Araripe et les petits bassins intracontinentaux voisins (N.E. du Brésil): formation et évolution dans le cadre de l'ouverture de l'Atlantique Equatorial. Comparaison avec les bassins ouest-Africains situés dans le même context. In: Simpósio sobre a Bacia do Araripe e Bacias Interiores do Nordeste, 1, Crato, pp 113–134

Beurlen K (1962) A geologia da Chapada do Araripe. An Acad Bras Ciênc 34(3):365–370

Bobco FER, Goldberg K, Bardola T (2017) Modelo deposicional do Membro Ipubi (Bacia do Araripe, Nordeste do Brasil) a partir da caracterização faciológica, petrográfica e isotópica dos evaporitos. Pesquisa em Geociências 44(3):431–451

Bom MHH, Ceolin D, Kochhan KGD, Krahl G, Fauth G, Bergue CT, Savian JF, Strohschoen Junior O, Simões MG, Assine ML (2021) Paleoenvironmental evolution of the Aptian Romualdo formation, Araripe Basin Northeastern Brazil. Glob Planet Chang 203:103528

Brito-Neves BB, Santos EJ, Van Schmus WR (2000) Tectonic history of the Borborema Province, Northeastern Brazil. In: Cordani UG, Milani EJ, Thomaz Filho A, Campos DA (eds) Tectonic evolution of South America, 31 International Geological Congress, p 151–182

Cerri RI, Warren LV, Spencer CJ, Varejão FG, Promenzio P, Luvizotto GL, Assine ML (2022) Using detrital zircon and rutile to constrain sedimentary provenance of Early Paleozoic fluvial systems of the Araripe Basin, Western Gondwana. J S Am Earth Sci 116:103821

Campos-Soto S, Cobos A, Caus E, Benito MI, Fernández-Labrador L, Suarez-Gonzalez P, Quijada IE, Mas R, Royo-Torres R, Alcalá L (2017) Jurassic coastal park: a great diversity of palaeoenvironments for the dinosaurs of the Villar del Arzobispo Formation (Teruel, eastern Spain). Palaeogeogr Palaeoclimatol Palaeoecol 485:154–177. https://doi.org/10.1016/j.palaeo.2017.06.010

Carvalho IS (2000) Geological environments of dinosaur footprints in the intracratonic basins from Northeast Brazil during South Atlantic opening (Early Cretaceous). Cretac Res 21:255–267. https://doi.org/10.1006/cres.1999.0194

Carvalho IS, Viana MSS, Lima Filho MF (1994) Dinossauros do Siluriano: um anacronismo crono-geológico nas bacias interiores do Nordeste? In: 38° Congresso Brasileiro de Geologia, Camboriú, vol 3. Boletim de Resumos Expandidos, Camboriú, Santa Catarina, SBG,pp 213–214

Carvalho IS, Viana MSS, Lima Filho MF (1995) Os icnofósseis de dinossauros da Bacia do Araripe (Cretáceo Inferior, Ceará–Brasil). An Acad Bras Ciênc 67(4):433–442

Carvalho IS, Freitas FI, Neumann V (2012) Chapada do Araripe. In: Hasui Y, Carneiro CD, Almeida FFM, Bartorelli A (eds) Geologia do Brasil. Beca, São Paulo, pp 510–513

Carvalho IS, Borghi L, Leonardi G (2013) Preservation of dinosaur tracks induced by microbial mats in the Sousa Basin (Lower Cretaceous), Brazil. Cretac Res 44:112–121

Carvalho IS, Borghi L, Fernandes ACS (2017) Microbial mediation in invertebrate trace fossil preservation in Sousa Basin (Early Cretaceous), Brazil. Cretac Res 69:136–146

Carvalho IS, Melo BGV, Borghi L, Rios-Netto AM, Andrade JAFG, Freitas FI (2018) Dinosaur tracks from the Aptian (Early Cretaceous) of Araripe Basin, Brazil. In: 4° Simpósio Latino-Americano de Icnología. Santa Marta, Colombia, Resúmenes, p 69

Carvalho IS, Rios-Netto AM, Borghi L, Leonardi G (2019a) Dinoturbation structures from the Aptian of Araripe Basin, Brazil, as tools for stratigraphic correlation. In: 3rd International Stratigraphic Congress, Strati 2019. Milano, Italy, p 262

Carvalho IS, Rios-Netto AM, Borghi L, Paula Freitas A, Leonardi G, Andrade JA (2019b) Dinosaur trampling from the Rio da Batateira Formation—Lower Cretaceous of Araripe Basin, Brazil. In: Paleo Falls, Livro de Resumos, Évora, Portugal, p 21

Carvalho IS, Leonardi G, Rios-Netto AM, Borghi L, Paula Freitas A, Andrade JA, Freitas, FI (2021a) Dinosaur trampling from the Aptian of Araripe Basin, NE Brazil, as tools for paleoenvironmental interpretation. Cretacous Res 117:104626. https://doi.org/10.1016/j.cretres.2020.104626

Carvalho IS, Leonardi G, Andrade JAFG, Freitas FI, Borghi L, Rios-Netto AM, Figueiredo SMD, Cunha PP (2021b) Dinoturbation in the Exu Formation (Cenomanian, Upper Cretaceous) from the Araripe Basin, Brazil. In: 3rd Virtual Paleontological Congress, 2021. Book of Abstracts 3rd Paleontological Virtual Congress, vol 1. pp 193–193

Carvalho IS, Neto, JX, Leonardi G, Andrade JAFG, Freitas FI, Borghi L, Rios-Netto AM, Figueiredo SMD, Cunha PP (2022). Estruturas de Dinoturbação do Cretáceo Superior da Bacia do Araripe (Formação Exu), Brasil. In: 27° Congresso Brasileiro de Paleontologia, Cuiabá, 2022. Paleontologia em Destaque, 37 (Ed. Especial), p 133. https://doi.org/10.4072/paleodest.2022.37.ed.especial

Carvalho IS, Silva DS, Andrade JAFG, Borghi L, Viana MSS, Lima Filho MF, Silva RS, Rios-Netto AM, Figueiredo SD, Costa GT (2023) A new dinosaur tracksite in the 'Early Paleozoic', Mauriti Formation, Araripe Basin, Brazil In: 4th Paleontological Virtual Congress, 2023, Espanha. Book of abstracts of 4th paleontological virtual congress, 2023, 1:162–162

Carvalho IS, Dias JJ, Silva DS, Borghi L, Andrade JAFG, Viana MSS, Lima Filho MF, Schmitt RS, Rios Netto AM (2024) A new dinosaur tracksite from the Araripe Basin (Brazil) and the Putative Early Paleozoic age for the Mauriti Formation. In: Taylor LH, Raynolds RG and Lucas SG 2024, Vertebrate Paleoichnology: A Tribute to Martin Lockley. New Mexico Museum of Natural History and Science Bulletin 96: 339–344

Catto B, Jahnert RJ, Warren LV, Varejão FG, Assine ML (2016) The microbial nature of laminated limestones: lessons from the Upper Aptian, Araripe Basin, Brazil. Sed Geol 341:304–315

Chagas DB, Assine ML, Freitas FI (2007) Fácies sedimentares e ambientes deposicionais da Formação Barbalha no Vale do Cariri, Bacia do Araripe, Nordeste do Brasil. Geociências 26:512–521

Custódio MA, Quaglio F, Warren LV, Simões MG, Fürsich FT, Perinotto JAJ, Assine ML (2017) The transgressive-regressive cycle of the Romualdo Formation (Araripe Basin): sedimentary archive of the Early Cretaceous marine ingression in the interior of Northeast Brazil. Sed Geol 359:1–15

Dias JJ, Batista DL, Corecco L, Carvalho IS (2022) Bacia do Araripe: Biotas do Cretáceo do Gondwana. In: Corecco L (ed) Paleontologia do Brasil: Paleoecologia e Paleoambientes. Interciência, Rio de Janeiro, pp 129–190

Dias JJ, Carvalho IS (2020) Remarkable fossil crickets' preservation from Crato Formation (Aptian, Araripe Basin), a Lagerstätte from Brazil. J S Am Earth Sci 98:102443

Dias JJ, Carvalho IS (2022) The role of microbial mats in the exquisite preservation of Aptian insect fossils from the Crato Lagerstätte, Brazil. Cretacous Res 130:105068

Dias JJ, Carvalho IS, Buscalioni ÁD, Umamaheswaran R, López-Archila AI, Prado G, Andrade JAFG (2023) Mayfly larvae preservation from the Early Cretaceous of Brazilian Gondwana: Analogies with modern mats and other Lagerstätten. Gondwana Res 124(2023):188–205

Falkingham PL, Hage J, Bäker M (2014) Mitigating the Goldilocks effect: the effects of different substrate models on track formation potential. R Soc Open Sci 1:140225. https://doi.org/10.1098/rsos.140225

Falkingham PL, Bates KT, Margetts L, Manning PL (2011) The 'Goldilocks' effect: preservation bias in vertebrate track assemblages. J R Soc Interface 8(61):1142–1154. https://doi.org/10.1098/rsif.2010.0634

Fambrini GL, Lemos DR, Tesser S Jr, Araújo JT, Silva-Filho WF, Souza BY, Neumann VH (2011) Estratigrafia, Arquitetura Deposicional e Faciologia da Formação Missão Velha (Neojurássico-Eocretáceo) na Área-Tipo, Bacia do Araripe, Nordeste do Brasil: Exemplo de Sedimentação de Estágio de Início de Rifte a Clímax de Rifte. Revista do Instituto de Geociências - USP 11(2):55–87

Fambrini GL, Neumann VH, Barros CL, Silva SM, Galm PC, Filho JA (2013) Análise estratigráfica da Formação Brejo Santo, Bacia do Araripe, Nordeste do Brasil: implicações paleogeográficas. Revista do Instituto de Geociências - USP 13(4):3–28

Fambrini GL, Silvestre DC, Barreto Junior AM, Silva-Filho WF (2020) Estratigrafia da Bacia do Araripe: estado da arte, revisão crítica e resultados novos. Geologia USP Série Científica 20(4):169–212

Fauth G, Kern HP, Villegas-Martín J, Mota MAL, Santos Filho MAB, Catharina AS, Leandro LM, Luft-Souza F, Strohschoen O Jr, Nauter-Alves A, Tungo EJF, Bruno MDR, Ceolin D, Baecker-Fauth S, Bom MHH, Lima FHO, Santos A, Assine ML (2023) Early Aptian marine incursions in the interior northeastern Brazil following the Gondwana breakup. Sci Rep 13:6728

Françolin JBL, Szatmari P (1987) Mecanismo de rifteamento da porção Oriental da margem Norte brasileira. Revista Brasileira de Geociências 17(2):196–207

Fürsich FT, Custódio MA, Matos AM, Hethke M, Quaglio F, Warren LV, Assine ML, Simões MG (2019) Analysis of a late Aptian (Cretaceous) high-stress ecosystem: the Romualdo Formation of the Araripe Basin, Northeastern Brazil. Palaeogeogr Palaeoclimatol Palaeoecol 95:268–296

García-Ramos JC, Piñuela L, Lires JL (2006) Atlas del Jurásico de Asturias. Oviedo. Ediciones Nobel.

Gaspary J, Anjos NFR (1964) Estudo hidrogeológico de Juazeiro do Norte–Ceará. Série Hidrogeologia 3, SUDENE, Grupo de Estudos do Vale do Jaguaribe, Departamento de Recursos Naturais, p 25

Goldberg K, Premaor E, Bardola T, Souza PA (2019) Aptian marine ingressions in the Araripe Basin: implications for paleogeographic reconstruction and evaporite accumulation. Mar Petrol Geol 107:214–221

Gomes JMP, Rios-Netto AM, Borghi L, Carvalho IS, Filho JGM, Sabaraense LD, Araújo BC (2021) Cyclostratigraphic analysis of the early Cretaceous laminated limestones of the Araripe Basin, NE Brazil: estimating sedimentary depositional rates. J S Am Earth Sci 112(1):103563

Guedes CB, Arena MC, Santos HN, Valle B, Santos JA, Favoreto J, Borghi L (2022) Sedimentological and geochemical characterization of microbial mats from Lagoa Vermelha (Rio de Janeiro, Brazil). J Sediment Res 92:591–600

Guerra-Sommer M, Siegloch AM, Degani-Schmidt I, Santos ACS, Carvalho IS, Andrade JAFG, Freitas FI (2021) Climate change during the deposition of the Aptian Santana Formation (Araripe Basin, Brazil): Preliminary data based on wood signatures. J S Am Earth Sci 111:103462

Hashimoto AT, Appi CJ, Soldan AL, Cerqueira JR (1987) O Neo-Alagoas nas bacias do Ceará, Araripe e Potiguar (Brasil): caracterização estratigráfica e paleoambiental. Revista Brasileira de Geociências 17(2):118–122

Heimhofer U, Ariztegui D, Lenniger M, Hesselbo SP, Martill DM, Rios-Netto AM (2010) Deciphering the depositional environment of the laminated Crato fossil beds (Early Cretaceous, Araripe Basin, North-eastern Brazil). Sedimentology 57:677–694

Iniesto M, Gutiérrez-Silva P, Dias JJ, Carvalho IS, Buscalioni AD, López-Archilla AI (2021) Soft tissue histology of insect larvae decayed in laboratory experiments using microbial mats: taphonomic comparison with Cretaceous fossil insects from the exceptionally preserved biota of Araripe, Brazil. Palaeogeogr Palaeoclimatol Palaeoecol 564:110156

Kellner AWA (1999) Short note on a new dinosaur (Theropoda, Coelurosauria) from the Santana Formation (Romualdo Member, Albian), Northeastern Brazil. Boletim do Museu Nacional 49:1–8

Kellner AWA, Campos DA (1996) First Early Cretaceous theropod dinosaur from Brazil. Newues Jahrbuch Für Geologie Um Paläontologie 199(2):151–166

Kroth M, Borghi L, Bobco FER, Araújo BC, Silveira LF, Duarte G, Ferreira LO, Guerra-Sommer M, Mendonça JO (2021) Aptian shell beds from the Romualdo Formation (Araripe Basin): Implications for paleoenvironment and paleogeographical reconstruction of the Northeast of Brazil. Sed Geol 426:106025

Leonardi G (1994) Annotated Atlas of South America Tetrapod Footprints (Devonian to Holocene) with an appendix on Mexico and Central America. CPRM (Serviço Geológico do Brasil), Brasília, Brasil

Leonardi G, Mietto P (2000) Le piste liassiche di dinosauri dei Lavini di Marco. In: Leonardi G, Mietto P (eds) Dinosauri in Italia. Accademia Editoriale, Pisa-Roma, pp 169–246

Leonardi G, Borgomanero G (1981) Sobre uma possível ocorrência de Ornithischia na Formação Santana, Chapada do Araripe (Ceará). Revista Brasileira de Geociências 11(1):1–4

Leonardi G, Spezzamonte M (1994) New tracksites (Dinosauria: Theropoda and Ornithopoda) from the Lower Cretaceous of the Ceará, Brasil. Studi Trentini di Scienze Naturali, Acta Geologica 69(1992):61–70

Lindoso RM, Carvalho IS (2021) The Cretaceous fishes of Brazil: a paleobiogeographic perspective. In: Pradel A, Denton JSS, Janvier P (eds) Ancient fishes and their living relatives: a Tribute to John G. Maisey, Verlag Dr. Friedrich Pfeil, Germany, pp 233–238

Lockley M, Conrad K (1989) The paleoenvironmental context, preservation and paleoecological significance of dinosaur tracksites in the Western USA. In: Gillette DD, Lockley MG (eds) Dinosaur Tracks and Traces. Cambridge University Press, Cambridge, pp 121–134

Lopes GLB, Barreto AMF (2021) Paleoecological and biomechanical inferences regarding the paleoichthyofauna of the Romualdo Formation, Aptian-Albian of the Araripe Basin, state of Pernambuco, northeastern Brazil. J S Am Earth Sci 111:103444

Mabesoone JM (1990) Problemas sedimentológicos-estratigráficos das bacias interiores do Nordeste. In: Simpósio sobre a Bacia do Araripe e Bacias Interiores do Nordeste, 1, Crato, pp 135–141

Maisey JG (2011) Northeastern Brazil: Out of Africa? In: Carvalho IS, Srivastava NK, Strohschoen Jr O, Lana CC (eds) Paleontologia: Cenários de Vida, Interciência, vol 4. Rio de Janeiro, p 515–529

Manning PL (2008) *T. rex* speed trap. In: Carpenter K, Larson PL (eds) *Tyrannosaurus rex*; The Tyrant King. Indiana University Press, Bloomington, pp 204–231

Marques FO, Nogueira FC, Bezerra FH, Castro DL (2014) The Araripe Basin in NE Brazil: an intracontinental graben inverted to a high-standing horst. Tectonophysics 630:251–264

Martill DM, Bechly G, Loveridge RF (2007) The Crato fossil beds of Brazil: window into an Ancient World. Cambridge University Press, Cambridge, p 675p

Martill DM, Cruickshank ARI, Frey E, Small PG, Clarke M (1996) A new crested maniraptoran dinosaur from the Santana Formation (Lower Cretaceous) of Brazil. J Geol Soc 153:5–8

Martill DM, Frey E, Sues HD, Cruickshank ARI (2000) Skeletal remains of a small theropod dinosaur with associated soft structures from the Lower Cretaceous Santana Formation of northeastern Brazil. Can J Earth Sci 37:891–900

Marty D (2008) Sedimentology, taphonomy, and ichnology of Late Jurassic dinosaur tracks from the Jura carbonate platform (Chenevez—Combe Ronde tracksite, NW Switzerland): insights into the tidal-flat paleoenvironment and dinosaur diversity, locomotion and palaeoecology. GeoFocus 21:1–278

Marty D (2011) Formation, taphonomy, and preservation of vertebrate tracks. In: Dinosaur Tracks–Obernkirchen, April 14–17, Abstracts, p 37

Marty D, Meyer CA, Billon-Bruyat J-P (2006) Sauropod trackway patterns expression of special behaviour related to substrate consistency? An example from the Late Jurassic of northwestern Switzerland. Hantekeniana 5:38–41. ISSN 1219-3933

Matos RMD (1992) The northeast Brazilian rift system. Tectonics 11:766–791

Milàn J, Bromley RG (2006) True tracks, undertracks and eroded tracks, experimental work with tetrapod tracks in laboratory and field: palaeogeography. Palaeoclim Palaeoecol 231:253–264. https://doi.org/10.1016/j.palaeo.2004.12.022

Milàn J, Clemmensen LB, Bonde N (2004) Vertical sections through dinosaur tracks (Late Triassic lake deposits, East Greenland)–undertracks and other subsurface deformation structures revealed. Lethaia 37:285–296. https://doi.org/10.1080/00241160410002036

Milàn J, Avanzini M, Clemmensen LB, García-Ramos JC, Piñuela L (2006) Theropod foot movement recorded by Late Triassic, Early Jurassic and Late Jurassic fossil footprints. New Mex Mus Nat Hist Sci Bull 37:352–364

Moratalla JJ, Lockley MG, Buscalioni AD, Fregenal-Martínez MA, Meléndez N, Ortega F, Pérez-Moreno BP, Pérez-Asensio E, Sanz JL, Schultz RJ (1995) A preliminary note on the first tetrapod trackways from the lithographic limestones of Las Hoyas (Lower Cretaceous, Cuenca, Spain). Geobios 28(6):777–782

Naish D, Martill DM, Frey E (2004) Ecology, systematics and biogeographical relationships of dinosaurs, including a new theropod, from the Santana Formation (?Albian, Early Cretaceous) of Brazil. Hist Biol 18:1–14. https://doi.org/10.1080/08912960410001674200

Neumann VH, Cabrera L (2002a) A Tendência Expansiva do Sistema Lacustre Aptiano-Albiano do Araripe durante sua evolução: Dimensões e Morfologia. IG. Série B, Estudos e Pesquisas, Recife-PE 11:176–188

Neumann VH, Cabrera L (2002b) Características hidrogeológicas gerais, mudanças de salinidade e caráter endorréico do sistema lacustre Cretáceo do Araripe, NE Brasil. Revista de Geologia (UFC). Fortaleza 15:43–54

Neumann VH, Borrego AG, Cabrera L, Dino R (2003) Organic matter composition and distribution through the Aptian-Albian lacustrine sequences of the Araripe Basin, northeastern Brazil. Int J Coal Geol 54(1–2):21–40

Osés GL, Petri S, Becker-Kerber B, Romero GR, Rizzutto MA, Rodrigues F, Galante D, Silva TF, Curado JF, Rangel EC, Ribeiro RP, Pacheco MLAF (2016) Deciphering the preservation of fossil insects: a case study from the Crato Member, Early Cretaceous of Brazil. PeerJ 4:1–28

Osés GL, Petri S, Voltani CG, Prado GMEM, Galante D, Rizzutto MA, Rudnitzki ID, Silva EP, Rodrigues F, Rangel EC, Sucerquia PA, Pacheco MLAF (2017) Deciphering pyritization-kerogenization gradient for fish soft-tissue preservation. Sci Rep 7:1–15

Paula Freitas ABL (2010) Análise estratigráfica do intervalo siliciclástico Aptiano da Bacia do Araripe (Formação Rio da Batateira). Dissertação (Mestrado em Geologia). Instituto de Geociências, Universidade Federal do Rio de Janeiro, Rio de Janeiro

Paula Freitas ABL, Borghi L (2011) Fácies sedimentares e sistemas deposicionais siliciclásticos aptianos da Bacia do Araripe. Geociências 30(4):529–543

Paula-Freitas ABL, Borghi L, Carvalho IS (2007) Âmbar na Formação Rio da Batateira, Bacia do Araripe (Cretáceo Inferior, Brasil). In: Carvalho, IS, Cassab, RCT, Scwanke, C, Carvalho, MA, Fernandes, ACS, Rodrigues, MAC, Carvalho, MSS, Arai, M, Oliveira, MEQ (Eds) Paleontologia: Cenários de Vida, Editora Interciência, 1, Rio de Janeiro, pp 169–176

Pereira PA, Cassab RCT, Barreto AMF (2017) Paleoecologia e paleogeografia dos moluscos e equinoides da Formação Romualdo, Aptiano-Albiano da Bacia do Araripe, Brazil. Anuário do Instituto de Geociências 40:180–198

Pérez-Lorente F (2015) Dinosaur footprints and trackways of La Rioja. Bloomington, Indiana, Indiana University Press. ISBN 9780253015150

Ponte FC, Appi CJ (1990) Proposta de revisão da coluna litoestratigráfica da Bacia do Araripe. In: 36 Congresso Brasileiro de Geologia. Natal, Brasil, pp 211–226

Prado G, Arthuzzi JCL, Osés GL, Callefo F, Maldanis L, Sucerquia P, Becker-Kerber B, Romero GR, Quiroz-Valle FR, Galante D (2021) Synchroton radiation in palaeontological investigations: examples from Brazilian fossils and its potential to South American palaeontology. J S Am Earth Sci 108:102973

Ribeiro AC, Ribeiro GC, Varejão FG, Battirola LD, Pessoa EM, Simões MG, Warren LV, Riccomini C, Pojato-Ariza FJ (2021) Towards an actualistic view of the Crato Konservat-Lagerstätte paleoenvironment: a new hypothesis as an Early Cretaceous (Aptian) equatorial and semi-arid wetland. Earth Sci Rev 216:103573

Rios-Netto AM, Regali MSP (2007) Estudo bioestratigráfico, paleoclimático e paleoambiental do intervalo Alagoas (Cretáceo Inferior) da bacia do Araripe, nordeste do Brasil (Poço 1-PS-11-CE). In: Carvalho IS, Cassab RCT, Schwanke, C, Carvalho MA, Fernandes ACS, Rodrigues MAC, Carvalho MSS, Arai M, Oliveira MEQ (eds) Paleontologia: Cenários de Vida, 1 edn, vol 2. Rio de Janeiro, Editora Interciência, pp 479–488. ISBN 9788571931855

Rios-Netto AM, Regali MSP, Carvalho IS, Freitas FI (2012) Palinoestratigrafia do Intervalo Alagoas da Bacia do Araripe. Revista Brasileira de Geociências 42(2):331–342

Santana W, Tavares M, Martins CAM, Melo JPP, Pinheiro AP (2022) A new genus and species of brachyuran crab (Crustacea, Decapoda) from the Aptian-Albian (Cretaceous) of the Araripe Sedimentary Basin, Brazil. J S Am Earth Sci 116:103848

Santos Filho MAB, Ceolin D, Fauth G, Lima FHO (2023) Ostracods from the Barbalha and Crato Formations, Aptian of the Araripe Basin, northeast Brazil. Zootaxa 5319(3):332–350

Santos VF, Callapez PM, Rodrigues NPC (2013) Dinosaur footprints from the Lower Cretaceous of the Algarve Basin (Portugal): New data on the ornithopod palaeoecology and palaeobiogeography of the Iberian Peninsula. Cretac Res 40:158–169. https://doi.org/10.1016/j.cretres.2012.07.001

Sayão JM, Saraiva AÁF, Brum AS, Bantim RAM, de Andrade RCIP, Cheng X, Lima FJ, Silva HP, Kellner AWA (2020) The first theropod dinosaur (Coelurosauria, Theropoda) from the base of the Romualdo Formation (Albian), Araripe Basin, Northeast Brazil. Sci Rep 10(1):1–15

Selden P, Nudds J (2012) Evolution of fossil ecosystems. Manson Publishing, London, 288p

Teixeira MC, Mendonça Filho JG, Oliveira AD, Assine ML (2017) Faciologia orgânica da Formação Romualdo (Grupo Santana, Cretáceo Inferior da Bacia do Araripe): caracterização da matéria orgânica sedimentar e interpretação paleoambiental. Geologia USP Série Científica 17(4):19–44

Tomé ME, Araripe R, Oliveira D, Barreto A, Prado L, Pedrosa F, Pereira P, Nascimento LR, NG C (2022) Early Cretaceous Ostracoda (Crustacea) from south-central Araripe Basin, Brazil, with descriptions of seven new species. Zootaxa 5159 (4): 535–557

Valenzuela M, García-Ramos JC, de Centi CS (1988) Las huellas de dinosaurios del entorno de Ribadesella. Central Lechera Asturiana. ISBN: 84-404-2845-6

Varejão FG, Warren LV, Simões MG, Fürsich FT, Matos SA, Assine ML (2019) Exceptional preservation of soft tissues by microbial entombment: insights into the taphonomy of the Crato Konservat-Lagerstätte. Palaios 34:331–348

Varejão FG, Warren LV, Simões MG, Buatois LA, Mangano MA, Bahniuk AMR, Assine ML (2021a) Mixed siliciclastic-carbonate sedimentation in an evolving epicontinental sea: Aptian record of marginal marine settings in the interior basins of north-eastern Brazil. Sedimentology. https://doi.org/10.1111/sed.12846

Varejão FG, Silva VR, Assine ML, Warren LV, Matos SA, Rodrigues MG, Fürsich FT, Simões MG (2021b) Marine of freshwater? Accessing the paleoenvironmental parameters of the Caldas Bed, a key marker bed in the Crato Formation (Araripe Basin, NE Brazil). Braz J Geol 51(1):e2020009

Warren LV, Varejão FG, Quaglio F, Simões MG, Fürsich FT, Poiré DG, Catto B, Assine ML (2017) Stromatolites from the Aptian Crato formation, a hypersaline lake system in the Araripe Basin, northeastern Brazil. Facies 63(3):1–19

Chapter 7
Walking in the Gondwanic Floodplains of Rio do Peixe Basins

Giuseppe Leonardi and Ismar de Souza Carvalho

7.1 Introduction

The Rio do Peixe basins are a set of four associated basins in the Northeastern region of Brazil: Sousa, Uiraúna-Brejo das Freiras (also known as Triunfo Basin), Pombal and Vertentes (or Icozinho Basin) (Fig. 7.1). The Sousa Basin comprises an area of ~675 km² (~60 km E-W, ~17 km N-S, E-W-oriented) and is located in the western part of the Paraíba State, in the counties of Aparecida, São João do Rio do Peixe, and Sousa (Fig. 7.2). This basin, when compared with the others, contains the largest number of dinosaur tracks.

The Triunfo Basin, is nearby to the Sousa Basin. It is located in the northwest of Paraíba State, in the counties of Uiraúna, Poço, Brejo das Freiras, Triunfo, and Santa Helena. It has a roughly triangular shape, NE-SW oriented, with an area of ~450 km². The Pombal Basin, the easternmost basin (NW–SE oriented), is a small one, with an area of about 81 km² (~27 km long and ~3 km wide). There are few outcrops, and the rocks are generally excessively coarse-grained to preserve dinosaur footprints. The Vertentes Basin (or Icozinho) is a small, long and very narrow basin, NE-SW oriented, with an area of about 74 km² (~37 km long and ~1 to 3 km wide). The total area of these four basins is estimated at ~1,280 km² (Gonzaga et al. 2022).

G. Leonardi (✉)
Istituto Cavanis, Dorsoduro 898, 30123 Venezia, Italy
e-mail: leonardigiuseppe879@gmail.com

CCMN/IGEO, Departamento de Geologia, Universidade Federal do Rio de Janeiro, Cidade Universitária, Ilha do Fundão, Rio de Janeiro, Estado do Rio de Janeiro 21949-900, Brazil

I. S. Carvalho
CCMN/IGEO, Departamento de Geologia, Universidade Federal do Rio de Janeiro, Cidade Universitária, Ilha do Fundão, Rio de Janeiro, Estado do Rio de Janeiro 21.910-200, Brazil
e-mail: ismar@geologia.ufrj.br

Centro de Geociências, Universidade de Coimbra, Rua Sílvio Lima, 3030-790 Coimbra, Portugal

© The Author(s), under exclusive license to Springer Nature Switzerland AG 2024
I. S. Carvalho and G. Leonardi (eds.), *Dinosaur Tracks of Mesozoic Basins in Brazil*,
https://doi.org/10.1007/978-3-031-56355-3_7

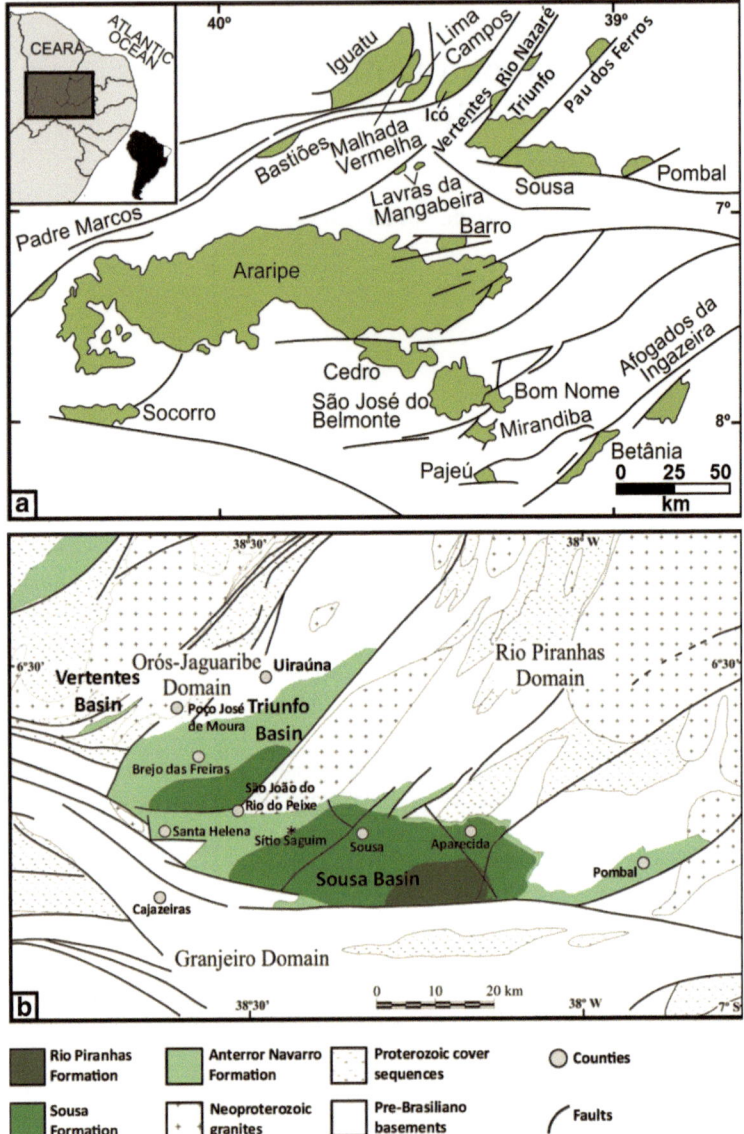

Fig. 7.1 Location map and geological framework of the Rio do Peixe basins area. **a** The intracratonic sedimentary basins of Northeastern Brazil. **b** Location and geological map of the four Rio do Peixe basins, western Paraíba, Brazil (modified from Carvalho 2000a and Castro et al. 2007)

Fig. 7.2 The Sousa Basin half-graben, from its northern margin. Overview of the Sousa Basin from Riachão dos Oliveira hill (**a**) and from Benção de Deus hill (**b**)

The origin of these basins is related to fault movements along preexisting structural trends of the Proterozoic basement during South and equatorial Atlantic Ocean opening (Ramos 2023). The Rio do Peixe basins, like many others in the region, are aborted rifts. In the clastic continental sediments, dinosaur footprints are the most abundant fossils. The main tetrapod ichnofauna comprises isolated footprints and trackways of large and small theropods, besides ornithopods, sauropods and ankylosaurs. Fish trails are found in the Sousa Basin (Muniz 1985; Leonardi and Muniz 1985). There are also invertebrate ichnofossils such as trails and burrows produced by arthropods and annelids (Fernandes and Carvalho 2001). The fossils are palynomorphs, plant fragments and some logs, ostracods, conchostracans and fish scales. Rare dinosaur bones (Ghilardi et al. 2014, 2016; Carvalho et al. 2017) and Crocodylomorpha bones (Carvalho and Nobre 2001) were also found. Sousa, Triunfo and Pombal basins comprise 42 ichnofossiliferous sites. The dinosaur ichnofaunas present a same stratigraphic-time-paleogeographical context, and represent parts of a widespread megatracksite. In these basins was recorded a great number of individual dinosaur tracks. These represent the passage of at least 629 individuals of the various dinosaur clades; there are also a large number of small chelonian half-swimming tracks and some Crocodylomorpha traces.

The environmental setting at that time was influenced by the initial development of the Gondwana breakup, with an endemic biota living nearby ephemeral rivers and shallow lakes under hot climate conditions. These were preserved in alluvial fans, braided, meandering rivers and shallow lake deposits of Berriasian to lower Barremian age.

After the first description of dinosaur tracks in the Sousa Basin (Moraes 1924) there was an increase of new ichnosites (Leonardi 2021) and, consequently, a relevant literature on the subject was produced (Leonardi 1979a, b, 1980a, b, 1981, 1984a, b, 1985, 1987a, b, 1989, 1994, 2008, 2011, 2021; Leonardi et al. 1987a, b, c; Leonardi and Carvalho 2000c, 2002, 2021; Godoy and Leonardi 1985; Campos et al. 2015; Carvalho 1989, 1996, 2000a, 2000b, 2004a, 2004b; Carvalho et al. 1993a, b; Carvalho et al. 2013a, b, c, 2016, 2017; Carvalho and Fernandes 1992; Carvalho and Leonardi 1992, 2007, 2021, 2023; Santos and Santos 1987a, 1987b, 1989; Carvalho and Carvalho 1990; Fernandes and Carvalho 2001, 2007; Leonardi and Santos 2006; Santos et al. 2016; Siqueira et al. 2011; Viana et al. 1993).

7.2 Geological Context

The Rio do Peixe basins are four sedimentary basins (Fig. 7.1): Sousa, Triunfo, Pombal and Vertentes located at the western extremity of Paraíba State and in the southeastern Ceará State, in the Proterozoic Borborema Province, Northeastern Brazil. They largely correspond to the current catchment area of the Peixe River, a tributary of the Piranhas River (Fig. 7.3).

These basins are located in the context of the Interior Rift System of Northeast Brazil. They are rift basins that evolved as a consequence of the Cretaceous Gondwana break-up, alongside earlier structural trends of the basement, during the South Atlantic Ocean opening (Matos 1992; Rapozo et al. 2021; Matos et al. 2021). As a result, there was the origin of several sedimentary basins throughout the normal and transcurrent fault movements within the Precambrian basement (Nogueira et al. 2015).

During the first rifting phase, the two major faults bordering the basins (Portalegre Fault and Malta Fault) caused normal displacement (Pichel et al. 2022) and left-lateral transtension on an E-W-striking major fault (Rapozo et al. 2019). The fault patterns controlled the E-W oriented Sousa half-graben and the SW oriented Triunfo, Vertentes and Pombal half-grabens. In the interior of these basins there are many minor faults, generally parallel to the major alignments, mainly to the Portalegre Fault (Araújo et al. 2019; Torabi et al. 2021; Nicchio et al. 2022; Oliveira et al. 2022; Pichel et al. 2022; Freitas et al. 2023). During the Lower Cretaceous (Berriasian to Hauterivian), under the same tectonic stress pattern, the basinal areas increased and during the last tectonic stage (early Barremian), there was a change in the tectonic

Fig. 7.3 The Sousa Basin and the Peixe River. **a** The Peixe River with its ciliary forest and its valley, corresponding in large part to the Sousa Basin, with the savannah and caatinga environments, partly cultivated with cotton; **b** The Peixe River during a long dry season, when the best outcrops of the Sousa Formation become visible

pattern, and sediment accumulation began to decline (Carvalho and Leonardi 2021). Later, in the post-rift phase, there would have been a reversal: the Rio do Peixe basins were subjected to a horizontal compression (ESE-WNW) from the Late Cretaceous onwards (Nogueira et al. 2015; Lima et al. 2017; Barbosa et al. 2021; Bezerra et al. 2017, 2023; Maciel et al. 2018).

The time interval of sedimentation, based on ostracods and palynomorphs, is characteristic of Berriasian to early Barremian stages (Early Cretaceous; Carvalho 2000a, 2004a; Lourenço et al. 2021). However, beneath the Lower Cretaceous succession (Fig. 7.4a, b), there are Lower Devonian (Eolochkovian–?Eopraguian) rocks identified through palynological analysis from drillings by Petrobras (Roesner et al. 2011). These sediments are the Santa Helena Group, divided in the Pilões Formation and Triunfo formations (Silva et al. 2014), indicating a multi-phase history of these basins (Silva et al. 2014; Carvalho and Leonardi 2021; Pichel et al. 2022). After the short Devonian deposition, during the Cretaceous, the Sousa Basin accumulated more than 2 km thick in the depocenter, and Triunfo Basin is a 2.5–3 km thick half-graben

Fig. 7.4 a Antenor Navarro Formation succession at Serrote do Letreiro; **b** Molds of large mud craks from the Sousa Formation; **c** The Moura Formation, a Cenozoic and Recent deposits of loose sediments. All images from the Sousa Basin

(Córdoba et al. 2007; Carvalho and Leonardi 2021). On the contrary, during the later phases of Cenozoic reactivation of the marginal coastal basins of Pernambuco and Paraíba (Neogene-Quaternary; Lima et al. 2017), it seems that there was not expressive reactivation of the fault system in the Rio do Peixe basins, aspect that is also confirmed by a low sediment supply (Moura Formation; Fig. 7.4c).

So, after an initial Paleozoic phase, these basins were filled with Lower Cretaceous reddish and greenish shales, mudstones, siltstones and sandstones of the Rio do Peixe Group (Fig. 7.5). This unit comprises three formations. The Antenor Navarro Formation on the border of these basins is interpreted as coalescing alluvial fans and braided fluvial systems. The Sousa Formation, which is essentially microclastic (fine

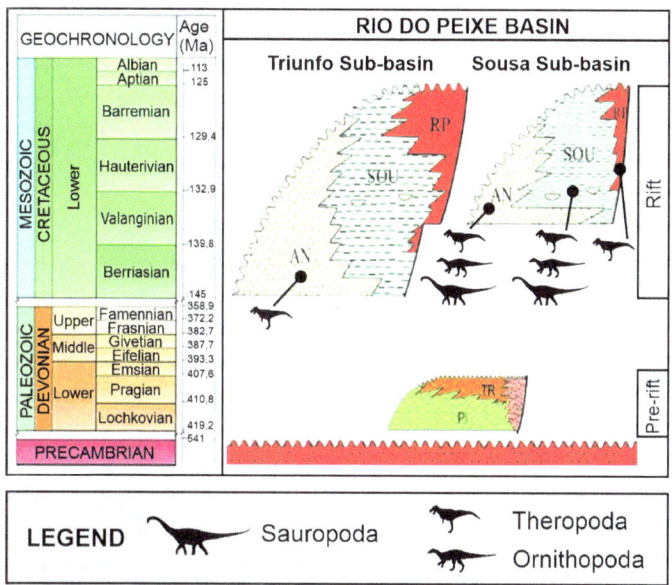

Fig. 7.5 Rio do Peixe Basins stratigraphical chart, and the lithostratigraphic units with dinosaur footprints. Abbreviations: Pl: Pilões Formation; TR: Triunfo Formation (Devonian Group Santa Helena); AN: Antenor Navarro Formation; SOU: Sousa Formation; RP: Rio Piranhas Formation. Modified from Rapozo et al. (2021)

sandstones, siltstones, argillites, marls), indicates lacustrine, swampy, and meandering fluvial environments with microbial influence (Carvalho et al. 2013a). The Rio Piranhas Formation is interpreted as alluvial fans and temporary and braided rivers.

These deposits demonstrate the direct control of sedimentation by the tectonic activity (Souza et al. 2021; Oliveira et al. 2022). Deposition along the faulted borders of the basins are alluvial fans, changing to a braided fluvial system more distally. A meandering fluvial system with a wide floodplain was established in the central region of the basins, where perennial and/or temporary lakes were established (Carvalho 2000a; 2004a; Leonardi and Carvalho 1992, 2021; Lourenço 2021).

The mudstones, siltstones and black shales of the Sousa Formation (locality Sítio Saguim, Sousa Basin) are prospective for hydrocarbon generation (low liquid hydrocarbon generation potential but a moderate gas potential, ANP - Agência Nacional de Petróleo 2008; Iemini 2009). However, a validation to hydrocarbon exploration in the basin is so far expected (Carvalho et al. 2013c; Muniz et al. 2017; Gonzaga et al. 2022; Fig. 7.6).

Fig. 7.6 Oil exudation from Sousa formation at Sítio Saguim, interior of the Sousa Basin

7.3 Footprints: Diversity and Paleobiological Interpretation

The Sousa and Triunfo basins have a paleontological significance due to the abundance of dinosaur ichnofaunas that are part of an extensive Early Cretaceous mega-tracksite (Viana et al. 1993; Leonardi and Carvalho 2000, 2002) established during the break-up of Gondwana and the early stages of the South Atlantic opening. After 48 years of field work, 42 tracksites (26 in Sousa Formation; 11 in Antenor Navarro Formation; 5 in Rio Piranhas Formation) and about 96 tracks-bearing levels of the Rio do Peixe basins (74 in Sousa Formation; 17 in Antenor Navarro Formation; 5 in Rio Piranhas Formation) were recognized, mainly in the Sousa Basin (Leonardi 2021; Leonardi and Carvalho 2021).

These tracksites and the correspondent ichnofossiliferous levels contain track-ways or isolated footprints assigned to different clades of dinosaurs: 395 large theropods (Fig. 7.7**a**); 31 smaller theropods with a third toe substantially longer than the other two toes; five additional, different kinds of small theropods; 16 medium-size theropods from Serrote do Letreiro (for a total of about 447 individual theropods); about 90 sauropods (Fig. 7.7**b**); 30 graviportal ornithopods (among them four quadrupedal and one sub-quadrupedal trackways, along with some isolated footprints, probably pertaining also to quadrupedal animals) (Fig. 7.7**c**); six small ornithopods; one ankylosaur (Fig. 7.7**d**); one small quadrupedal Thyreophora (alto-gether 36 ornithopods, 38 ornithischians); and at least 53 indeterminate dinosaur tracks. In total, the number of identifiable individual dinosaurs is 576, and the total number of individual dinosaurs, including the indeterminate tracks is at least 629; there are, in addition, some representatives of the mesofauna. There are also four possible dinosaur tail impressions (Leonardi and Carvalho 2000, 2021). These numerical data are updated with respect to those previously provided (Leonardi and Carvalho 2021) both for the increase of the discovered locations and for a revision of some classification cases.

The meso-ichnofauna, very rare in these basins, is represented by just one set of batrachopodid prints; some crocodilian traces (tracks and a body imprint in the mudstone) (Fig. 7.8a); one isolated lacertoid footprint (Fig. 7.8b); and a very large number of small chelonian swimming tracks (Leonardi and Carvalho 2000, 2021; Fig. 7.8c). The absence, for now, of pterosaur tracks is odd. The mammals seem to have left no traces, although an occurrence (of poor quality), cannot be excluded, with an eventual trackway at Riacho do Cazé (Sousa County), Antenor Navarro Formation, Sousa Basin (Fig. 7.8d). The detailed description and classification of the ichnofossiliferous sites of Sousa and Triunfo basins (and the few material found in the two smaller basins of Pombal and Vertentes) is presented by Leonardi and Carvalho (2021). Some new ichnosites (Serrote do Mocó Fig. 7.9a and Araçá-Rio Novo, Fig. 7.9b, Leonardi 2021; Pereiros, Carvalho and Leonardi 2023 (Fig. 7.9c); Engenho Novo 3rd, Leonardi 2021; (Fig. 7.9d) and the new site Buscapé, Fazenda Abóbora (unpublished) were recently discovered in the Sousa Basin, Sousa Forma-tion. Table 7.1 summarizes the data from these ichnosites and Table 7.2 gives the geographical coordinates, to aid the tracksites location.

7.3.1 Fossil Tracks Complete or Replace Bones in Dinosaur Documentation

One of the advantages of fossil tracks is that in many paleoenvironments and conti-nental stratigraphic units, where body fossils are poor or absent, good quality tracks complement or sometimes completely replace the documentation on the existence of dinosaurs and other animals. This is for example the case of the western Paraíba

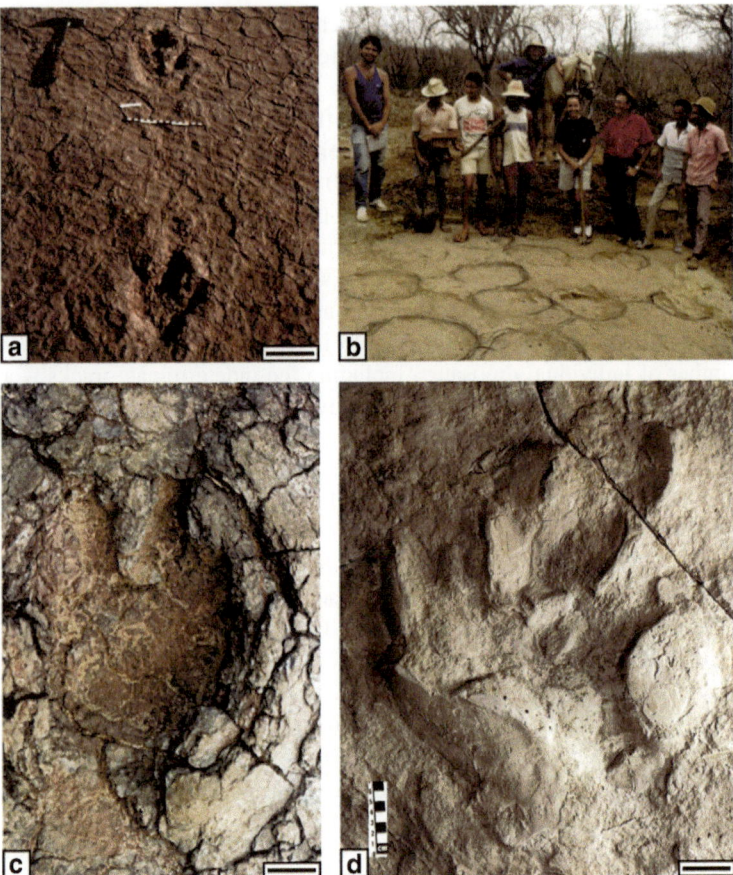

Fig. 7.7 The Rio do Peixe Basins contain trackways or isolated footprints assigned to different clades of dinosaurs: **a** Theropods; **b** Sauropods. Photo by Franco Capone; c Ornithopods; **d** Cast of an ankylosaurian hand-foot set (SOES 7), from Serrote do Pimenta, Antenor Navarro Formation. Photo by M. de Fátima C. F. dos Santos. Graphic scales: **a** = 10 cm; **b** = the diameter of the footprints of sauropods in excavation is 40 to 80 cm; **c** = 8 cm; **d** = 5 cm

State, but also other areas of the Brazilian northeast: the Mesozoic skeletons and bones are rare, with the exception of the Araripe Basin.

The fossil tracks present a true bonanza in the Rio do Peixe basins and especially in the Sousa Basin, with more than 600 dinosaurs recorded by their footprints in such a small area, and a very high diversity index. The body fossils, on the contrary, are reduced so far to a few bones corresponding to two individuals of titanosaurids (Ghilardi et al. 2014, 2016; Carvalho et al. 2017) and one or two notosuchian crocodyliformes (Carvalho and Nobre 2001). In the Sousa Basin, even if only a titanosaurid fibula had been found so far (Ghilardi et al. 2014, 2016), from

Fig. 7.8 Non-dinosaur tracks in the Sousa Basin. **a** Crocodilian body imprint at Tapera, Sousa Formation; **b** The single lacertoid footprint in the Rio do Peixe Basins, from Serrote do Pimenta ichnosite, Antenor Navarro Formation; **c** Footprints of swimming chelonians and a theropod, from the Piau locality, Sousa Formation; **d** A possible occurrence of mammal tracks, from Riacho do Cazé, Sousa County, Antenor Navarro Formation. Graphic scales: **a** = people as a graphic scale; **B** = 2 cm; **C** = 20 cm; **D** = 10 cm

the fossil footprints, we can estimate the existence of about 608 dinosaur individuals of distinct clades. In the Triunfo Basin, the unique formally described dinosaur species is the sauropod *Triunfosaurus leonardii* Carvalho et al. 2017 (Fig. 7.10), despite the presence of theropod footprints. Among the tracks from these basins it is possible to estimate at least 447 individual theropods divided into large predators, mostly abelisaurids, but without excluding spinosaurs (340 individuals, represented by at least five different forms; Fig. 7.11); 31 small theropods, probably noasaurids or velocisaurids (Fig. 7.12), with long and slender feet, producing tracks similar to those of the *Grallator-Eubrontes* plexus of the Laurasian continents; 16 medium-sized theropod individuals, different, but attributable to the same plexus and forming a single population (Fig. 7.13); not to mention the about 60 swimming theropods, that one cannot classify correctly. The sauropods are less frequent in these formations. In the Sousa Basin their fossil footprints indicate 67 individuals divided into five great groups that marched in herds, with evident gregariousness; and 23 isolated individual tracks (about 90 sauropods, all together). Sauropod tracks are often of

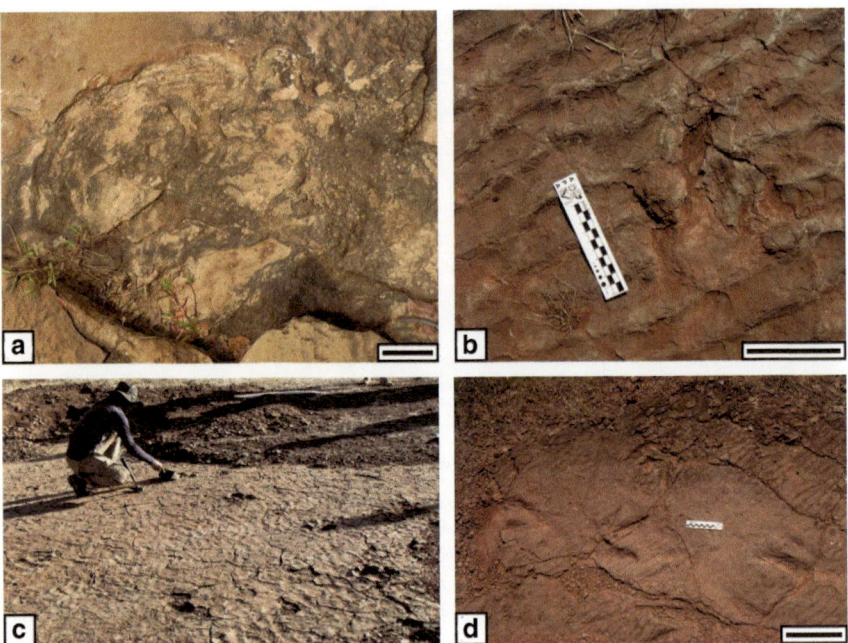

Fig. 7.9 Tracksites from the Sousa Basin. **a** Serrote do Mocó, with sauropod tracks; Photo by Luiz Carlos Gomes da Silva; **b** Theropod tracks and ripple marks at Araçá Rio Novo; **c** The Pereiros ichnosite, with three theropod trackways; **d** Engenho Novo 3rd, with several theropod and sauropod tracks. Graphic scales: **a** = 15 cm; **b** = 10 cm; **c** = Ismar Carvalho as reference scale; **d** = 20 cm

poor quality, and typically very large (Fig. 7.14**a**). The largest, with a diameter of 120 cm, is located in the locality Piau-Caiçara.

The ornithopods were rarer, and their trackways are partially bipedal (Fig. 7.14**b**), partly quadruped (Fig. 7.14**c–d**) and partly semi-quadruped. Most are isolated and therefore not gregarious individuals, usually graviportal animals of large dimensions (about 30 individuals). The main ones have been assigned to three ichnogenera (Leonardi 1979a, 1984a), attributed to iguanodontids, probably of African affinity, without excluding other types. There are also some specimens of small ornithopod tracks (about 6 specimens). One of the most important discoveries in the Sousa Basin was that of a hand-foot pair of an ankylosaurian, probably a nodosaurid (SOES 7; Leonardi 1984a, 1994). The discovery of it in 1979 was the first indication of the presence of ankylosaurs in South America (Fig. 7.15**a–b**). Another particularly interesting specimen is a rather enigmatic short trackway, of difficult interpretation because it is an underprint. It is the track SOPP 15 (Leonardi 1994, 58; Leonardi and Carvalho 2021), of four hand-foot sets, found in the locality Passagem das Pedras, level 1, perhaps a Thyreophoroidea trackmaker. Together they are 38 Ornithischia (6.60% of the total individual tracks classifiable); and 537 Saurischia (93.23% of the total individual tracks classifiable).

Table 7.1 Ichnosites and ichnofaunas of the Rio do Peixe Basins. Ank = Ankylosauria; Chel = Chelonians; Croc = Crocodylomorph; IS = Dinosauria; uncertain classification; Juv = juvenile dinosaur; Liz = Lizard-like track; Mam = ?Mammal tracks; Ornith = quadruped ornithischian; Triunfo = Triunfo Basin

Code	Ichnosite	Basin	Formation	Theropods	Sauropods	Ornithopods	Others dinos	IS	Total dinos	Levels
Sousa Formation										
SOAB	Abreu	Sousa	Sousa	1					1	1
ANAC	Araçá de Cima	Sousa	Sousa	1					1	1
ANAN	Araçá Rio Novo	Sousa	Sousa	>5					>5	?
ANBD	Barragem do Domício	Sousa	Sousa	3					3	3
SOBP	Baixio do Padre	Sousa	Sousa	1		1			2	2
SOBU	Buscapé, ex Abobora	Sousa	Sousa	1					1	1
ANEN1	Engenho Novo1	Sousa	Sousa	24	1	1		2	28	?
ANEN2	Engenho Novo2	Sousa	Sousa	2					2	1
ANEN3	Engenho Novo3	Sousa	Sousa	>9	>10				19	2
ANJU	Juazeirinho	Sousa	Sousa	4					4	4
SOMA	Matadouro	Sousa	Sousa	4					4	2
SOPP	Passagem das Pedras	Sousa	Sousa	14		4	1 Juv 1 ornith	~20	~40	8
SOPE	Pedregulho	Sousa	Sousa	2					2	2
SJPE	Pereiros	Sousa	Sousa	3					3	2
SOCA	Piau-Caiçara	Sousa	Sousa	~170	9	7	(1 croc Chel)	~23	~209	25
SOPU	Piau 2	Sousa	Sousa	2					2	2
SOPI	Piedade	Sousa	Sousa	4		2		1	7	3
SOPM	Poço do Motor	Sousa	Sousa	8					8	3

(continued)

Table 7.1 (continued)

Code	Ichnosite	Basin	Formation	Theropods	Sauropods	Ornithopods	Others dinos	IS	Total dinos	Levels
SOPV	Poço da Volta	Sousa	Sousa	1					1	1
ANSM	Serrote do Mocó	Sousa	Sousa		~1				1	1
SOSA	Sítio Saguim	Sousa	Sousa	4		1			5	1
APTA	Tapera	Sousa	Sousa	>5	2	1	(1croc)		>8	3
APVR1	Varzea dos Ramos 1	Sousa	Sousa	>60	1	?1			~62	2
APVR2	Varzea dos Ramos 2	Sousa	Sousa	1					1	1
APVR3	Varzea dos Ramos 3	Sousa	Sousa	10					10	2
ANZO	Zoador	Sousa	Sousa	1					1	1
Antenor Navarro Formation										
UIAR	Arapuã	Triunfo	Ant.Nav		1				1	1
ANAR	Aroeira	Sousa	Ant.Nav		1				1	1
UIBA	Baleia	Triunfo	Ant.Nav	2					2	1
ANCA	Cabra Assada	Sousa	Ant.Nav	7		2		4	13	?
POGR	Grotão	Pombal	Ant.Nav					1	1	1
UIPO	Pocinho	Triunfo	Ant.Nav	4					4	2
SORC	Riacho do Cazé	Sousa	Ant.Nav	~5	~5		(mam?)		~10	2
SORP	Riacho dos Oliveira	Sousa	Ant.Nav			1			1	?
SOSL	Serrote do Letreiro	Sousa	Ant.Nav	>24	~20	1		2	~47	3
SOES	Serrote do Pimenta	Sousa	Ant.Nav	~32	8	3	ank, (liz)		~44	3
SOFB	Floresta dos Borba	Sousa	Ant.Nav	>4	~20	>2			~26	3

(continued)

Table 7.1 (continued)

Code	Ichnosite	Basin	Formation	Theropods	Sauropods	Ornithopods	Others dinos	IS	Total dinos	Levels
Rio Piranhas Formation										
SOCV	Curral Velho	Sousa	Rio Pir	7		5			12	1?
SOLF 1	Lagoa do Forno 1	Sousa	Rio Pir	6	8				14	1
SOLF 2	Lagoa do Forno 2	Sousa	Rio Pir	1					1	1
SOMD	Mãe d'Água	Sousa	Rio Pir	>2	>2	4			>8	>1
SOFP	Fazenda Paraíso	Sousa	Rio Pir	13	1				14	>1
Total ichnofaunas of Sousa Formation										
26	26	–	26	>340	>24	>18	2	46	>430	>74
Total ichnofaunas of Antenor Navarro Formation										
11	11	–	11	>78	55	>9	1	7	>150	>17
Total ichnofaunas of Rio Piranhas Formation										
5	5	–	5	>29	>11	9	0	0	>49	>5
Total ichnofaunas of Rio do Peixe Basins										
42	42	–	42	>447	>90	>36	3	53	>629	>96

Table 7.2 The ichnosites of the Rio do Peixe Basins and geographic coordinates, datum WGS 84 (Siqueira et al. 2011; Leonardi and Carvalho 2021). The coordinates were collected at different times, over 48 years, using different methods and devices (or maps)

Code	Ichnosite	Basin	Formation	Coordinates
SOAB	Abreu	Sousa	Sousa	06°44′25″S/ 38°19′00″W
ANAC	Araçá de Cima	Sousa	Sousa	06°44.995S/ 38°24.673W
ANAN	Araçás Rio Novo	Sousa	Sousa	06°44′41.2″S/ 38°25′08.6″W
ANBD	Barragem do Domício	Sousa	Sousa	06°44.165 S/ 38°26.288W
SOBP	Baixio do Padre	Sousa	Sousa	06°45.113 S/ 38°19.993W
SOBU	Buscapé, ex Abobora	Sousa	Sousa	–
ANEN1	Engenho Novo1	Sousa	Sousa	06°42′51.7″S/ 38°24′43.0″W
ANEN2	Engenho Novo2	Sousa	Sousa	06°42′51.7″S/ 38°24′43.0″W
ANEN3	Engenho Novo3	Sousa	Sousa	06°42.896S/ 38°24.752″W
ANJU	Juazeirinho	Sousa	Sousa	06°44.685 S/ 38°25.144W
SOMA	Matadouro	Sousa	Sousa	06°45′06.93″S/ 38°13′41.72″W
SOPP	Passagem das Pedras	Sousa	Sousa	06°44′00.51″S/ 38°15′41.57″W
SOPE	Pedregulho	Sousa	Sousa	06°45′22.6″S/ 38°20′56.7″W
SOCA	Piau-Caiçara	Sousa	Sousa	06°44′24.9″S/ 38°19′54.9″W
SOPU	Piau 2	Sousa	Sousa	06°43′52″S/ 38°19′37″W
SOPI	Piedade	Sousa	Sousa	06°44′55.4″S/ 38°20′56.5″W
SOPM	Poço do Motor	Sousa	Sousa	06°44′18.139″S/ 38°15′28.947″W
SOPV	Poço da Volta	Sousa	Sousa	06°45′20.10″S/ 38°24′43.9″W
ANSM	Serrote do Mocó	Sousa	Sousa	06°41.7965/ 38°24.974″W
SJSP	Sítio Pereiros	Sousa	Sousa	06°47′18.71″S/ 38°29′11.81″W
SOSA	Sítio Saguim	Sousa	Sousa	06°43′24.3″S/38° 20′15.5″W

(continued)

Table 7.2 (continued)

Code	Ichnosite	Basin	Formation	Coordinates
APTA	Tapera	Sousa	Sousa	06°46.188 S/38° 06.695W
APVR1	Varzea dos Ramos 1	Sousa	Sousa	06°46′09.0″S/38° 06′40.5″W
APVR2	Varzea dos Ramos 2	Sousa	Sousa	06°46′38″S/ 38°05′42″W
APVR3	Varzea dos Ramos 3	Sousa	Sousa	06°46′38″S/ 38°06′15″W
ANZO	Zoador	Sousa	Sousa	06°45.301S/38 24.595W
UIAR	Arapuã	Triunfo	Ant.Navarro	06°34′45″S/ 38°25′40″W
ANAR	Aroeira	Sousa	Ant.Navarro	06°41′44″S/ 38°22′06″W
UIBA	Baleia	Triunfo	Ant.Navarro	06°12′10″S/ 38°25′13″W
POGR	Grotão	Pombal	Ant.Navarro	06°45′32″S/ 37°54′40″W
UIPO	Pocinho	Triunfo	Ant.Navarro	06°35′13″S/ 38°25′15″W
SORC	Riacho do Cazé	Sousa	Ant.Navarro	06°43.153″S/ 38°13.14.548W
SORP	Riacho dos Oliveira	Sousa	Ant.Navarro	06°43.347″S/ 38°14.636W
SOSL	Serrote do Letreiro	Sousa	Ant.Navarro	06°41′36.89″S/ 38°18′29.72″W
SOES	Serrote do Pimenta	Sousa	Ant.Navarro	06°43′18.8″S/ 38°11′44.1″W
SOFB	Floresta dos Borba	Sousa	Ant.Navarro	06°41′055″S/ 38°20′733″W
ANCA	Cabra Assada	Sousa	Ant.Navarro	06°49′53.8″S/38° 23′50.3″W
SOCV	Curral Velho	Sousa	Rio Piranhas	06°49′47.474″S/ 38°12′9.812″W
SOLF 1	Lagoa do Forno	Sousa	Rio Piranhas	06°48.066 S/ 38°10.039W
SOLF 2	Lagoa do Forno	Sousa	Rio Piranhas	06°48.563 S/ 38°10.492W

(continued)

Table 7.2 (continued)

Code	Ichnosite	Basin	Formation	Coordinates
SOMD	Mãe d'Água	Sousa	Rio Piranhas	06°48′58.5″S/ 38°12′41.5″W
SOFP	Fazenda Paraíso	Sousa	Rio Piranhas	06°48.793S/ 38°09.857W

Fig. 7.10 The sauropod *Triunfosaurus leonardii* Carvalho, Salgado, Lindoso, Araújo Jr., Nogueira & Agnelo, 2017 from the Triunfo Basin. In these basins the fossil bone are very rare. Graphic scale = 1 m. Art by Deverson Silva

Based on the numerical data, according to their characteristics (Leonardi 2021) there are 128 trackways assigned to herbivorous dinosaurs (22.26% of individual trackways and isolated footprints) and 447 trackways attributed to theropods (77.74% of the identifiable individual trackways and isolated footprints). The ratio of herbivorous to theropod individual trackways in this ichnofauna is 1 : 3.47. However, probably not all theropods were carnivorous and predatory; some were necrophagous; other forms of that clade could be herbivorous or omnivorous rather than strictly carnivorous. There were also piscivorous and insectivorous animals. This is particularly likely for small to midsize theropods.

There are at least 99 quadrupedal trackways (about 90 sauropods, and nine ornithischians, correspondent to about seven quadrupedal ornithopods and two quadrupedal thyreophorans; 17.22% of the identifiable individual trackways and isolated footprints) and 476 bipedal trackways (82.78% of the identifiable individual tracks). The ratio of quadrupedal to bipedal tracks is 1 : 4.81.

The relationship between youth and adult tracks is also interesting. In the Rio do Peixe basins, the former tracks are very rare, consequently, little can be said about the age-class structure of the trackmakers. The only footprint in Sousa Basin that is, almost certainly, that of a juvenile is an isolated tridactyl track on Passagem das Pedras site, which is the smallest dinosaur track discovered so far in these basins (footprint length = 5.6 cm). There are no other very small dinosaur individuals (hind-foot prints shorter than 12 cm). This phenomenon might indicate very heavy

Fig. 7.11 Diversity of tracks among the 447 individual theropods found in the Sousa Basin, Sousa Formation. **a** A trackway of *Moraesichnium barberenae* Leonardi 1979a, b **b** A theropod track from Caiçara-Piau locality; **c** Theropod footprint from Sítio Saguim; **d** An anomalous theropod footprint, pertaining to a normal theropod trackway, but that seems to belong to an ornithopod. It is a footprint with infilling material more coarse from the adjacent top layer. Graphic scale: **a** = the average stride is 197.2 cm; **b** = the length of the footprint is 20 cm; **c** and d = 5 cm

mortality on the part of very young individuals (Leonardi 1981). It is also possible to be an artifact of preservation, where small sized dinosaurs were not heavy enough to leave footprints because of the substrate firmness.

In a total of 42 ichnosites in the Rio do Peixe basins, against what one would expect in theory, those in which meat eaters outnumber plant eaters (31 sites out of 42; that is 73.81% of all sites) are more abundant than those in which the opposite occurs, (7 sites, 16.67%). There are three sites where the parity between carnivores

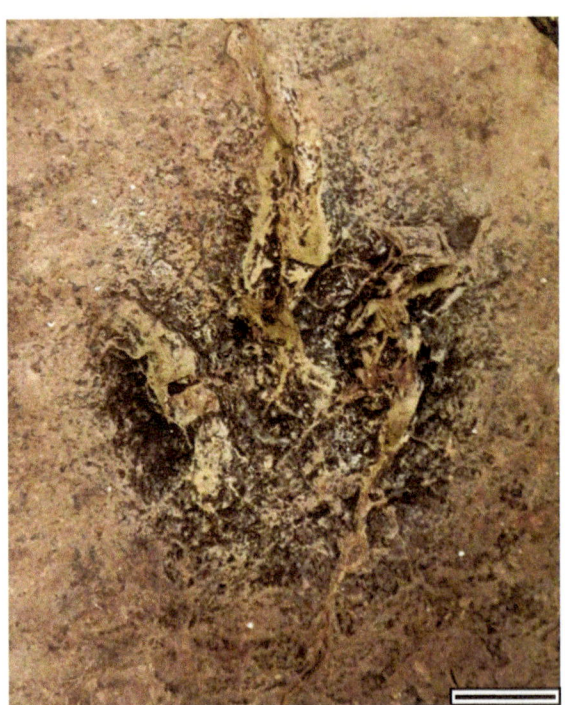

Fig. 7.12 A noasaurid track, 9 cm long, from the Caiçara-Piau locality, Sousa Formation, SOCA 1321. Photo by M. de Fátima C.F. dos Santos. Graphic scale = 1.3 cm

and herbivores is reached (7.14%) and a site with a track not classifiable (2.38%). Besides, ichnosites in which there are only theropod tracks, are rather numerous in the Rio do Peixe basins. They are 19, out of 42 ichnosites, and it corresponds to the 45.24% out of all 42 ichnosites. This high number of sites with apparent exclusive presence of theropods is well explained by Pérez-Lorente (2015, p. 325): "Because theropod footprints are the most abundant, so are outcrops with theropod footprints".

In addition to these 19 ichnosites where the theropods are represented only by tracks (45.24% out of 42 sites), there are: 1 ichnosite with tracks of theropods, sauropods, ornithopods and one quadrupedal Thyreophora (four clades, 2.38%); 1 ichnosite with tracks of theropods, ornithopods and one quadrupedal Thyreophora (three clades, 2.38%); 7 ichnosites with tracks of theropods, sauropods, ornithopods (three clades, 16.67%); 4 ichnosites with tracks of theropods and sauropods (two clades, 9.52%); 5 ichnosites with tracks of theropods and ornithopods (two clades, 11.90%); 3 sites with only sauropod tracks (7.14%); 1 site with only ornithopod tracks (2.38%); 1 site with only large unclassifiable herbivore tracks (2.38%). There are also few sites where rare tracks of animals of the mesofauna are also recorded, they are 4 (9.52% of all the 42 sites). These last localities are Caiçara-Piau, Riacho do Cazé, Serrote do Pimenta, and Tapera.

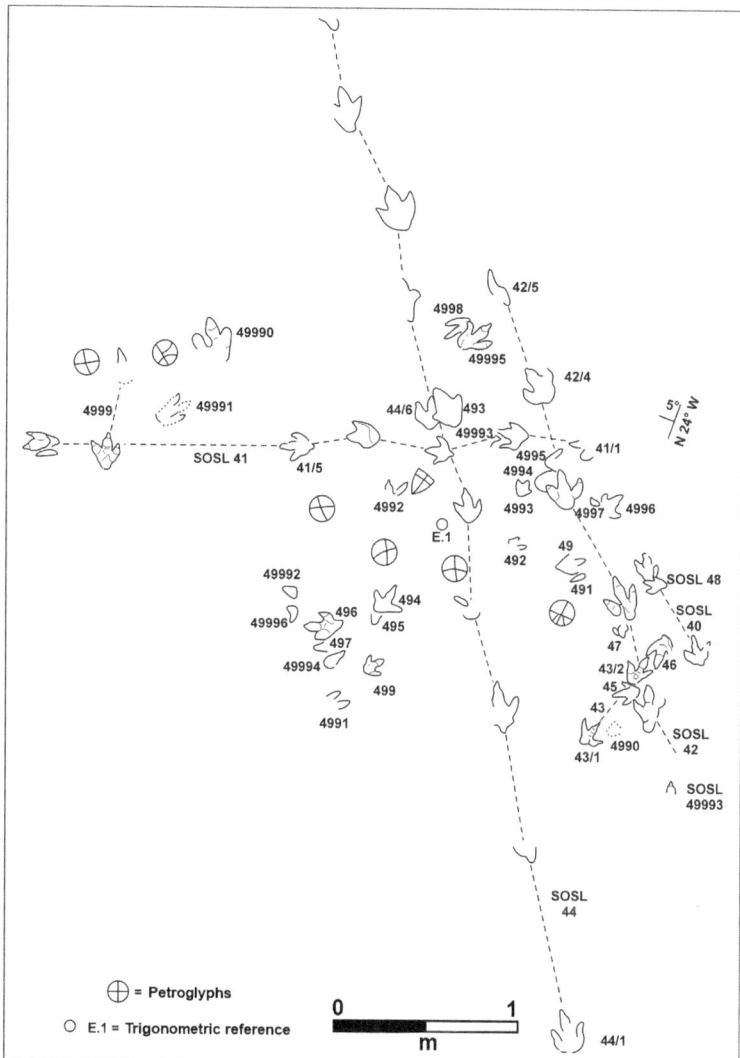

Fig. 7.13 An original drawing, by G. Leonardi, of the bedding surface of the Antenor Navarro Formation, at the Serrote do Letreiro tracksite, Sousa Basin. All individuals probably belonged to the same population. Fossil footprints are associated with later prehistoric petroglyphs

7.3.2 Behavior of the Rio do Peixe Dinosaurs

The study of the fossil tracks also is the most important and unparalleled (Gatesy and Ellis 2016) method for making inferences about the behavior of the track-makers. Seventy-eight trackways were sufficiently long and conveniently measur-able, permitted estimation of trackmaker speeds. The result was clear: the speed of

Fig. 7.14 Tracks of plant eating dinosaurs. **a** Manus-pes pair of a sauropod from Engenho Novo ichnosite; São João do Rio do Peixe County, Sousa Formation, Sousa Basin; **b** A trackway of a bipedal graviportal ornithopod from Baixio do Padre, Sousa County, Sousa Formation, Sousa Basin. It pertain to the ichnogenus *Caririchnium*, but the gait is herein bipedal; **c** The main trackway of the Passagem das Pedras ichnosite, *Sousaichnium pricei*, a semi-bipedal iguanodontid. Sousa County, Sousa Formation, Sousa Basin; **d** Holotype of *Caririchnium magnificum*, pertaining to a graviportal quadrupedal ornithopod. Serrote do Letreiro, Sousa County, Antenor Navarro Formation, Sousa Basin. Graphic scales: **a**= 15 cm; **b**, **c** and **d**: the average widths of the hind-footprints are respectively: 51.2; 35.7: 47.8 cm

fifty-nine of these trackways (75.64% of the sample) was estimated between 3 and 7 km/h. The trackmakers were, therefore, traveling with a walking gait. Seven trackways show a slower estimated speed (≤2 km/h; 8.98%); four of these are sauropods, three are ornithopods. Twelve trackways (15.38% of the sample) point to a speed between 8 and 23 km/h. Of these, eight (10.25%) have calculated speeds of 8–13 km/h; another four (5.12%) are distributed over a range between 13 and 23 km/h. These

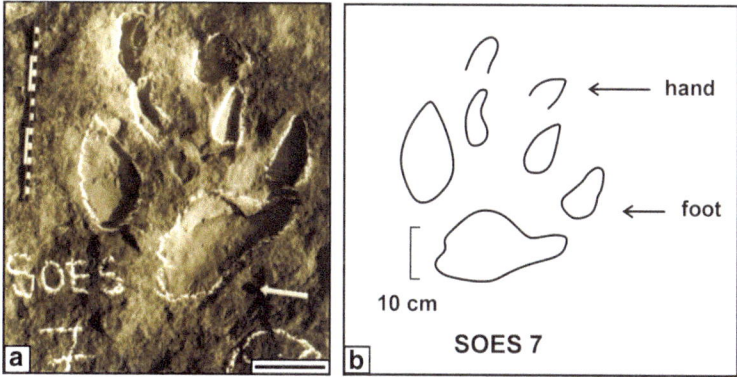

Fig. 7.15 a–b The discovery (1979) of this hand-foot set attributed to an ankylosaurian, probably a nodosaurid, at the Serrote do Pimenta tracksite (SOES 7; Leonardi 1984a; 1994), was the first evidence of ankylosaurs in South America. Sousa Basin, Antenor Navarro Formation. Graphic scale = 10 cm

last four trackways, all belonging to medium to large theropods (Fig. 7.16a), corresponds to the fastest runners of the Rio do Peixe ichnofauna (Leonardi et al. 1987a, b, c). It is important to be cautious when estimating or calculating speed from fossil track records (Lockley and Meyer 1999), despite the fact that we can calculate that dinosaurs in the Rio do Peixe basins in general kept a walking pace, and only rarely took a running gait. The quadrupeds always moved slowly or very slowly. The bipeds, including theropods, did not run very often or very quickly; the calculated maximum speed that was found in the Rio do Peixe Basins is about 23 km/h (Leonardi et al. 1987a, b, c; Fig. 7.16b). This situation is common for non-avian dinosaurs (Leonardi and Mietto 2000). A similar, more recent statement on low dinosaur speeds can be found, for example, in Xing et al. (2014).

The general evidence that dinosaurs have a high degree of metabolism is not doubted here. However, it is not so evidently reinforced by the known ichnological record, and it ought to be better examined on the basis of many detailed, extensive and statistic studies on their trackways (Leonardi et al. 1987a, b, c; Molnar and Farlow 1990), rather than on the basis of some isolated, biased, and/or unchecked information. When the latter occurs, it guides the huge racing dinosaurs of Bakker (1986a, b) and Paul (1987a, b).

The bearings or directions of the footprints from the Rio do Peixe Group of the Sousa Basin (Leonardi and Carvalho 2021), based on 386 individual trackways, point to a rather tetramodal model, with two main modes in the NE and SW quadrants, and two secondary modes in the other two quadrants. There seem, then, to be four associations of dinosaurs, moving along preferential paths, on different levels, and in diverse times and occasions. As earlier described for the locality Piau-Caiçara' tracks (Godoy and Leonardi 1985), and at least for the Sousa Basin tracks in general, most of these tracks are parallel or nearly parallel to the ridges of the ripple marks. These crests, in turn, indicate the dominant orientation of the water's edge, which

Fig. 7.16 On the behavior of dinosaur. **a** This very straight and narrow trackway corresponds to one of the fastest runners of the Rio do Peixe ichnofauna: a large theropod track (SOES 1) at Serrote do Pimenta (Antenor Navarro Formation), with about 22 km/h. Graphic scale = 20 cm; **b** Another theropod track (SOPP 3) at Passagem das Pedras (Sousa Formation), with an estimated speed of 23 km/h. **b** Photograph by Franco Capone

is often parallel to the regional faults that gave rise to the Rio do Peixe basins. It is clear, therefore, that the directions of movements were strongly conditioned by the local and regional morphology of the territory, in particular by the bodies of water and, indirectly, by the regional tectonic patterns (Godoy and Leonardi 1985; Leonardi 1989; Leonardi and Carvalho 2021). It would be possible that they could easily reach Africa afoot (nearly 450 km from Sousa), by following one of these paths, specifically along the Patos-Garoua fault system, possibly reaching the current territory of Cameroon.

All tracks in the Rio do Peixe basins, including those of sauropods, are rather narrow, attesting to an entirely erect position of the trackmaker. All sauropods were clearly quadrupeds. Theropods, both large and small, were all bipedal (Molnar and Farlow 1990), with very narrow trackways, in contrast to the old model of large theropods giving a Cossack dance show (Molnar and Farlow 1990; Wade 1989). Ornithopods, in the Sousa Basin, were bipeds, quadrupeds or, in one case, semi-quadrupeds. In some cases, the tail mark is perhaps preserved in the Sousa Basin. The rarity of tail drags or marks is habitual for dinosaurs. It is evident that most of the dinosaurs in the Rio do Peixe basins, including both bipedal and quadruped ones, kept their tail away from touching the ground.

Dinosaur tracks of the Rio do Peixe basins produced by bipeds (probably 476 individual tracks, or ~82.78% of the 575 classified dinosaur tracks) heavily surpass those of quadrupeds (about 99 individuals, or ~17.22% of the serviceable sample). The ratio of biped tracks to quadruped tracks is thus 4.81: 1. We stated earlier that walking gaits surpass almost entirely running gaits. There are cases of semi-bipedal or semi-quadruped animals; especially ornithopods. No hopping, galloping or sprawling (except for a single lizard-like footprint) gaits are represented at these basins. A particular case is that of the trackway SOPP 1, of an iguanodontid, quadruped or semi quadruped in a different way, because in its very long trackway it leaned slightly on the ground with the right hand only, and not with the left. It can be interpreted as a taphonomic aspect or even an abnormal behavior. A rather usual manner of gait in the Sousa Basin (~10.43%) is that of dinosaurs, mainly theropods, which, swimming (Fig. 7.17a) and perhaps searching for fish or other food in shallow water, pushed with their feet on the bottom of a shallow lake bed, and produced what are called swimming-tracks or more correctly, half-swimming-tracks (Leonardi 1987). Altogether, there are about 59 individual theropod half-swimming tracks and a single probable ornithopod half-swimming track in these basins (~60 cases vs. 1).

Except for sauropods, which almost always lived in herds, most dinosaurs in the Rio do Peixe basins were lonely animals. The gregarious behavior is attested by clusters of sauropod tracks of at least 7–20 individuals (Fig. 7.17b; Leonardi 1989, 1994; Carvalho 2000b; Leonardi and Santos 2006); the number of animals in these herds could have been higher, because some tracks were probably destroyed by erosion, and some have yet to be found out and/or excavated. Theropods and ornithopods, instead, ordinarily traveled as individual animals. There are, however, three exceptions among the theropods: the population of small and medium-sized theropods (~16 individuals at Serrote do Letreiro; Fig. 7.17c); the assemblage of some long-heeled theropods of the ichnogenus *Moraesichnium* Leonardi 1979a, b at Passagem das Pedras (Fig. 7.17d); and perhaps the nearly 30 theropods of Piau-Caiçara farm on the "rainy" level 13/2.

7.4 Paleogeographic Distribution of the Footprints

The South American (and, in general, Gondwanan) dinosaurs are very different from those of the northern continents. Some integration and coexistence between the dinosaur species of the two supercontinents, Gondwana and Laurasia, occurred much later with the phenomenon that we could name: Dinosaur American Biotic Interchange (DABI), which occurred towards the end of the Cretaceous, due to the junction between the two American continents, a phenomenon analogous with the GABI, the Great American Biotic Interchange (Cione et al. 2015).

The diversity of the dinosaur faunas between Northern and Southern America depends on the probable total biogeographic isolation of South American and, more generally, Gondwanan faunas, from those of boreal continents (Laurasia) during Middle and Late Jurassic and almost all of the Cretaceous, a typical case of endemism (Bonaparte 1986, 2007). There is, instead, a notable affinity between South American dinosaur faunas and those of the other Gondwanan plates: Africa, Madagascar,

Fig. 7.17 a Typical imprint of a swimming theropod, pushing on the bottom with the toetips. Piau-Caiçara locality, Sousa County, Sousa Basin, Sousa Formation; **b** The gregarious behavior is attested by clusters of sauropod tracks of at least 7–20 individuals, which proceed in parallel herd, here preserved on the bottom of the rivulet Riacho do Pique, at Serrote do Letreiro, Sousa County. Antenor Navarro Formation, Sousa Basin. Photo by Franco Capone; **c** Gregarious (rare) behavior of a theropod population in Serrote do Letreiro tracksite, Sousa County. Antenor Navarro Formation, Sousa Basin; **d** An assemblage of several long-heeled theropods of the ichnogenus *Moraesichnium* at Passagem das Pedras, Sousa County. Sousa Formation, Sousa Basin. Photograph by Franco Capone. Graphic scale: a = 4 cm; **c** = 20 cm

India, Antarctica, and Australia. This applies even more to the present territory of Northeastern Brazil, which in the Early Cretaceous was probably still attached, at least partially, to Africa, but detached from the rest of what would later have been South America.

Indeed, the dinosaur fauna of the oldest portion of the Early Cretaceous (Berriasian to lower Barremian, Rio da Serra-Aratu stages) represented by their tracks in the Rio do Peixe basins, as well as the few other representatives of the meso-ichnofauna, had to be a very special fauna. As a matter, it ought to be more similar to that of West and Central Africa, rather than that of other regions of present-day Brazil and South America, since the easternmost area of the Brazilian territory (Rio Grande do Norte, Paraíba, at least part of the Ceará and Pernambuco states) was isolated by an epicontinental sea during the Early Cretaceous.

The landscape of the mentioned basins, during Early Cretaceous, were mainly flat surfaces, elongated narrow valleys, between chains of low mountains of mostly Proterozoic rocks, with a possible Paleozoic or Early Mesozoic cover, now disappeared. Low mountain or hill ranges, flanked the sinking basin. One side of the valley, controlled by a fault, usually a strike-slip fault, was steeper, rockier, and consequently less covered by vegetation. The other side could be less inclined and with a less inclined slope. The mountains on the sides of the basins could be covered on the top by woods or coniferous groves, mostly Araucariaceae and with all an undergrowth of tree ferns, Cycadoidea, like *Podozamites* and Conipherophyta, like *Otozamites*. Along the borders of the valleys ran ephemeral streams, generally of low flow, at the base on alluvial fans of Precambrian polymictic material, partially reworked, consisting of gravel and coarse immature sands. These torrential streams often originated anastomosed ephemeral channels in the largest basins.

The fauna could pass from the upland areas in the lowland prairies and vice-versa, both to graze, if they were plant-eaters, both to reach water points, easier to gush than in the highlands. If they were meat-eaters, just in those points that somehow served as a necessary gathering point of the fauna, especially in the seasons and other periods of dryness and aridity.

The sediments, transported by water, became gradually finer: coarse sand, fine-grained sand, silt, and finally mud. The interior of these basins, especially in those of greater area, were often abundant in water and low vegetation. At the depocenter there were lakes, shallow and ephemeral, with rather warm and alkaline waters.

7.5   Paleoenvironmental and Paleoclimatic Contexts

At the beginning of the Mesozoic, the southern hemisphere had a warm and arid climate. This is clearly observed by the generalized presence of wind deposits along the Brazilian and African intracratonic basins (Lima 1983). The connection of South America and Africa as a single continental block did not allow for greater humidity in what was then the continental hinterland. A greater degree of humidity was allowed due the fragmentation of the Gondwana supercontinent and the creation of a lake and river systems along the rift valleys. This suggests a link between climate change and the tectonic events that led to the separation of South-American and African continents. Although low, this humidity contributed to increased local rainfall and led to the growth of abundant vegetation in the region (Lima 1983; Carvalho 2000a).

During the Early Cretaceous, warm climate was widespread. According to Petri (1983, 1998) and Lima (1983), in the beginning of Cretaceous the climate was more humid in regions located to the south of the tropical domain (the Recôncavo-Tucano-Jatobá basins). Despite the tropical domain's hotter and drier climate, the existence of fresh-water lakes is suggested by invertebrate fossils, as the large conchostracans *Palaeolimnadiopsis reali* in some lacustrine facies of the Sousa Basin that locally provided more humid conditions (Carvalho 1989; Carvalho and Carvalho 1990).

At that time, the southern continents were still amalgamated in the Gondwana supercontinent, and the Atlantic Ocean was in its initial developing phase. In northeastern Brazil, across an area of hundreds of square kilometers, ephemeral rivers and shallow lakes constituted important environments for an abundant endemic biota in many basins (Lima Filho et al. 1999; Mabesoone et al. 1979, 2000). The rarity of levels with rain-drop marks (only one recognized in the Sousa Basin, none in the others; Fig. 7.18), the scarcity of plant remains, logs and fossils suggest a relatively arid to semi-arid climate.

Another aspect is that dinosaur footprints are so strongly distributed in the Sousa Basin, compared to other basins (Fig. 7.19). Although sedimentological and taphonomic aspects may well control this disparity, it also could depend on different microclimates, at least due the presence or not of lakes and the pH of their waters. One might think that the valley of the Sousa Basin was similar to, in the present time,

Fig. 7.18 The stratigraphic level 13/2 at the Piau-Caiçara tracksite, Sousa Formation, Sousa Basin. The upper surface of the mudstone layer was covered by footprints of about thirty theropods of the same population, and the small craters produced by the rain, fall after the passage of that pack of meat-eater dinosaurs. Here, in this representative specimen, a footprint and some other theropod digits. Photo by Ragnhild Borgomanero. The width of the track is 14 cm

Fig. 7.19 Environmental reconstruction of the Sousa Basin (Antenor Navarro Formation) in the Early Cretaceous (Berriasian-Berramian). One ankylosaurian and titanosaurians crossing a fluvial fan deposit on the North margin of the basin. In the background to the left (South), behind the two titanosaurids, one notices the straight line of Proterozoic hills that depend on the Malta fault (E-W). Art by Guilherme Gehr

the great oasis of Faiyum in the Libyan-Nubian desert, located in Egypt, southwest of Cairo, so fertile and luxuriant, compared to the situation of aridity surrounding, and also so rich in crocodiles.

7.6 Conclusions

The four Rio do Peixe basins, and notably the Sousa Basin, present an impressive amount of Early Cretaceous dinosaur tracks in sediments dated from the Berriasian to the lower Barremian. Data on the presence of different groups of dinosaurs and other coeval animals, their correlations, the numerical value and percentage of their presence in the overall sample were herein provided. Several aspects of their behavior have been deduced: speeds, manners of gaits, directions, posture, individual and social behavior. The importance of quantifying the ichnological material, by means of a large number of data, and to its statistical study, has also been underlined.

The presence of dinosaur tracks in the Rio do Peixe basins induced the scientific tourism, which enabled job positions allowing the economic flourishment of the region. It would however be important and urgent to carry out a new vulnerability diagnosis of these ichnosites and the establishment of proposals for geoheritage protection.

References

ANP - Agência Nacional de Petróleo. Brasil Round 9 (2008) Rio do Peixe Basin. Nona Rodada de Licitações. Cid Queiroz Fontes. Bid Area Department

Araújo REB, Bezerra FHR, Nogueira FCC, Balsamo F, Carvalho BRBM, Souza JAB, Sanglard JCD, Castro DL, Melo ACC (2019) Basement control on fault formation and deformation band damage zone evolution in the Rio do Peixe Basin, Brazil. Tectonophysics 745:117–131. https://doi.org/10.1016/j.tecto.2018.08.011

Bakker RT (1986a) The dinosaur heresies. Zebra Books, Kensington, New York, p 480. ISBN: 978-0140100556

Bakker RT (1986b) The return of the dancing dinosaurs. In: Czerkas SJ, Olson EC (eds) Dinosaurs past and present, vol 1. Washington University Press, Seattle, pp 38–69. ISBN: 978-0938644248

Barbosa ABS, Maia RP, Pontes CCC, Nogueira FCC, Bezerra FHR (2021) Conditioning of relief along fault zones with deformation bands in the Rio do Peixe Sedimentary Basin, Brazil. Revista Brasileira de Geomorfologia 22(2):385–406. https://doi.org/10.20502/rbg.v22i2.1948 http://www.lsie.unb.br/rbg/

Bezerra FH, Araujo R, Maciel I, Nogueira FC, Balsamo F, Storti F, Souza JA, Carvalho B (2017) The role of major rift faults in the evolution of deformation bands in the Rio do Peixe Basin, Brazil, European Geosciences Union General Assembly 2017, Wien, Austria, 23–28 April 2017. Poster. https://ui.adsabs.harvard.edu/abs/2017EGUGA..19.5438H/abstract

Bezerra FH, Marques FO, Vasconcelos DL, Rossetti DF, Tavares AC, Maia RP, Castro DL, Nogueira FCC, Fuck RA, Medeiros WE (2023) Analysis of mechanisms, timing and effects on structures and relief. J S Am Earth Sci 126(2023):104356. https://doi.org/10.1016/j.jsames.2023.104356

Bonaparte JF (1986) History of the terrestrial Cretaceous vertebrates of Gondwana. In: 4° Congreso Argentino de Paleontología Y Bioestratigrafía, Mendoza, Argentina 2, pp 63–95

Bonaparte JF (2007) Dinosaurios y pterosaurios de América del Sur. Albatros, Buenos Aires, Argentina, p 224

Campos HBH, Bonde N, Leal ME, Dantas MA (2015) A new dinosaur ichnosite from the early Cretaceous Sousa Formation, Northeastern Brazil. PeerJ PrePrints 3:e1413v1. https://doi.org/10.7287/peerj.preprints.1413v1

Carvalho IS (1989) Icnocenoses continentais: bacias de Sousa, Uiraúna-Brejo das Freiras e Mangabeira. MS thesis, Universidade Federal do Rio de Janeiro, Rio de Janeiro, Brazil [unpublished]

Carvalho IS (1996) As pegadas de dinossauros da bacia de Uiraúna-Brejo das Freiras (Cretáceo Inferior, estado da Paraíba). In: 4° Simpósio Sobre o Cretáceo do Brasil, Boletim, Rio Claro, São Paulo, UNESP, Brasil, 4, pp 115–121

Carvalho IS (2000a) Geological environments of dinosaur footprints in the intracratonic basins of Northeast Brazil during the Early Cretaceous opening of the South Atlantic. Cretac Res 21(2000):255–267

Carvalho IS (2000b) Huellas de saurópodos Eocretácicas de la cuenca de Sousa (Serrote do Letreiro, Estado da Paraíba, Brasil). Ameghiniana 37(3):353–362

Carvalho IS (2004a) Bacias cretáceas interiores do Nordeste. Fundação Paleontológica Phoenix 70:1–4

Carvalho IS (2004b) Dinosaur footprints from Northeastern Brazil: taphonomy and environmental setting. Ichnos 11:311–321

Carvalho IS, Carvalho MGP (1990) O significado paleoambiental dos conchostráceos da Bacia de Sousa In: Campos DA, Viana MSS, Brito PM, Beurlen G (eds) Simpósio Sobre a Bacia do Araripe e Bacias Interiores do Nordeste, Crato, 1, Crato, Ceará, Sociedade Brasileira de Paleontologia, Brazil, pp 329–333

Carvalho IS, Fernandes ACS (1992) Os icnofósseis da Bacia de Mangabeira, Cretáceo do Ceará. In: 2° Simpósio Sobre as Bacias Cretácicas Brasileiras, Boletim de Resumos Expandidos, Rio Claro, São Paulo, UNESP, Brazil, p 105–106. Carvalho IS, Fernandes ACS (eds) (2007) Icnologia. Sociedade Brasileira de Geologia, Serie Textos, 3. São Paulo, Brazil, p 117

Carvalho IS, Leonardi G (1992) Geologia das bacias de Pombal, Sousa, Uiraúna-Brejo das Freiras e Vertentes (Nordeste do Brasil). An Acad Bras Ciênc 64:231–252

Carvalho IS, Leonardi G (2007) The dinosaur valley natural monument: dinosaur tracks from Rio do Peixe Basins (Lower Cretaceous, Brazil). In: 5ª Reunión Argentina de Icnología, y 3era Reunión de Icnología del Mercosur, Ushuaia, Argentina, 28–30 March, 2007, p 51

Carvalho IS, Leonardi G (2021) Fossil footprints as biosedimentary structures for paleoenvironmental interpretation: examples from Gondwana. J S Am Earth Sci 106:102936. https://doi.org/10.1016/j.jsames.2020.102936

Carvalho IS, Leonardi G (2023) Dinosaur tracks from the Sítio Pereiros ichnosite, Triunfo Basin (Lower Cretaceous) and the dinosaur diversity in the Rio do Peixe basins, Northeastern Brazil. Cretac Res 1444:105446. https://doi.org/10.1016/j.cretres.2022.105446

Carvalho IS, Nobre PH (2001) Um Crocodylomorpha (?Notosuchia) da Bacia de Uiraúna (Cretaceo Inferior), Nordeste do Brasil. Revista Brasileira de Paleontologia 2:123–124

Carvalho IS, Borghi L, Leonardi G (2013a) Preservation of dinosaur tracks induced by microbial mats in the Sousa Basin (Lower Cretaceous), Brazil. Cretac Res 44(2013):112–121

Carvalho, IS, Leonardi, G, Santos, WFS (2013b) Vale dos Dinossauros: a relevância das pegadas fósseis da Bacia de Sousa como patrimônio geológico. In: GeoBRheritage, 2° Simpósio Brasileiro de Patrimônio Geológico, Ouro Preto, MG, 24 a 28 de setembro de 2013, Anais

Carvalho IS, Mendes JC, Costa T (2013c) The role of fracturing and mineralogical alteration of basement gneiss in the oil exhsudation in the Sousa Basin (Lower Cretaceous), Northeastern Brazil. Cretac Res 47(2013):47–54

Carvalho IS, Viana MSS, Lima Filho MF (1993a) Bacia de Cedro: a icnofauna cretácica de vertebrados. An Acad Bras Ciênc 65:459–460

Carvalho IS, Viana MSS, Lima Filho MF (1993b) Os icnofósseis de vertebrados da bacia do Araripe (Cretáceo Inferior, Ceará-Brasil). An Acad Bras Ciênc 65:459

Carvalho IS, Viana MSS, Lima Filho MF (1994) Dinossauros do Siluriano: um anacronismo crono-geológico nas bacias interiores do Nordeste? In: 38° Congresso Brasileiro de Geologia. Boletim de Resumos Expandidos, Sociedade Brasileira de Geologia, Camboriú, Santa Catarina, Brazil, 3, pp 213–214

Carvalho IS, Araújo-Junior HI, Nogueira FC, Soares JA, Salgado L, Lindoso RM, Leonardi G (2016) Taphonomic and paleoenvironmental aspects of the Lower Cretaceous Rio Piranhas formation (Triunfo Basin, Northeastern Brazil) Based on faciological and paleontological data. In: 48° Congresso Brasileiro de Geologia, Sociedade Brasileira de Geologia, Porto Alegre, RS, Brazil. Abstracts and programs.

Carvalho IS, Leonardi G, Rios-Netto AM, Borghi L, Paula Freitas A, Andrade JA, Freitas FI (2021) Dinosaur trampling from the Aptian of Araripe Basin, Brazil, as tools for stratigraphic correlation. Cretac Res 117:104626. https://doi.org/10.1016/j.cretres.2020.104626

Carvalho IS, Salgado RM, Lindoso JR, Araújo Júnior HI, Nogueira FCC, Soares JA (2017) A new basal titanosaur (Dinosauria, Sauropoda) from the Lower Cretaceous of Brazil. J S Am Earth Sci 75:74–84. https://doi.org/10.1016/j.jsames.2017.01.010

Castro DL, Oliveira DC, Castelo Branco RMG (2007) On the Tectonics of the Neocomian Rio do Peixe Rift Basin, NE Brazil: lessons from gravity, magnetics, and radiometric data. J S Am Earth Sci 24:184–202

Cione AL, Gasparini GM, Soibelzon E, Soibelzon LH, Tonni EP (2015) The great American biotic interchange. A South American Perspective. Springer, Heidelberg, p 117. https://doi.org/10. 1007/978-94-017-9792-4

Córdoba VC, Antunes AF, Sá FFJ, Lins FAPL (2007) Stratigraphic and structural analysis of the Rio do Peixe Basin, Northeastern Brazil: Integration based on the pioneer seismic survey 0295-RIO-DO-PEIXE-2D. Boletim de Geociências da Petrobras 16(1):53–68

Fernandes, ACS, Carvalho IS (2001) Icnofósseis de invertebrados da Bacia de Sousa (Estado da Paraíba, Brasil): a localidade de Serrote do Letreiro. In: 1° e 2° Simpósios Sobre a Bacia do Araripe e Bacias Interiores do Nordeste, Crato, Ceará. Comunicações 2001. Coleção Chapada do Araripe, Brazil, 1, pp 147–155

Fernandes, ACS, Carvalho IS (2007). As Pegadas de Dinossauros da Bacia do Rio do Peixe: Elementos de Transformação Cultural em Souza, Paraíba – Brasil. In: Quinta Reunión Argentina de Icnología, Tercera Reunión de Icnología del Mercosul, p 57

Freitas RBRM, Nogueira FCC, Vasconcelos DL, Honório GB, Nicchio MA, Stohler RC, Souza JAB (2023) 3D topological analysis in deformation bands: insights for structural characterization and impact on permeability. J Struct Geol 176:104959

Gatesy SM, Ellis RG (2016) Beyond surfaces: a particle-based perspective on track formation. In: Falkingham PL, Marty D, Richter A (eds) Dinosaur tracks: the next steps. Indiana University Press, Bloomington, chapter 5, pp 82–91

Ghilardi AM, Aureliano T, Duque RRC, Fernandes MA, Barreto AMF, Chinsamy A (2016) A new titanosaur from the Lower Cretaceous of Brazil. Cretac Res 16(2016):16–24. https://doi.org/10. 1016/j.cretres.2016.07.001

Ghilardi AM, Aureliano T, Duque RRC et al (2014) An Early Cretaceous dinosaur from Sousa Basin, Brazil. In: 4th international paleontological congress, abstract volume, Mendoza, Argentina, p 327

Godoy LC, Leonardi G (1985) Direções e comportamento dos dinossauros da localidade de Piau-Caiçara, Sousa, Paraíba (Brasil), Formação Sousa (Cretáceo Inferior). Departamento Nacional da Produção Mineral, Coletânea de Trabalhos Paleontológicos. Serie "Geologia", 27. Seção Paleontologia e Estratigrafia, Brazil 2:65–73

Gonzaga FAS, Santos Filho JIS, Santos JSI, Teles GS, Oliveira HBL (2022) Aspectos petrofísicos e mineralógicos de rochas sedimentares da formação Sousa, bacia do Rio do Peixe (PB). Revista de Geociências do Nordeste 8(2):91–101. https://doi.org/10.21680/2447-3359.2022v8n2ID19514

Iemini JA (2009) Fácies orgânicas de uma sucessão sedimentar cretácea da Bacia de Sousa, PB, Brasil. Programa de Pós-Graduação em Geologia, Universidade Federal do Rio de Janeiro. Dissertação de Mestrado, Brazil [Unpublished]

Leonardi G (1979a) Nota Preliminar Sobre Seis Pistas de Dinossauros Ornithischia da Bacia do Rio do Peixe (Cretáceo Inferior) em Sousa, Paraíba, Brasil. An Acad Bras Ciênc 51(3):501–516

Leonardi G (1979b) New archosaurian trackways from the Rio do Peixe Basin, Paraíba, Brazil. Annali dell'Università di Ferrara, N.S., S. IX 5(14):239–249

Leonardi G (1980a) *Isochirotherium* sp.: Pista de um gigantesco Tecodonte na Formação Antenor Navarro (Triássico), Sousa, Paraíba, Brasil. Revista Brasileira de Geociências 10(4):186–190

Leonardi G (1980b) Dez novas pistas de Dinossauros (Theropoda Marsh, 1881) na Bacia do Rio do Peixe, Paraíba, Brasil. In: 1er Congreso Latinoamericano de Paleontología, Actas, Buenos Aires, 1978, Argentina, 1, pp 243–248

Leonardi G (1981) Ichnological data on the rarity of young in North East Brazil dinosaurian populations. An Acad Bras Ciênc 53(2):345–346

Leonardi G (1984a) Le impronte fossili di dinosauri. In: Bonaparte JF, Colbert EH, Currie PJ, de Ricqles AJ, Leonardi G et al (eds) Sulle orme dei dinosauri. Venezia-Mestre, Erizzo, 1984. (Esplorazioni e ricerche, IX), pp 161–186

Leonardi G (1984b) Rastros de um mundo perdido. Ciência Hoje 2(15):48–60

Leonardi G (1985) Vale dos dinossauros: uma janela na noite dos tempos. Revista Brasileira De Tecnologia 16(1):23–28

Leonardi G (ed) (1987a) Glossary and Manual of Tetrapod Footprint Palaeoichnology. Brasília, DNPM (Serviço Geológico do Brasil), 20 plates, 20 pages of tables, p 117

Leonardi G (1987b) Pegadas de dinossauros (Carnosauria, Coelurosauria, Iguanodontidae) na Formação Piranhas da Bacia do Rio do Peixe, Sousa, Paraíba, Brasil. In: 10° Congresso Brasileiro de Paleontologia, Anais Sociedade Brasileira de Paleontologia, Rio de Janeiro, Brazil, 1, pp 337–351, 3 plates

Leonardi G (1989) Inventory and Statistics of the South American Dinosaurian Ichnofauna and its Paleobiological Interpretation. In: Gillette DD, Lockley MG (eds) Dinosaur Tracks and Traces. Cambridge University Press, New York, pp 165–178

Leonardi G (1994) Annotated Atlas of South America tetrapod footprints (Devonian to Holocene) with an appendix on Mexico and Central America. Companhia de Pesquisa de Recursos Minerais, Brasília, Brazil, p 248, 35 plates

Leonardi G (2008) Trinta e três anos a procura de pegadas fósseis nas bacias brasileiras. In: 44° Congresso Brasileiro de Geologia, Anais, Curitiba, Paraná, Brazil, pp 1–3

Leonardi G (2011) What do the dinosaur tracks of the Rio do Peixe Basins (Paraíba, Brasil) point at? In: Carvalho IS, Srivastava NK, Strohschoen O, Lana CC (eds) Paleontologia: Cenários de Vida. Rio de Janeiro: Editora Interciência, Brazil, 3, pp 669–680

Leonardi G (2021) Main results of 45 years of ichnological research on the dinosaur tracks of the Rio do Peixe basins (Paraíba, Brazil, Early Cretaceous). Annali del Museo Civico di Rovereto, Sez. Archeologia, Storia, Scienze Naturali 37:159–182

Leonardi G, Carvalho IS 2000 As pegadas de dinossauros das bacias Rio do Peixe, PB. In: Schobbenhaus C, Campos DA, Queiroz ET, Winge M, Berbert-Born M (eds) Sítios Geológicos e Paleontológicos do Brasil. Published on Internet at the address http://www.unb.br/ig/sigep/sitio026/sit io026.htm. p 15

Leonardi G, Carvalho IS (2002) Icnofósseis da Bacia do Rio do Peixe, PB. O mais marcante registro de pegadas de dinossauros do Brasil. In: Schobbenhaus C, Campos DA, Queiroz ET, Winge M, Berbert-Born M (eds) Sítios geológicos e paleontológicos do Brasil. Brasília. Brazil, Departamento Nacional de Produção Mineral, pp 101–111

Leonardi G, Carvalho IS (2021) Dinosaur tracks from Brazil: a lost world of Gondwana. Indiana University Press, Bloomington, Indiana, USA. xv + p 456

Leonardi G, Mietto P (eds) (2000) Dinosauri in Italia. Le orme giurassiche dei Lavini di Marco (Trentino) e gli altri resti fossili italiani. Accademia Editoriale, Pisa-Roma, p 494

Leonardi G, Muniz GCB (1985) Observações icnológicas (Invertebrados e Vertebrados) no Cretáceo continental do Ceará (Brasil), com menção a moluscos dulçaquícolas. In: 9° Congresso Brasileiro de Paleontologia, Fortaleza, Sociedade Brasileira de Paleontologia, Resumo das Comunicações, Fortaleza, Ceará, Brazil, p 45

Leonardi G, Santos MFCF (2006) New dinosaur tracksites from the Sousa Lower Cretaceous basin (Paraíba, Brazil). Studi Trentini di Scienze Naturali, Acta Geologica 81(2004):5–21

Leonardi, G, Lima CV, Oliveira FHL (1987a) Os dados numéricos relativos às pistas (e suas pegadas) das Icnofaunas dinossaurianas do Cretáceo Inferior da Paraíba, e sua interpretação estatística. I - Parâmetros das pistas. In: 10° Congresso Brasileiro de Paleontologia, Anais da Sociedade Brasileira de Paleontologia, Rio de Janeiro, Brazil, 1, pp 377–394

Leonardi, G, Lima CV, Oliveira FHL (1987b) Os dados numéricos relativos às pistas (e suas pegadas) das Icnofaunas dinossaurianas do Cretáceo Inferior da Paraíba, e sua interpretação estatística. II - Parâmetros das pegadas. In: 10° Congresso Brasileiro de Paleontologia, Anais da Sociedade Brasileira de Paleontologia, Rio de Janeiro, Brazil, 1, pp 395–417

Leonardi G, Lima CV, Oliveira FHL (1987c) Os dados numéricos relativos às pistas (e suas pegadas) das icnofaunas dinossaurianas do Cretáceo Inferior da Paraíba, e sua interpretação estatística. III – Estudo estatístico. In: 10° Congresso Brasileiro de Paleontologia, Anais da Sociedade Brasileira de Paleontologia, Rio de Janeiro, Brazil, 1, pp 419–444

Leonardi G, Santos MFCF, Barbosa FHS (2021) First dinosaur tracks from the Açu Formation, Potiguar Basin (mid-Cretaceous of Brazil). An Acad Bras Ciênc 93:e20210635. https://doi.org/10.1590/0001-3765202120210635

Lima JCF, Bezerra FHR, Rossetti DF, Barbosa JA, Medeiros WE, Castro DL, Vasconcelos DL (2017) Neogene–Quaternary fault reactivation influences coastal basin sedimentation and landform in the continental margin of NE Brazil. Quat Int 438(Part A):92–107. https://doi.org/10.1016/j.quaint.2016.03.026

Lima MR (1983) Paleoclimatic reconstruction of the Brazilian Cretaceous based on palynology data. Revista Brasileira de Geociências 13:223–228

Lima Filho MF, Mabesoone JM, Viana MSS (1999) Late Mesozoic history of sedimentary basins in NE Brasilian Borborema Province before the final separation of South America and Africa 1: Tectonic-sedimentary evolution. In: 5° Simpósio Sobre o Cretáceo do Brasil, Boletim, UNESP Rio Claro, Brazil, pp 605–611

Lockley MG, Meyer C (1999) Dinosaur tracks and other fossil footprints of Europe. Columbia University Press, New York, p 360

Lourenço MCM, Jardim de Sá EF, Córdoba VC, Pichel LM (2021) Multi-scale tectono-stratigraphic analysis of Pre- and Synrift sequences in the Rio do Peixe Basin, NE Brazil. Mar Pet Geol 130:105127. https://doi.org/10.1016/j.marpetgeo.2021.105127

Mabesoone JM, Lima PJ, Ferreira EMD (1979) Depósitos de cones aluviais antigos, ilustrados pelas formações Quixoá e Antenor Navarro (Nordeste do Brasil). In: 9° Simpósio de Geologia do Nordeste, Anais, Recife, Sociedade Brasileira de Geologia/Núcleo Nordeste, Brazil, 7, pp 225–235

Mabesoone JM, Viana MSS, Neumann VH (2000) Late Jurassic to Mid-Cretaceous Lacustrine Sequences in the Araripe-Potiguar Depression of Northeastern Brasil. In: Gierlowski-Kordesch EH, Kelts KR (eds) Lake basins through space and time, AAPG studies in geology, 46, pp 197–208

Maciel IB, Dettori D, Balsamo F, Bezerra FHR, Vieira MM, Nogueira FCC, Salvioli-Mariani E, Sousa JAB (2018) Structural control on clay mineral authigenesis in faulted arkosic sandstone of the Rio do Peixe Basin, Brazil. Minerals 8(9):408. https://doi.org/10.3390/min8090408www.mdpi.com/journal/minerals

Matos RMD (1992) The Northeast Brazilian rift system. Tectonics 11:766–791

Matos RMD, Krueger A, Norton I, Casey K (2021) The fundamental role of the Borborema and Benin-Nigeria provinces of NE Brazil and NW Africa during the development of the South Atlantic Cretaceous Rift system. Mar Pet Geol 127:104872

Molnar RE, Farlow JO (1990) Carnosaur paleobiology. In: Weishampel DB, Dodson P, Osmolska H (eds) The dinosauria. University of California Press, Berkeley, pp 210–224

Moraes LJ (1924) Serras e montanhas do Nordeste. In: Inspectoria de Obras Contra as Seccas. Geologia. Rio de Janeiro. Ministério da Viação e Obras Públicas. (Serie I. D. Publ. 58). 2nd ed. Coleção Mossoroense, 35(1). Fundação Guimaraes Duque, Rio Grande do Norte, Brazil, pp 43–58

Muniz GCB (1985) Cochlichnus sousensis, icnoespécie da Formação Sousa, Grupo Rio do Peixe, no Estado da Paraíba. In: Coletânea de Trabalhos Paleontológicos. DNPM, Brasília, Brazil, pp 239–242

Muniz YM, Fernandes YL, Costa Jr NJR (2017) Avaliação do Potencial Gerador da Formação Sousa, Bacia do Rio do Peixe, Utilizando Dados de Pirólise Rock-Eval e Carbono Orgânico Total. In: 10th Simpósio Sul Brasileiro de Geologia. http://ssbg2017anais.siteoficial.ws/ST1/ST102.pdf

Nicchio MA, Balsamo F, Cifelli F, Nogueira FCC, Aldega L, Bezerra FHR, Vasconcelos DL, Souza JAB (2022) An integrated structural and magnetic fabric study to constrain the progressive extensional tectonics of the Rio do Peixe Basin, Brazil. Tectonics 41:e2022TC007244. https://doi.org/10.1029/2022TC007244

Nogueira FCC, Marques FO, Bezerra FHR, Castro DL, Fuck RA (2015) Cretaceous intracontinental rifting and post-rift inversion in NE Brazil: insights from the Rio do Peixe Basin. Tectonophysics 644–645:92–107. https://doi.org/10.1016/j.tecto.2014.12.016

Oliveira LSB, Nogueira FCC, Vasconcelos DL, Balsamo F, Bezerra FHR (2022) Pérez YAR (2022) Mechanical stratigraphy influences deformation band pattern in arkosic sandstones, Rio do Peixe Basin, Brazil. J Struct Geol 155:104510

Paul GS (1987a) Predation in the meat-eating dinosaurs. In: Currie PJ, Koster EH (eds) Symposium of Mesozoic Terrestrial Ecosystems, Drumheller, Alberta, Canada: Short Papers Tyrrell Museum Palaeontology, pp 171–176

Paul GS (1987b) The Science and Art of Restoring the Life Appearance of Dinosaurs and Their Relatives. In: Czerkas SJ, Olson EC (eds) Dinosaurs Past and Present, Los Angeles County: Natural History Museum, vol. II: 5–49.

Pérez-Lorente F (2015) Dinosaur footprints and trackways of La Rioja. Indiana University Press, Bloomington, p 376. ISBN: 9780253015150

Petri S (1983) Brazilian Cretaceous paleoclimates: evidence from clay-minerals, sedimentary structures and palynomorphs. Revista Brasileira de Geociências 13(4):215–222

Petri S (1998) Paleoclimas da era Mesozóica no Brasil - evidências paleontológicas e sedimentológicas. Revista da Universidade de Guarulhos 6:22–38

Pichel LM, Antunes AF, Fossen H, Rapozo BF, Finch E, Córdoba VC (2022) The interplay between basement fabric, rifting, syn-rift folding and inversion in the Rio do Peixe Basin, NE Brazil. Basin Res 35(1):1–25. https://doi.org/10.1111/bre.12704

Ramos GV, Castro, DL Vasconcelos DL, Nogueira FCC, Bezerra FHR, Nicchio MA (2023) Architectural rift geometry of the Rio do Peixe Basin (Brazil): implications for its tectonic evolution and Precambrian heritage. Tectonophysics (2023).https://doi.org/10.1016/j.tecto.2023.230173

Rapozo BF, Antunes AF, Córdoba VC (2019) Interpretação Sismoestrutural e Sismoestratigráfica da Porção SE do Semi-graben de Brejo das Freiras, Bacia do Rio do Peixe, NE do Brasil. In: XVII Simpósio Nacional de Estudos Tectônicos/ XI International Symposium on Tectonics At: Bento Gonçalves/ RS, Brazil, May 2019

Rapozo BF, Córdoba VC, Antunes AF (2021) Tectono-stratigraphic evolution of a Cretaceous intracontinental rift: example from Rio do Peixe Basin, north-eastern Brazil. Mar Pet Geol 126:104899

Roesner EH, Lana CC, Herisse AL, Melo JHG (2011) Bacia do Rio do Peixe (PB): novos resultados biocronoestratigráficos e paleoambientais. In: Carvalho IS, Srivastava NK, Strohschoen Jr O, Lana CC (eds) Paleontologia: Cenários de Vida. Rio de Janeiro: Interciência, Brazil, 3, pp 135–141. ISBN 978–85–7193–273–9

Santos MFCS, Santos CLA (1987a) Sobre a ocorrência de pegadas e pistas de dinossauros na localidade de Engenho Novo, Antenor Navarro, Paraíba (Grupo Rio do Peixe, Cretáceo Inferior), In: 10º Congresso Brasileiro de Paleontologia, Anais, Rio de Janeiro, Brazil, 1, pp 353–366

Santos MFCS, Santos CLA (1987b) Novas pegadas de dinossauros retiradas de uma cerca de pedras no sítio Cabra Assada, Antenor Navarro, Paraíba (Grupo Rio do Peixe, Cretaceo Inferior). In: 10º Congresso Brasileiro de Paleontologia, Anais, Rio de Janeiro, 1987, Brazil, 1, pp 367–376

Santos MFCS, Santos CLA (1989) Alguns parâmetros relativos as pegadas de dinossauros em Várzea dos Ramos, município de Sousa, Paraíba. In: 11º Congresso Brasileiro de Paleontologia, Anais, Sociedade Brasileira de Paleontologia, Curitiba, Paraná, Brazil, 1, pp 373–80

Santos WFS, Carvalho IS, Brilha JB, Leonardi G (2016) Inventory and assessment of palaeontological sites in the Sousa Basin (Paraíba, Brazil): preliminary study to evaluate the potential of the area to become a Geopark. Geoheritage 8:315–332. https://doi.org/10.1007/s12371-015-0165-9

Silva JGD, Córdoba VC, Caldas LHO (2014) Proposta de novas unidades litoestratigráficas para o Devoniano da Bacia do Rio do Peixe, Nordeste do Brasil - proposal of new lithostratigraphic units for the Devonian of the Rio do Peixe Basin, Northeast of Brazil. Braz J Geol 44(4):561–578

Siqueira LMP, Polck MAR, Hauch ACG et al (2011) Sítios Paleontológicos das Bacias do Rio do Peixe: Georreferenciamento, Diagnóstico de Vulnerabilidade e Medidas de Proteção. Anuário Do Instituto De Geociências 34(1):9–21

Souza DHS, Nogueira FCC, Vasconcelos DL, Torabi A, Souza JAB, Nicchio MA, Pérez YAR, Balsamo F (2021) Growth of cataclastic bands into a fault zone: a multiscalar process by microcrack coalescence in sandstones of Rio do Peixe Basin, NE Brazil. J Struct Geol 146:104315

Torabi A, Balsamo F, Nogueira F, Souza JAB (2021) Variation of thickness, internal structure and petrophysical properties in a deformation band fault zone in siliciclastic rocks. Mar Pet Geol 133(10):105297. https://doi.org/10.1016/j.marpetgeo.2021.105297

Viana MSS, Lima-Filho MF, Carvalho IS (1993) Borborema Megatracksite: uma base para correlação dos "arenitos inferiores" das bacias intracontinentais do Nordeste do Brasil. Simpósio de Geologia do Nordeste, Sociedade Brasileira de Geologia/núcleo Nordeste, Boletim 13:23–25

Wade M (1989) The stance of dinosaurs and the cossack dancer syndrome. In: Gillette DD, Lockley MG (eds) Dinosaur tracks and traces. Cambridge University Press, New York, pp 73–82

Xing L, Lockley MG, Zhang J, Klein H, Persons WS IV, Dai H (2014) Diverse Sauropod-, Theropod-, and Ornithopod-Track Assemblages and a New Ichnotaxon *Siamopodus xui* ichnosp. nov. from the Feitianshan Formation, Lower Cretaceous of Sichuan Province, Southwest China. Palaeogeogr Palaeoclimatol Palaeoecol 414:79–97

Chapter 8
Tracking Dinosaurs During the Equatorial and South Atlantic Opening

Giuseppe Leonardi, Maria de Fátima C. F. dos Santos, and Fernando Henrique de Souza Barbosa

8.1 Introduction

The breaking of Gondwana continent and the subsequent opening of the equatorial and southern Atlantic Ocean during the Cretaceous resulted in a number of marginal and intracratonic basins, due to the fault reactivation of the Precambrian shield. This great tectonic event gave origin to a number of marginal sedimentary basins as well as several small intracratonic basins in the northeastern region of Brazil, notably in the states of Ceará, Rio Grande do Norte, Paraíba, Pernambuco, Sergipe, Alagoas and Bahia (Fig. 8.1). The sediments of these basins reveal with some frequency the presence of dinosaurs in the continent that was emerging, especially through the record that comes from the discovery of their fossil tracks. These basins assemble the paleoceanographic, paleoclimatic, and biotic record changes during the late Barremian–Albian (Luft-Souza et al. 2021).

Apart from the well-known larger sedimentary basins, these smaller basins have aroused great interest among geoscientists, particularly in the field of tetrapod Ichnology, with a focus on dinosaurs. Some of these, located in the interior of

G. Leonardi (✉)
Istituto Cavanis, Dorsoduro 898, 30123 Venezia, Italy
e-mail: leonardigiuseppe879@gmail.com

CCMN/IGEO, Departamento de Geologia, Universidade Federal do Rio de Janeiro, Cidade Universitária, Ilha do Fundão, Rio de Janeiro, Estado do Rio de Janeiro 21949-900, Brazil

M. de Fátima C. F. dos Santos
Universidade Federal do Rio Grande do Norte, Museu Câmara Cascudo, Tirol, Natal, Estado do Rio Grande do Norte 59020-650, Brazil

F. H. de Souza Barbosa
Escola Normal Superior, Universidade do Estado do Amazonas, Manaus, Estado do Amazonas 69050-010, Brazil

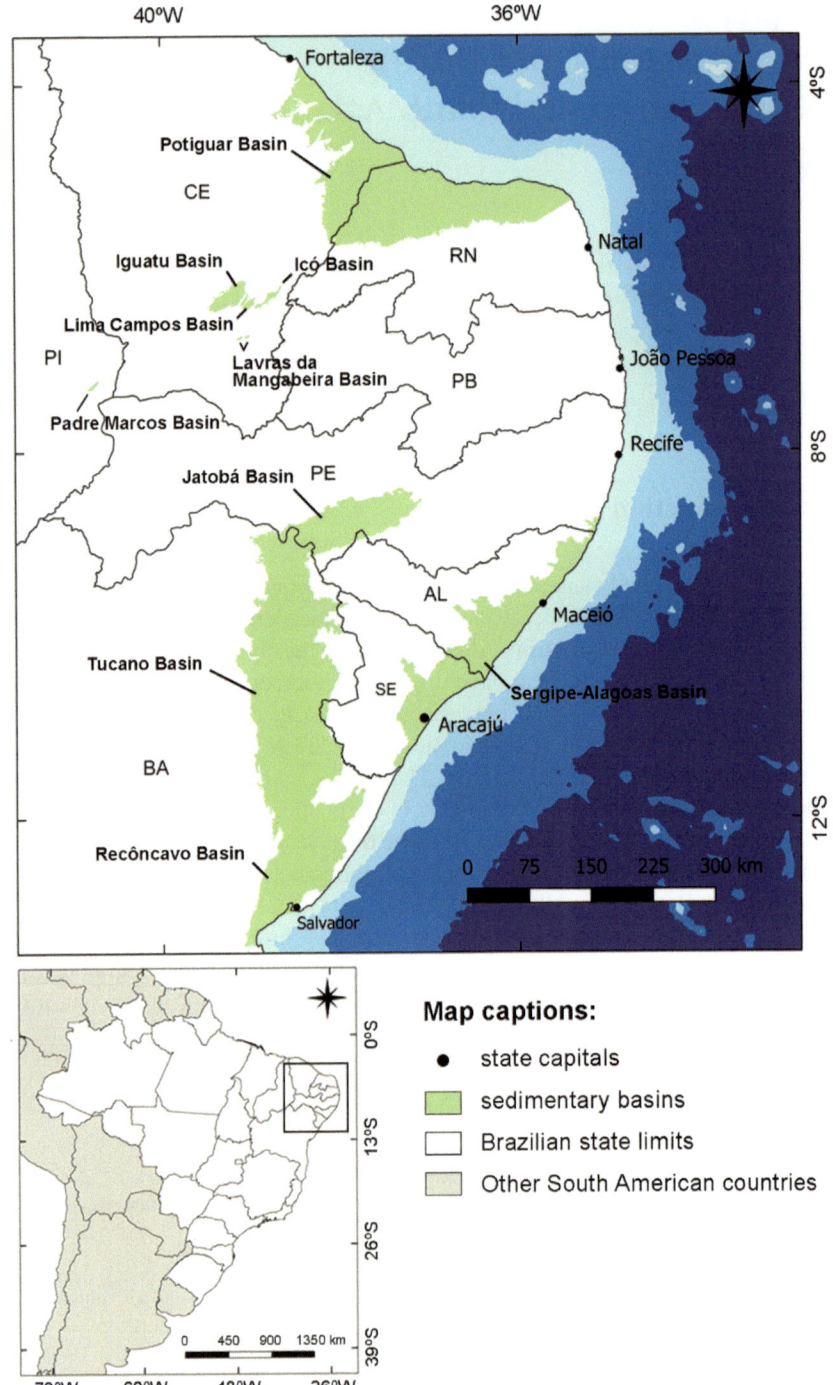

◀**Fig. 8.1** Geological map of the Brazilian northeastern sedimentary basins with the main occurrences of dinosaur tracks. Abbreviations: AL, Alagoas; BA, Bahía; CE, Ceará; PB, Paraíba; PE, Pernambuco; PI, Piauí; RN, Rio Grande do Norte; SE, Sergipe

Northeastern Brazil, presents a large number of fossil specimens and diversity have yielded interesting ichnosites (Leonardi and Muniz 1985; Leonardi and Spezzamonte 1994; Carvalho et al. 2020). In southeastern Ceará there is another group of small and medium-sized basins, controlled by the normal and transcurrent faults of the shear alignments Portalegre (~SW-NE) and Patos and Pernambuco (~W-E): Iguatu; Malhada Vermelha (also known as Palestina, or Igarói); Lima Campos; and Icó basins. The area of the Iguatu Basin is approximately 780 km^2, the Malhada Vermelha Basin, 65 km^2, Lima Campos Basin, 105 km^2 and Icó Basin, 120 km^2. To this assembly of basins, it can be joined the Rio dos Bastiões Basin (southern Ceará, about 40 km SW of Iguatu), which occur in a SW-NE small and narrow elongated depression. Fossil tracks of dinosaurs have been found in two of these basins: Lima Campos and Malhada Vermelha. The small basin of Icozinho (or Vertentes) is associated rather to the basins of the Rio do Peixe (Carvalho and Leonardi 1992). Little to the east of the basins is found another assemblage of small basins: Lavras de Mangabeira; Mangabeira; Iborepi in the municipality of Lavras da Mangabeira, which area does not exceed 63 km^2.

Other basins with dinosaur tracks include the Potiguar Basin, covering approximatelly an area of 48,000 km^2. It extends beyond the offshore, which is about 21,500 km^2, for about 200 km in the continental shelf to the north, beneath the ocean (Pessoa Neto et al. 2007). Additionally, there is the aborted rift of Jatobá, Tucano and Recôncavo with 500 km length and an average width of about 70 km (Dantas et al. 2019) and the narrow SW-NE marginal basin (300 km long and about 30 km wide) of Sergipe-Alagoas (Carvalho and Souza-Lima 2023).

8.2 Geological Context

During the Cretaceous South America and Africa breakup, the northern segment of the Atlantic Rift System was formed by the East Brazilian Rift System (EBRIS) (Chang et al. 1992).

The small intracratonic basins of the central region of northeastern Brazil (Fig. 8.2) are tectonic depressions mainly filled by Lower Cretaceous sediments, showing great similarities in their origin, shape and evolution (Ponte 1992; Mabesoone 1994; Valença et al. 2003). They resulted from mostly normal and transcurrent faults, (Carvalho and Melo 2012). They are mostly located in the western region of Paraíba and Rio Grande do Norte states, and in the southern part of the Ceará State. They present a great variety of invertebrates and vertebrates ichnofossils and, especially, dinosaur tracks (Leonardi 1979, 1989, 1994; Carvalho 1989; Carvalho et al. 1993a, b, 1994; Leonardi and Spezzamonte 1994), which demonstrates that despite the rarity of skeletons in this area, the dinosaur faunas of Brazil were rich and diverse.

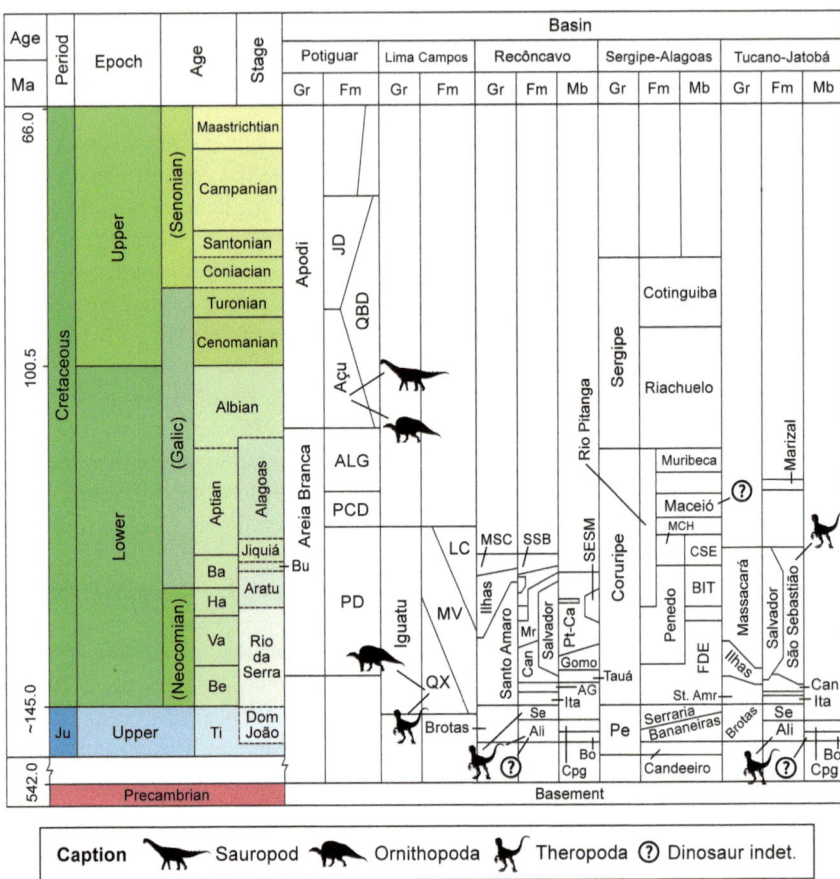

Fig. 8.2 Simplified stratigraphic sections and lithostratigraphic units with dinosaur footprints of the Potiguar Basin, Lima Campos Basin, Recôncavo Basin, Sergipe-Alagoas Basin, and Tucano-Jatobá Basin. Abbreviations: AG, Água Grande Formation; ALG, Alagamar Formation; Ali, Aliança Formation; Ba, Barremian; Be, Berriasian; BIT, Barra de Itiúba Formation; Bo, Boipeba Member; Bu, Buracica Stage; Can, Candeias Formation; Cpg, Capianga Member; CSE, Coq. Seco Formation; FDE, Feliz Deserto Formation; Fm, Formation (litostratigraphic unit); Gr, Group (litostratigraphic unit); Ha, Hauterivian; Ita, Itaparica Formation; JD, Jandaíra Formation; Ju, Jurassic; LC, Lima Campos Formation; Ma, million years; Mb, Member (litostratigraphic unit); MCH, Morro do Chaves Formation; Mr, Maracangalha Formation; MSC, Massacará Group, MV, Malhada Vermelha Formation; PCD, Pescada Formation; PD, Pendência Formation; Pe, Perucaba Group; Pt-Ca, Pitanga-Caruaçu Member; QBD, Quebradas Formation; QX, Quixoá Formation; Se, Sergi Formation; SESM; Sesmaria Member; SSB; São Sebastião Formation; St. Amr, Santo Amaro Group; Ti, Tithonian; Va, Valanginian (modified from Costa et al. 2007; Pessoa-Neto et al. 2007)

The Iguatu, Malhada Vermelha, Lima Campos and Icó grabens or semi-grabens, are neighboring basins, strongly controlled by the tectonic structures of the Protero-zoic basement, related to the ~SW-NE Portalegre shear zone. In these basins, clastic rocks are distributed mainly nearside the shearbundles that limit and control the basins. In more central areas, or away from active faults during sedimenta-tion, occur fine sandstones, siltstones, shales, mudstones, limestones and marls (Carvalho 2000a, b). The lithostratigraphy of these basins consist in the Iguatu Group (about 3,000 m thick), with the Quixoá (coarse sandstones with conglom-erate intercalations); Malhada Vermelha (fine sandstones, shales and marls) and Lima Campos (conglomeratic sandstones and fine sandstones) formations. Srivas-tava (1990) suggested that the lithostratigraphic units of these basins should follow the same terminology of the Rio do Peixe basins (Antenor Navarro, Sousa and Rio Piranhas formations), due the lithological and paleoenvironmental similarities (Carvalho 2000a, b).

The small basins of Lavras de Mangabeira, Mangabeira and Iborepi show a E-W orientation, controlled by the Portalegre shear zone. The basal sediments are the Antenor Navarro Formation, with coarse sandstones, breccia and polymictic conglomerates. There are also (in the locality Melancias) siltstones and mudstones, reddish color, similar to the deposits of the Sousa Formation (Carvalho 1989, 1993). Another small basin is the Padre Marcos, located in Cariri Valley rift system, in the Padre Marcos County, Piauí State, between the Parnaíba and Araripe basins. There are polymictic conglomerates and coarse sandstones, near the faulted margins. In the depocenter there are fine greenish sandstones, siltstones and shales. The lithostratigraphic context is similar to that of the Rio do Peixe basins.

The Potiguar Basin is one of the basins that is located in the Equatorial Margin. It was created during the Early Cretaceous, with the reactivation of the Precambrian basement shear zones, that culminated in the opening of the Atlantic Ocean (Pessoa Neto et al. 2007; Araújo et al. 2023). The sedimentary sequence of the Potiguar Basin is composed of more than 9 km thick and fills asymmetric grabens during the South American/Africa breakup (Pessoa Neto et al. 2007; Araújo et al. 2023). It is represented by three supersequences: (i) Rift Supersequence (Rift I e Rift II; Early Cretaceous); (ii) Pos-rift Supersequence (Alagoas Stage); and (iii) Drift Superse-quence (Albian to Recent). The fossil record of the Potiguar Basin is predominantly composed by marine invertebrates and vertebrates associated with the Jandaíra and Açu formations (Cassab 2003; Pereira et al. 2018, 2020a, b; Veiga et al. 2019; Dantas et al. 2021).

The long, aborted rift system Recôncavo-Tucano-Jatobá resulted from the crustal extension during the fragmentation of the supercontinent Gondwana. Apart from a Silurian-Devonian sequence, there is a Pre-rift Supersequence (Aliança and Sergi formations), interpreted as fluvial-eolian and lacustrine (Dom João stage; Tithonian, Upper Jurassic); and the Rift Supersequence (Candeias, São Sebastião and Salvador formations) dated as Berriasian-Barremian up to early Aptian, Lower Cretaceous (Silva et al. 2007; Costa et al. 2007).

The Sergipe-Alagoas marginal basin is oriented SW-NE and together with the Potiguar Basin, are the only Brazilian Cretaceous basins with a complete sedimentary record since the Upper Jurassic Pre-rift phase to the uppermost drift Cretaceous marine sediments (Cassab and Santos 1994). It is remarkable for its fossil-rich units, with several macrofossil and microfossil groups, and sedimentary sequences representing all phases encompassed by the Gondwana break-up and South Atlantic Ocean opening (Luft-Souza et al. 2021).

8.3 Footprints: Diversity and Paleobiological Interpretation

8.3.1 Iguatu, Malhada Vermelha, Lima Campos and Icó Basins

In the Malhada Vermelha Basin, in the Cabeça de Negro locality (Leonardi and Spezzamonte 1994; 06° 21′ 52″ S; 39° 04′ 17″ W) in a remarkable series of outcrops and loose slabs, which represents a transition between the Quixoá and Malhada Vermelha formations, one footprint was found (ORCN 1; Figs. 8.3 and 8.4). It was attributed to a medium-sized theropod.

In the Lima Campos Basin are found two ichnosites. The first one is the richest locality with dinosaur trackways in this region, and is located at the county of Orós, in the São Romão farm, 3 km east of lgarói, along the road linking this town to the CE-113 highway (6° 21′ 52″ S; 39° 01′ 18″ W). The track-bearing surface is located in the basal portion of the Quixoá (or Antenor Navarro) Formation. It is a 15 m^2 area and it shows nine dinosaur short trackways, all bipedal, and some isolated footprints, belonging to bipedal dinosaurs; there is a total of fourteen individuals. This association is constituted of fourteen bipedal animals, including about seven individuals and three different forms of theropods (ORSR 1, ORSR 2, ORSR 3, ORSR 4, ORSR 11, ORSR 12, ORSR 14; Figs. 8.3 and 8.4) and perhaps three individuals and two different forms of medium-sized ornithopods (ORSR 8, ORSR 9 and ORSR 13). The small tracks (ORSR 5, ORSR 6 and ORSR 7) are too small to be classifiable in this brittle material. The track ORSR 10 in not classifiable too. This classification is different from the one given in the work cited above (Leonardi and Spezzamonte 1994).

Most of the footprints are well discernible, some are very deep, with high displacement rims in front or behind, with some sliding marks. The trackways and separated footprints show the directions subparallel to SW-NE (n = 8) or the contrary (n = 3), parallel to that of the main Portalegre fault direction; only two tracks (ORSR 1 and ORSR 4) cross the others in a SE-NW direction. Particularly interesting are the trackways pertaining to three tiny sized dinosaurs of the same kind, which are very rare in the Rio do Peixe Group basins and in almost all of the northern and northeastern Brazil (Leonardi and Carvalho 2021). In these two trackways (ORSR 5 and ORSR 6), the footprints have an average length of 100 mm and 110 mm, respectively; and

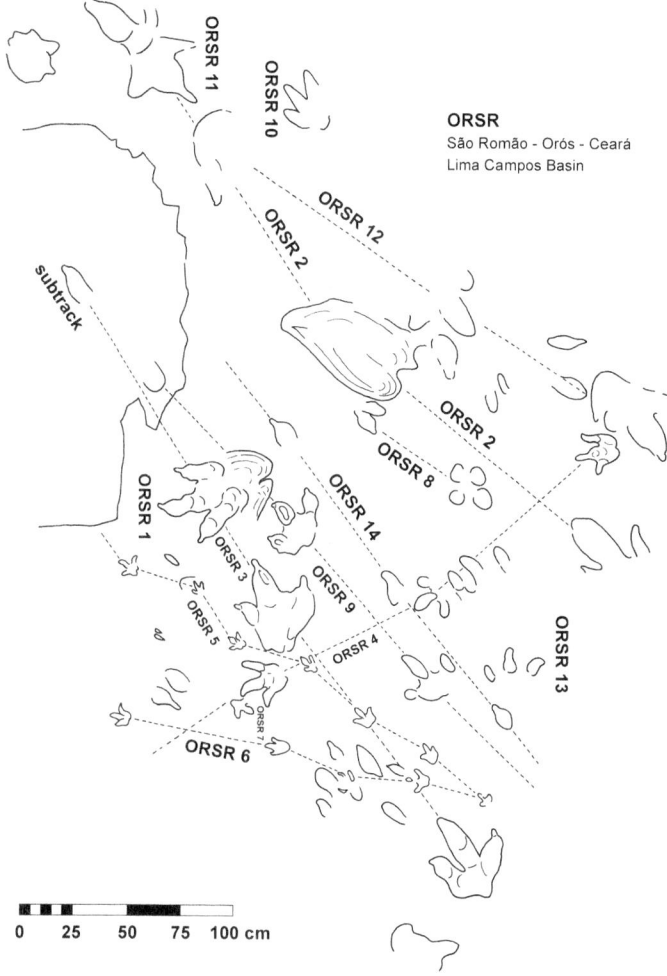

Fig. 8.3 Map of the theropods and ornithopods trackways in a rocky pavement of the Quixoá Formation (Lower Cretaceous, Lima Campos Basin) at the tracksite of São Romão, Orós, State of Ceará (from Leonardi and Spezzamonte 1994)

a width of 85 and 102 mm, respectively. These dimensions are similar to these of the isolated footprint ORSR 7. Another isolated track was found in the tracksite of the Tijuca Farm, Orós County, 0.5 km W of the highway CE-113, N of the town of Lima Campos. On the coarse sandstones of the Quixoá (or Antenor Navarro) Formation, there is an isolated and incomplete footprint (LCTI 1), rather shallow but very well imprinted in an isolated flat boulder, and attributed to a large theropod (Fig. 8.4).

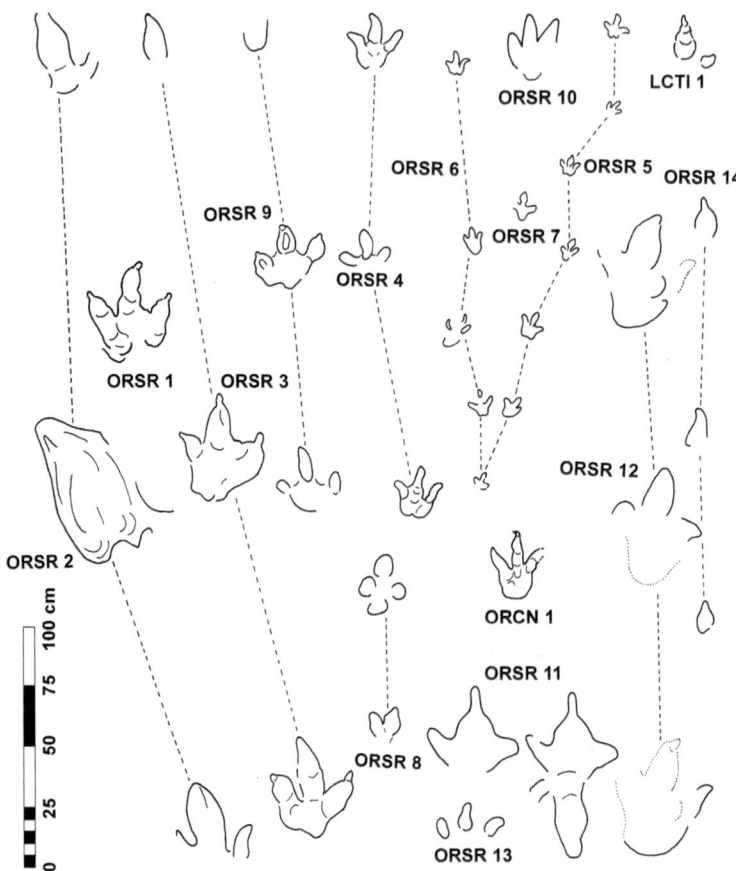

Fig. 8.4 The same dinosaur trackways of the Fig. 8.3 and those of the locality Tijuca (LCTI 1), and of the site Cabeça de Negro in the Malhada Vermelha Basin (ORCN 1), both attributed to theropods (from Leonardi and Spezzamonte 1994)

8.3.2 The Mangabeira and Padre Marcos basins

In the small basins of Lavras de Mangabeira, Mangabeira and Iborepi, the ichno-logical record is represented by dinosaur footprints of poor-quality, in the bed of the Rosario Creek (unpublished material), vertical pedotubules and horizontal and vertical invertebrate tubes, possibly assigned to *Skolithos* Haldemann, 1840 (Carvalho and Fernandes 1992). The latter were on the edge of a rural road, which leads to Quintaús, on an outcrop of red-coloured sandstone, which seems to correspond to a hard ground.

In the Padre Marcos Basin, fossil tracks of dinosaurs were identified in the Sítio tracksite (Juazeiro do Quitó, Jaicós county), and attributed to theropods (Carvalho and Viana 1996; Carvalho 2001).

8.3.3 Potiguar Basin

Dinosaur skeletal remains were found at the western end of the Potiguar Basin, in the municipality of Quixeré, Ceará State, in Açu Formation (Albian-Cenomanian). These materials come from the Açu-4 informal stratigraphic unity (Cenomanian). They were classified as carcharodontosaurids, abelisaurids and sauropods (Santos et al. 2005; Pereira et al. 2020a, b). Later on, an association of dinosaur tracks has been described in this basin, also in the Açu Formation (Leonardi et al. 2021). These footprints are located at the site known as Pingos Farm (5°34'10 S, 37°02'20''W, datum WGS 84; Fig. 8.5). This ichnofauna includes four individual tracks. Three of them (AÇPI 1, AÇPI 2 and AÇPI 3) are assigned to large sauropod hind-footprints. The AÇPI 2 and AÇPI 3 are poor quality footprints, with no anatomical details, exhibiting only a clear, although low, displacement rim, and an oval outline. AÇPI 1 (Fig. 8.5c), on the other hand, is a track deeply imprinted on a sandstone cracked surface, featuring a large and high displacement rim, particularly proeminent in front and some morphological details. The footprint is wider in the front portion and narrower in the rear, and it has the characteristic outline of a bell; the heel outline is roundish; on the front margin, one can observe at least four claw impressions. Including the displacement rims, the track measures about 100 cm in width and 140 cm in lenght, a probable large titanosaurid trackmaker. This footprint is part of a short trackway with at least three hind footprints, while the footprints of the manus, as in the first one that was reported, are probably covered by the displacement rim of the pedes. Several other sauropod tracks, probably of the same population, were discovered on the last visit (July 2021) and deserve a further detailed study. A preliminary analysis indicates that the sauropod trackway seems, in this way, to be associated with the same kind of titanosaurid track found there (AÇPI 1). It is not uncommon when it comes to sauropods, animals that frequently lived and moved gregariously in herds. This apparent appearing and disappearing of tracks probably depends on the fact that there is an accelerated weathering, with loss of structures and outlines. The last footprint (AÇPI 4) is attributed to an ornithopod. It is a very deep, left hind-footprint, longer than it is wide, with three roundish and short hooves and a rounded (monolobed) interdigital pad (Fig. 8.5d). The footprint presents a large displacement rim, 15–20 cm wide. The hooves have a blunted distal end, and they are short, especially toes II and IV, which are much shorter than toe III. The III digit is spatulate. Digit II is separated from the heel pad by a rather typical notch or incision. Based on comparisons with other ornithopod tracks, it is possible that it could have been imprinted by some kind of iguanodontid.

Fig. 8.5 The Pingos tracksite in the Potiguar Basin and its surroundings. A **a** The plateau of sandstone Açu 3, on which lies the ichnosite of Pingos; **b** The cave, called *Gruta dos Pingos*, is a good reference mark to find the rocky pavement with the dinosaur tracks, which is located about fifteen meters above the cavern. Photograph by Leonardo Menezes; **c** A fine and large sauropod hind-footprint (AÇPI 1), with a large and high displacement rim, especially large in front of the track; in a trackway of three consecutive footprints; **d** Deep ornithopod footprint (AÇPI 4), probably a left one. Scale bar: 15 cm (in **c**) and 20 cm (in **d**)

8.3.4 *Tucano and Recôncavo Rift and Sergipe-Alagoas Basin*

There is an uncertain theropod footprint at an outcrop of the Aliança Formation (Tithonian, Lower Jurassic) in the locality of Penedo, Bahia (Tucano North Basin), near the right margin of the São Francisco River.

In the Recôncavo Basin several load- and fluidization structures were found in deposits of the Dom João Stage (Upper Jurassic, middle and late Tithonian), which have been interpreted as dinoturbation processes (Carvalho and Borghi 2008). In the

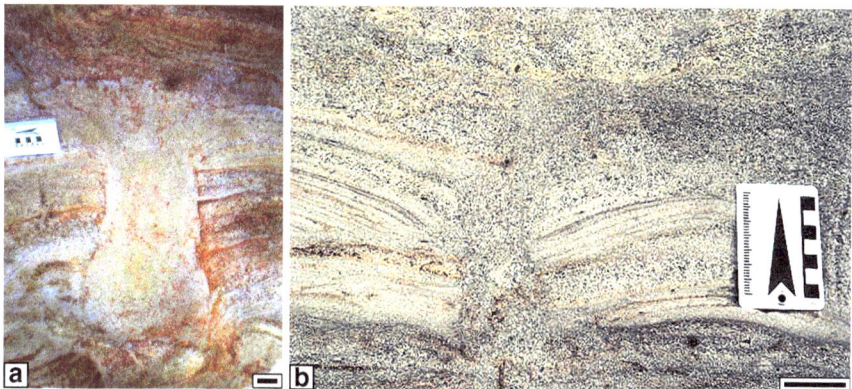

Fig. 8.6 **a** Cross section of dinosaur deep footprints presenting fragmentation of layers partially solidified, partly with fluidization in the internal surrounding matrix; **b** Outcrop of the Maceió Formation (Sergipe-Alagoas Basin) where cross section tracks are found in coarse grained sandstones. Scale bars: 5 cm. Photographs by Ismar de Souza Carvalho

Aliança Formation (Boipeba Member), they observed load structures with a more pronounced concavity at the base. In the Sergi Formation (Uppermost Jurassic), Carvalho and Borghi (2008) discovered similar deformation structures, with larger dimensions, possessing verticalized walls, in the shape of a tube, denoting a deformation of up to 30 cm in lamination depth (Fig. 8.6). These footprints are interpreted as load and fluidization structures that could result from the weight exerted by the autopods of large tetrapods in unconsolidated sediments. As the structures are vertical, sectioning the lamination, the details of the trackmaker autopodia are not visible. These can then be classified only as specimens of the order Dinosauria, without a more detailed classification.

Recently, Dantas et al. (2019) preliminarily communicate the finding of three isolated dinosaur footprints, on the surface of layers of the Lower Cretaceous São Sebastião Formation, from the North Tucano Basin, ichnosite of Canindé de São Francisco, State of Sergipe, not far from the right bank of the São Francisco River. They are isolated tridactyl footprints, well preserved. All three are assigned to small theropods; they are accompanied by bird tracks, and therefore by avian-theropods (Dantas, MAT, personal information). It seems likely that the age of those footprints is early Aptian. Another three dinosaur tracks were discovered by Carvalho and Souza-Lima (2023) in the Sergipe-Alagoas Basin. These tracks were found in sandstones of the Maceió Formation, in cross-section on seashore bars, on the beach of Bicingui, in the municipality of Japaratinga (State of Alagoas), in the NE extremity of the Sergipe-Alagoas Basin. In these sandstones occur phenomena of fluidizations and apparent load-casts, which are interpreted by the aforementioned authors and by us as resulting surely as a load-cast, but this was produced by a load exerted by the autopods of large and heavy vertebrates. These are features of dinoturbation, preserved in Aptian sandstones and they comprise absolutely the first evidence of dinosaurs in the Sergipe-Alagoas Basin. These tracks reach 20 cm of maximum width and an approximate 46 cm in length. The cross-section footprints do not allow to easily recognize their trackmaker in detail. On the other hand, during the Albian, the only animals with feet of this size were, as far as is known, dinosaurs, and probably non-theropod dinosaurs.

8.4 Paleogeographical Distribution of the Footprints

Between the Late Jurassic and Early Cretaceous, the Gondwana supercontinent began to break apart. By the Early Jurassic, the extant South American continent had separeted from Florida and the Gulf area, and the Central Atlantic Ocean was initiating its opening. However, it is from the Callovian (late Middle Jurassic; Matos 1992) that the rupture between the Western Gondwana, now corresponding to South America, and the eastern part becomes more evident, affecting the Brazilian Northeastern region. The Atlantic equatorial margin had settled. Now, a complex of tensions and stresses lead to crustal stretch, rifting and the break of West and East Gondwana, resulting in the opening of the Central and South Atlantic (Cainelli and Mohriak 1999). This complex of tensions opened a deep rift valley in the Recôncavo Basin and the Tucano and Jatobá basins during the Dom João Stage (Upper Jurassic) (Carvalho and Borghi 2008). Around the same time, in the Araripe Basin there was the deposition of the Brejo Santo and Missão Velha formations (Assine 2007) and in the Sergipe-Alagoas Basin the Serraria and Bananeiras formations (Campos Neto et al. 2007).

Later, in the Potiguar Basin at the beginning of the Cretaceous, the Pendência, Pescada and Alagamar formations (Rio da Serra-Aratu stages; late Berriasian-Aptian) record the rift conditions (Assine 2007; Pessoa Neto et al. 2007). Also, at the beginning of the Cretaceous (Berriasian), throughout the northeast opened other small, narrow and elongated basins, from SW to NE, almost parallel to the current Atlantic coastline, and more rarely have a W-E orientation. Although the names of the lithostratigraphic units changes, they generally exhibit similar sequences of sediments. The Rio do Peixe Group corresponds to these basins with its three formations: Antenor Navarro, Sousa and Rio Piranhas.

Later began the opening of the marginal basins of Paraíba-Pernambuco during Aptian (Córdoba et al. 2007). During the Albian-Cenomanian begin the deposition of the psammitic Açu Formation in the Potiguar Basin. The dinosaur track assemblage at Fazenda Pingos probably needs to be located in the informal subunit Açu 3 (Early Cenomanian, Vasconcelos et al. 1990; Leonardi et al. 2021). The lithostratigraphic unit of Açu Formation represents the initial phase of transgression during the Early Cretaceous resulting in vertical stacking of river systems. This includes braided systems at the base, transitioning to coarse meanders, fine meanders and estuarine at the top. This deposition is linked to the drift phase of the Potiguar Basin evolution, before the implementation of a tide-dominated carbonate platform/ramp (Jandaíra Formation; Turonian-Campanian; Pessoa-Neto et al. 2007).

There is the presence of different groups of dinosaur in almost all of these basins, as well as in the surrounding territory (Fig. 8.7). The tectonic events, from Callovian to the late Lower Cretaceous, ends with the definitive opening of the central and southern Atlantic Ocean, and with the beginning of the oceanic crust, approximately 110 million years ago, at the Albian-Cenomanian. The origin and evolution of the South Atlantic was a long and complex process, during which the rocks of the basement of Borborema and Benin-Nigeria provinces acted as an "obstacle" to the complete South Atlantic opening. Seawater initially invaded the continental areas as an epicontinental sea in the northeastern region of Brazil crust (Carvalho 2022). The ocean floor spreading was a diachronic phenomenon, in three distinct moments (approximately 130 Ma, 113 Ma and 110 Ma) starting from south to north. Only after the end of the Early Cretaceous (about 110 million years) did a continuous crust of oceanic nature be established (Darros de Matos et al. 2021a, b; Carvalho 2022). The South American continent was thus separated and individualized from Africa. So, the late Aptian crustal separation led to the evolution of the Brazilian Atlantic margin (Chang et al. 1992). Its fauna from this moment on, had to evolve separately from that of Africa.

8.5 Paleoenvironmental and Paleoclimatic Contexts

The ancient landscape in today's Brazilian northeast, particularly in the area between the Potiguar and Araripe basins, was mostly mountainous and arid at the beginning of the Cretaceous. It comprised mountains and low ranges of rocky hills, primarily formed from Proterozoic fault rises and eroded over an extended period. The large plateaus, such as the distinctive Chapada do Araripe and others had not yet formed. It was difficult to humidity reach the middle of an immense megacontinent, resulting in rare rainy seasons. Several formations from the Early Cretaceous indicate the presence of aeolian sands and silts. Bedding surfaces with impressions of raindrops are rare. Plant remains are not abundant in the Cretaceous formations of the region, especially when compared with the underlying Jurassic formations. The overall climate was likely warm and arid, prevailing greenhouse conditions (Luft-Souza et al. 2021). The rivers, at the bottom of the valleys, were stony streams, usually dry, flowing only after the rare flash floods and during the limited time of the seasonal rains.

If vegetation was indeed scarce, the fauna was undoubtedly influenced by this, and in regions lacking humid microenviroments with lakes, it could not have be abundant. In addition, the fauna had to consist of animals adapted to life in an arid or semi-arid environment. The absence (so far at least) of both fossil bones or footprints in about half of the basins mentioned in this chapter, seems to confirm this general picture of the environment, landscape and climate.

However, with the beginning of the crust stretching and the consequent break-up of the region, the landscape favored the formation of fluvial valleys, lakes, and more widespread the presence of small and sometimes large lakes (Petri 1983). This change

Fig. 8.7 Environmental reconstruction of life in the intracratonic basins of the Brazilian northeast during the Early Cretaceous. A herd of sauropods passes from one basin to another, over the crest of a horst. On the shore there is a herd of ouranosaurid-like ornithopods, and a few abelisaurids at rest. At the base of the mountains to the west runs a transcurrent fault. Art by Guilherme Gehr

is particularly evident in the case of the Sousa Basin, as well as in the Potiguar, Lima Campos, and Malhada Vermelha basins.

As the climate became progressively wetter and less arid with the Gondwana breakup, small rift basins with a lacustrine and fluvial evironments were settled. This is concurrent with the progressive separation of South America and Africa. The presence of freshwater lakes with alkaline pH, high temperature and high amount of nutrients is shown in many basins by the numerous specimens of the giant conchostracean *Paleolimnadiopsis reali* (Carvalho and Srivastava 1996). Consequently, the environment became suitable for the life and concentration of tetrapods and especially dinosaurs, from diverse clades. Other minor evidences of these faunas, especially dinosaurian, are also found in the areas of Mangabeira, Padre Marcos and Rio Nazaré basins.

The final rupture of Gondwana and the beginning of spreading of an oceanic crust barred further passageway for non-flying continental tetrapods, and even contributed to the total change of marine currents in the region and important wind system changes (Scherer et al. 2020). We must also consider the great importance of microclimates or local climates for the biota, depending on factors such as altitude, proximity to small or large bodies of water, especially the sea and its currents, and the relationships between flora and fauna and between different animal groups.

8.6 Conclusions

The presence of a number of ichnosites with dinosaur tracks in the basins along the Brazilian Atlantic coast, or slightly more inland, indicates a remarkable diversity index. This diversity is compared with the dinosaurian fauna on the western shore of Africa. These ichnosites increase the record provided by the tetrapod body fossils, which are very rare in the area, and contribute to the understanding of the paleoenvironments, paleogeography, and paleoclimate of the Brazilian Northeast.

Dinosaur tracks have been identified in several basins, initially revealing the presence in the Early Cretaceous (Berriasian-Albian) of many clades of non-avian dinosaurs. These include sauropods (probably titanosaurids and Diplodocoidea), a majority of theropods (likely Abelisauroidea or spinosaurids), ornithopods, both graviportals (iguanodontids) and small-sized ones, and avian theropods. Alongside conventional tracks, some others were discovered in cross-section. In total, about 32 dinosaur individuals have been found in these 17 basins so far. Their presence has often assisted in dating stratigraphic units, confirming them as Mesozoic and particularly Cretaceous. It has also contributed to a better understanding of environments and climate, explaining the relationship between the presence and movements (including possible migrations) of dinosaur faunas and the tectonic structure of the region and recalling the connection of NE Brazil with the African continent.

Considering that the sediments of several of these basins closely resemble those of the Rio do Peixe basins, and given that some of them have not been visited by ichnologists for decades due to their isolated locations, it is suggested that further periodic research should be conducted. Erosion may reveal interesting new material, but it also poses a risk of destruction.

References

Araújo LN, Mello CL, Anderson Moraes A, Silva AT (2023) Numerical modeling of the Cenozoic tectonic stress fields and associated structural patterns in the northwestern portion of the Potiguar Basin. Geologia USP Série científica 23(3):14–167

Assine ML (2007) Bacia do Araripe. Boletim de Geociências da Petrobrás 15(2):371–389

Cainelli C, Mohriak WU (1999) Some remarks on the evolution of sedimentary basins along the eastern Brazilian continental margin. Epis J Int Geosci 22(3):206–216

Campos Neto OPA, Lima WS, Cruz FEG (2007) Bacia de Sergipe-Alagoas. Boletim de Geociências da Petrobras 15(2):405–415

Carvalho IS (1989) Icnocenoses continentais: bacias de Sousa, Uiraúna-Brejo das Freiras e Mangabeira. MS thesis, Universidade Federal do Rio de Janeiro, Rio de Janeiro, Brazil [unpublished]

Carvalho IS (1993) Os Conchostráceos Fósseis das Bacias Interiores do Nordeste do Brasil. Universidade Federal do Rio de Janeiro, Instituto de Geociências, PhD Thesis [unpublished]

Carvalho IS (2000a) Geological environments of dinosaur footprints in the intracratonic basins from Northeast Brazil during South Atlantic opening (Early Cretaceous). Cretac Res 21(2000):255–267

Carvalho IS (2000b) Huellas de saurópodos Eocretácicas de la cuenca de Sousa (Serrote do Letreiro, Estado da Paraíba, Brasil). Ameghiniana 37(3):353–362

Carvalho IS (2001) Conchostráceos da Bacia de Padre Marcos (Cretáceo Inferior) Estado do Piauí, Brasil. Acta Geologica Leopoldensia XXIV (52.53):349–357

Carvalho IS (2022) Paleogeografia. Cenários da Terra. 1 ed. Rio de Janeiro. Editora Interciência. 450 pp ISBN: 9786589367482

Carvalho IS, Fernandes ACS (1992) Os icnofósseis da Bacia de Mangabeira, Cretáceo do Ceará. In: 2º Simpósio Sobre as Bacias Cretácicas Brasileiras, Boletim de Resumos Expandidos, Rio Claro, São Paulo, UNESP, Brazil, pp 105–106

Carvalho IS, Borghi L (2008) Estruturas de Dinoturbação na Bacia do Recôncavo: Implicações Paleoambientais. In: 44º Congresso Brasileiro de Geologia. Curitiba, Paraná, Anais, p 815

Carvalho IS, Leonardi G (1992) Geologia das bacias de Pombal, Sousa, Brejo das Freiras e Vertentes (Nordeste do Brasil). An Acad Bras Ciênc 64(3):231–252

Carvalho IS, Melo JHG (2012) Bacias Interiores do Nordeste. In: Hasui Y, Carneiro CDR, Almeida FFM, Bartorelli A (eds) Geologia do Brasil, Beca, São Paulo.

Carvalho IS, Srivastava KN (1996) Conchostráceos paleolimnadiopsídeos da Bacia do Rio Nazaré (Cel. João Pessoa), Rio Grande do Norte. In: 4º Simpósio sobre o Cretáceo do Brasil. Boletim do 4º Simpósio sobre o Cretáceo do Brasil, Águas de São Pedro, São Paulo, pp 151–155

Carvalho IS, Viana MSS (1996) A bacia de Padre Marcos (Cretáceo Inferior, estado do Piauí) e sua icnofauna dinossauriana. In: 39º Congresso Brasileiro de Geologia, Salvador. Anais, SBG, 2, pp 265–267

Carvalho IS, Viana MSS, Lima Filho MF (1993a) Bacia de Cedro: a icnofauna cretácica de vertebrados. An Acad Bras Ciênc 65:459–460

Carvalho IS, Viana MSS, Lima Filho MF (1993b) Os icnofósseis de vertebrados da bacia do Araripe (Cretáceo Inferior, Ceará-Brasil). An Acad Bras Ciênc 65:459

Carvalho IS, Viana MSS, Lima Filho MF (1994) Dinossauros do Siluriano: um anacronismo cronogeológico nas bacias interiores do Nordeste? In: 38º Congresso Brasileiro de Geologia, Camboriú. Boletim de Resumos Expandidos, Camboriú, Santa Catarina, Sociedade Brasileira de Geologia, 3, pp 213–214

Carvalho IS, Leonardi G, Rios-Netto AM, Borghi L, Freitas AP, Andrade JA, Freitas FI (2020) Dinosaur trampling from the Aptian of Araripe Basin, Brazil, as tools for stratigraphic correlation. Cretac Res 117:104626. https://doi.org/10.3301/IJG.2020.24

Carvalho IS, Souza-Lima W (2023) Processos de dinoturbação na Formação Maceió (Cretáceo Inferior), Bacia de Sergipe-Alagoas. In: VI Simpósio Brasileiro de Paleontologia de Vertebrados, Ribeirão Preto. Boletim de Resumos, pp 61–62

Carvalho IS, Viana MSS (1996) A bacia de Padre Marcos (Cretáceo Inferior), estado do Piauí e sua icnofauna dinossauriana, In: Anais do Congresso Brasileiro de Geologia, 39(2), pp 265–267

Cassab, RCT (2003). Paleobiologia dos Gastrópodos da Formação Jandaíra, Cretáceo Superior da Bacia Potiguar. Tese (Doutorado em Geologia). Instituto de Geociências, Universidade Federal do Rio de Janeiro, Rio de Janeiro

Cassab RCT, Santos MEM (1994) The Sergipe-Alagoas Basin. In: Beurlen G, Campos DA, Viviers MC (eds) Stratigraphic range of Cretaceous mega- and microfossils of Brazil. Universidade Federal do Rio de Janeiro, Centro de Ciências Matemáticas e da Natureza, Instituto de Geociências, pp 161–231

Chang HK, Kowsmann RO, Figueiredo AMF, Bender AA (1992) Tectonics and stratigraphy of the East Brazil Rift system: an overview. In: Ziegler PA (ed) Geodynamics of rifting, Volume II. Case history studies on rifts: North and South America and Africa, Tectonophysics, 213, pp 97–138

Córdoba VC, Antunes AF, Sá FFJ, Lins FAPL (2007) Stratigraphic and structural analysis of the Rio do Peixe Basin, Northeastern Brazil: integration based on the pioneer seismic survey 0295-RIO-DO-PEIXE-2D. Boletim de Geociências da Petrobras 16(1):53–68

Costa IP, Milhomem PD, Bueno GV, Lima e Silva HS, Kosin MD (2007) Sub-bacia do Tucano Norte e Bacia de Jatobá. Boletim de Geociências da Petrobras 15, pp 445–453

Dantas MAT, Teixeira, FAP, Santos, DB et al (2019) Dinosaur footprints from the lower cretaceous (Aptian, Tucano Basin) of Canindé de São Francisco, Sergipe, Brazil. In: XXVI Congresso Brasileiro de Paleontologia. Uberlândia, pp 269–270

Dantas EP, Medeiros VC, Cavalcante R (2021) Mapa Geológico do Estado do Rio Grande do Norte. Serviço Geológico do Brasil - CPRM, Recife.1 mapa color. 132,72cm x 85,45cm. Escala 1:500.000

Darros de Matos RM, Krueger A, Norton I, Casey K (2021a) Fundamental role of the Borborema and Benin-Nigeria provinces of NE and NW Africa during the development of the South Atlantic Cretaceou System. Mar Pet Geol 127:104872

Darros de Matos RM, Almeida MWE, CB et al (2021b) A solution to the Albian fit challenge between the South American and African plates based on key magmatic and sedimentary events late in the rifting phase in the Pernambuco and Paraíba basins. Mar Petrol Geol 128:105038

Leonardi G (1979) Nota Preliminar Sobre Seis Pistas de Dinossauros Ornithischia da Bacia do Rio do Peixe (Cretáceo Inferior) em Sousa, Paraíba, Brasil. An Acad Bras Ciênc 51(3):501–516

Leonardi G (1989) Inventory and statistics of the South American dinosaurian ichnofauna and its Paleobiological interpretation. In: Gillette DD, Lockley MG (eds) Dinosaur tracks and traces. Cambridge University Press, New York, pp 165–178

Leonardi G (1994) Annotated Atlas of South America Tetrapod footprints (Devonian to Holocene) with an appendix on Mexico and Central America. CPRM (Serviço Geológico do Brasil), Brasília, Brasil

Leonardi G, Muniz GCB (1985) Observações icnológicas (Invertebrados e Vertebrados) no Cretáceo Continental do Ceará (Brasil), com menção a moluscos dulçaquícolas. In: IX Congresso Brasileiro de Paleontologia, Resumo de Comunicações, Fortaleza, p 45

Leonardi G, Spezzamonte M (1994) New tracksites (Dinosauria: Theropoda and Ornithopoda) from the Lower Cretaceous of the Ceará, Brasil. Studi Trentini di Scienze Naturali, Acta Geologica 69:61–70

Leonardi G, Santos MFCF, Barbosa FHS (2021) First dinosaur tracks from the Açu Formation, Potiguar Basin (mid-Cretaceous of Brazil). An Acad Bras Ciênc 93:e20210635

Leonardi G, Carvalho IS (2021) Dinosaur Tracks of Rio do Peixe Basins, Brazil: a Lost World of Gondwana. Indiana University Press, Bloomington, U.S.A

Luft-Souza F, Fauth G, Bruno MDR, Mota MAL, Vásquez-García B, Santos-Filho MAB, Terra GJS (2021) Sergipe-Alagoas Basin, Northeast Brazil: a reference basin for studies on the early history of the South Atlantic Ocean. Earth Sci Rev 229:104034

Mabesoone JM (1994) Sedimentary basins of Northeast Brasil. Federal University Pernambuco, Geology Department, Special Publication, p 2

Matos RMD (1992) The Northeast Brazilian rift system. Tectonics 11:766–791

Pereira PVLGC, Marinho TD, Candeiro CR, Bergqvist LP (2018) A new titanosaurian (Sauropoda, Dinosauria) osteoderm from the Cretaceous of Brazil and its significance. Ameghiniana 55(6):644–650

Pereira PVLGC, Ribeiro TB, Brusatte SL, Candeiro CRA, Marinho TS, Bergqvist LP (2020a) Theropod (Dinosauria) diversity from the Açu Formation (mid-Cretaceous), Potiguar Basin, Northeast Brazil. Cretac Res 114:104517

Pereira PVLGC, Veiga IM, Ribeiro TB, Cardozo RH, Candeiro CRA, Bergqvist LP (2020b) The path of giants: a new occurrence of Rebbachisauridae (Dinosauria, Diplodocoidea) in the Açu Formation, NE Brazil, and its paleobiogeographic implications. J S Am Earth Sci 100:102515

Pessoa-Neto OC, Soares UM, Silva JGF (2007) Bacia Potiguar. Boletim de Geociências da Petrobras 15(2):357–369

Petri S (1983) Brazilian Cretaceous paleoclimates: evidence from clay-minerals, sedimentary structures and palynomorphs. Revista Brasileira de Geociências 13(4):215–222

Ponte FC (1992) (1992) Origem e evolução das pequenas bacias cretácicas do interior do Nordeste do Brasil. 2° Simpósio sobre as Bacias Cretácicas Brasileiras, Rio Claro. São Paulo, Universidade Estadual Paulista, Resumos expandidos, pp 55–58

Santos MFCF, Florêncio CP, Reyes-Perez YA, Bergqvist LP, Porpino KO, Uchoa AF, Lima-Filho FP (2005) Dinossauros na Bacia Potiguar: o registro da primeira ocorrência. In: Boletim de resumos expandidos do XXI Simpósio de Geologia do Nordeste "A Geologia e a Sociedade", Recife, pp 325–328

Scherer CM, Mello RG, Ferronatto JP, Amarante FB, Reis AD, Souza EG, Goldberg K (2020) Changes in prevailing surface-palaeowinds of western Gondwana during Early Cretaceous. Cretac Res 116:104598

Silva RC, Carvalho IS, Schwanke C (2007) Vertebrate dinoturbation from the Caturrita Formation (Late Triassic, Paraná Basin), Rio Grande do Sul State, Brazil. Gondwana Res 11:303–310

Srivastava NK (1990) Aspectos geológicos e sedimentológicos das bacias de Iguatu, Lima Campos e Malhada Vermelha (Ceará). In: Simpósio sobre a Bacia do Araripe e bacias interiores do Nordeste, Campos DA, Vianna MSS, Brito PM, Beurlen G. (eds), 1, Crato, pp 209–222

Valença LMM, Neumann VH, Mabesoone JM (2003) An overview on Callovian-Cenomanian intracratonic basins of Northeast Brazil: onshore stratigraphic record of the opening of the Southern Atlantic. Geol Acta 1(3):261–275. https://doi.org/10.1244/105.000001614

Vasconcelos EP, Lima Neto FF, Roos S (1990) Unidades de correlação da Formação Açu, Bacia Potiguar. In: XXXVI Congresso Brasileiro de Geologia, 36, Natal, pp 227–240

Veiga IMMG, Bergqvist LP, Brito PM (2019) The fish assemblage of the Cretaceous (?Albian-Cenomanian) Açu Formation, Potiguar Basin, Northeastern Brazil. J S Am Earth Sci 93:162–173

Chapter 9
Equatorial Dinosaurs During the Opening of Atlantic Ocean: The São Luís Basin Footprints

Ismar de Souza Carvalho(iD) and Rafael Matos Lindoso(iD)

9.1 Introduction

In the northeastern Brazil, the origin of South Atlantic Ocean led to the formation of vast subaerial environments suitable to generation of ichnological record. Such record includes mainly dinosaur trackways and isolated footprints commonly found also in the Sousa, Triunfo, Cedro, Malhada Vermelha, Lima Campos, Potiguar, Araripe basins, as well as in São Luís Basin, from Berriasian to Cenomanian (Leonardi 1980a, 1994; Leonardi and Spezzamonte 1994; Carvalho 2000, 2004; Carvalho et al. 2021; Leonardi and Carvalho 2021; Leonardi et al. 2021). Some of these ichnological records in Brazil shed light on most of our comprehension concerning the climate and paleoecological aspects during the break-up of Western Gondwana (Carvalho 2004; Carvalho et al. 2013, 2021; Leonardi and Carvalho 2021).

In the São Luís Basin (Fig. 9.1), northern Brazil, dinosaur isolated footprints and trackways have been attributed to small and large theropods, sauropods and ornithopods in six localities of São Luís and Alcântara counties (Carvalho and Gonçalves 1994; Carvalho 1994b, 1995, 2001; Carvalho and Araújo 1995; Carvalho and Pedrão 1998). The São Luís Basin ichnocenoses are considered to compose a megatracksite, and the most of these footprint-bearing strata are associated to Cenomanian tidal flat deposits in an estuarine environmental context (Rossetti 1997, 1998;

I. S. Carvalho (✉)
CCMN/IGEO, Departamento de Geologia, Universidade Federal do Rio de Janeiro, 21.910-200 Cidade Universitária, Ilha do Fundão, Rio de Janeiro, Estado do Rio de Janeiro, Brazil
e-mail: ismar@geologia.ufrj.br

Centro de Geociências, Universidade de Coimbra, Rua Sílvio Lima, 3030-790 Coimbra, Portugal

R. M. Lindoso
Departamento Acadêmico de Biologia. Av. Getúlio Vargas, Instituto Federal de Educação, Ciência e Tecnologia do Maranhão, Estado do Maranhão, 4 Monte Castelo, São Luís 65030-005, Brazil
e-mail: rafael.lindoso@ifma.edu.br

© The Author(s), under exclusive license to Springer Nature Switzerland AG 2024
I. S. Carvalho and G. Leonardi (eds.), *Dinosaur Tracks of Mesozoic Basins in Brazil*,
https://doi.org/10.1007/978-3-031-56355-3_9

Fig. 9.1 Location map of
São Luís Basin and its
tectonic relationship with the
Parnaíba Basin, Northeastern
Brazil (modified from
Pedreira da Silva et al. 2003)

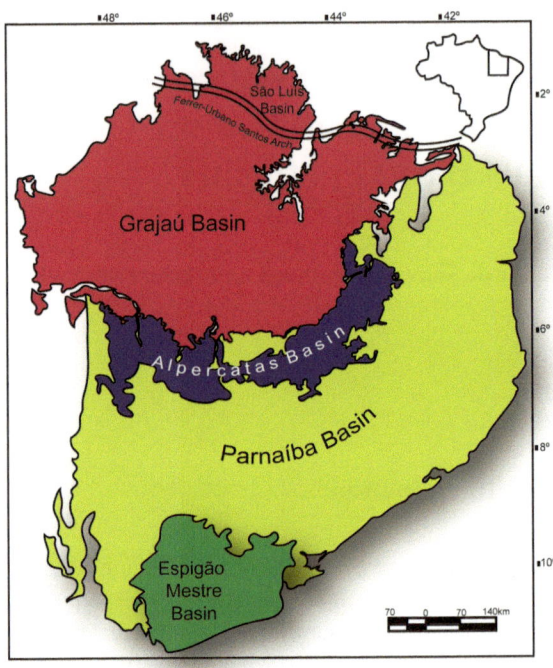

Carvalho and Pedrão 1998; Carvalho 2001; Medeiros and Schultz 2001; Medeiros et al. 2014).

The São Luís Megatracksite, whose ichnosites are distributed in the borders of São Marcos Bay (23,600 km^2), occurs in the Alcântara Formation (Rossetti and Truck-enbrodt 1997) and includes small and large theropods, sauropods and ornithischians (bipedal and quadrupedal) (Carvalho 2001). The potential trackmakers are physically represented by body fossils found in the region, in particular by isolated remains of *Spinosaurus* sp., *Carcharodontosaurus* sp., noasaurids and Unenlagiinae (Lindoso et al. 2012; Medeiros et al. 2014; Medeiros 2006; Letizio et al. 2022). Ornithischian remains are not yet represented in the São Luís Basin, however, those of sauropods include Titanosauridae and Rebbachisauridae (Medeiros et al. 2014; Medeiros and Schultz 2001, 2002; Lindoso et al. 2013, 2019). Other fossils found in the Alcân-tara Formation are palynomorphs, plants (angiosperms), invertebrate ichnofossils, mollusks (Mytilidae, Inoceramidae, Pectinidae, Plicatulidae, Limidae, Ostreidae, Trigonidae and Matricidae), fishes (Dipnoi, Elasmobranchii and Actinopterygii), reptiles (Crocodyliforms, Mosasauria, Chelonia and Pterosauria) (Oliveira 1958; Klein and Ferreira 1979; Carvalho and Pedrão 1998; Arai 2001; Castro et al. 2004; Elias et al. 2007; Lindoso et al. 2011; Medeiros et al. 2014; Moraes-Santos et al. 2001; Vilas Bôas and Carvalho 1999).

We present an update of the dinosaur ichnological record from the Alcântara Formation, São Luís Basin, northern Brazil, and we emphasize their paleoecological and paleoenvironmental significance on the Western Gondwana context.

9.2 Geological Context

The São Luís Basin (Maranhão State) is a rift marginal basin of 18,000 km^2, whose evolution is related to the origin of the Brazilian equatorial margin (Fig. 9.1). The initial rifting occurred during the Aptian, by the simple shear stress and lithospheric thinning. The depositional pre-Cretaceous history is related to the Parnaíba Basin (Carvalho and Pedrão 1998). The sedimentary thickness is 4,500 m (Aranha et al. 1990) and the Cretaceous outcrops where the dinosaur footprints occur are named as Alcântara Formation (Cenomanian). These rocks consist of reddish sandstones, siltstones, shales and mudstones, with some interbedded carbonates composed of marls and limestones. The main sedimentary structures are channel and planar cross-stratification, ripple-marks, liquefaction structures, mud-cracks, herring-bone cross-stratification and hummocky cross-stratification.

The footprints of Alcântara Formation are found in fine-grained quartzose sandstones distributed in the São Luís and Alcântara counties. This set of footprints, in the outcrops surrounding the São Marcos Bay, is temporally chronocorrelated and occur in the same paleoenvironmental setting. The footprints from these localities are in the context of the São Luís Megatracksite, a wide coastal plain where theropods, ornithischians and sauropods lived and were the main trackmakers.

The region where many dinosaur communities lived comprises an estuary in a low gradient coastal plain and nearshore environments submitted to tidal currents (Rossetti 1996a, b, c; Carvalho 2000). The environmental interpretation points out to estuarine, nearshore and shallow marine environments affected by both tide and storm processes (Klein and Ferreira 1979; Rossetti 1994, 1996a) under a hot and dry climate.

In the Cenomanian deposits that outcrops in the São Luís Basin there are two depositional intervals (Fig. 9.2). The lower succession consists of well sorted and fine-grained sandstones interpreted as a regressive interval—an upward transition from seaward to landward settings of upper shoreface, foreshore, tidal channel, and lagoon-washover environments. Such deposits revealed a prograding, barred coast probably formed on the seaward portion of a wave-dominated estuarine system (Rossetti 1996a). The footprint bearing-strata are found in the upper succession (Rossetti 1996b) that consists of tidal-dominated deposits attributed to channel, sand flat, delta, and bay fill depositional settings of an estuary. Rossetti (1996b) considered that the lower and upper successions are part of two incised valley fills. The lower succession was deposited at a time of slow rise in relative sea-level, meanwhile, the upper succession records the transition from the transgressive to the highstand systems tract of a younger incised valley.

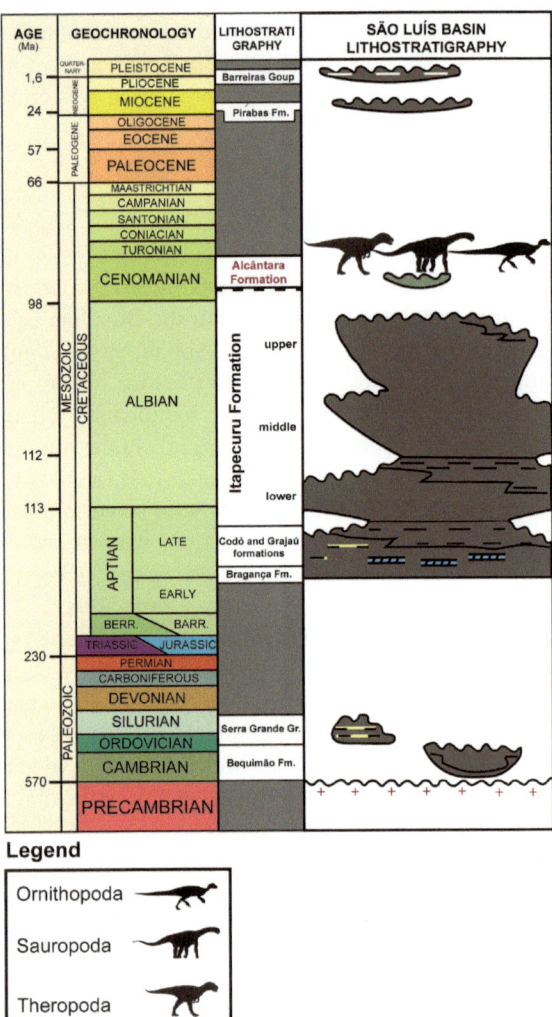

Fig. 9.2 Stratigraphic chart of São Luís Basin and the occurrence of dinosaur tracks (modified from Carvalho and Pedrão 1998)

9.3 Footprints: Diversity and Paleobiological Interpretation

Alongside the cliffs surrounding the São Luís Bay there are six ichnosites (Fig. 9.3) named as Ponta da Guia, Ponta do Farol, Praia do Boqueirão, Ilha do Medo, Praia da Baronesa and Praia Prefeitura de Alcântara (Carvalho and Gonçalves 1994; Carvalho 1995, 2001; Carvalho and Araújo 1995).

The localities of Ponta da Guia and Praia da Baronesa show the best-preserved footprints. In Ponta da Guia, the most striking footprints are the large-sized ones, distributed in seven short trackways, four of them parallel. Other footprints are found

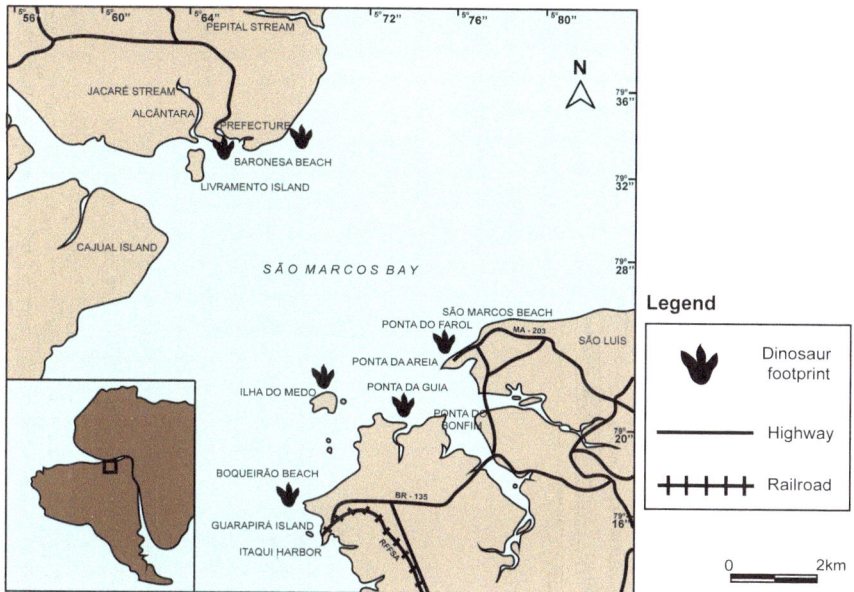

Fig. 9.3 The São Luís Megatracksite, Alcântara Formation, constituted of six ichnosites named as Ponta da Guia, Ponta do Farol, Praia do Boqueirão, Ilha do Medo, Praia da Baronesa and Praia Prefeitura de Alcântara (Baronesa Beach and Prefecture), and its paleogeographical position during the Cenomanian (lower left corner) (modified from Carvalho 2001)

in Praia da Baronesa, generally as isolated imprints, although there is a short trackway with three consecutive footprints.

9.3.1 Ponta da Guia Ichnosite

Seven short trackways (SLPG-A, SLPG-B, SLPG-C, SLPG-D, SLPG-E, SLPG-F and SLPG-G) four of them parallel, with tridactyl and mesaxonic footprints (Fig. 9.4). They are preserved as concave epirelief with 40–43 cm in width and 43–50 cm in length. The digits present the same size, and they are pointed or show claws. The trackways and footprints found in this ichnosite are described below; their codification SL means São Luís Basin and PG is the locality of Ponta da Guia.

SLPG-A is a trackway constituted of four consecutive tridactyl footprints (SLPG-A02, SLPG-A03, SLPG-A04 and SLPG-A05). The step angle is obtuse (165°) and the oblique pace presents an average value of 245 cm. The footprints are preserved as concave epirelief, with pointed digits (SLPG-A02 and SLPG-A05) or slightly rounded (SLPG-A03 and SLPG-A04), all almost the same size. The rear borders of the footprints are rounded and the preservation differences are related to the erosive surface where the footprints are found.

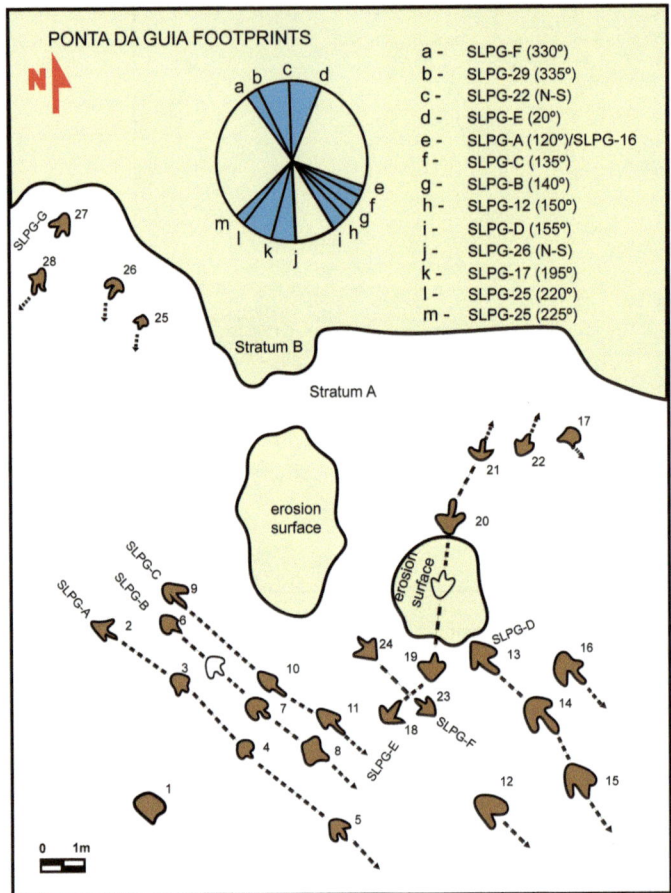

Fig. 9.4 Distribution map of the Ponta da Guia ichnosite with the theropod tracks (Carvalho 1994b)

SLPG-B trackway presents four consecutive footprints (SLPG-B06, SLPG-B s/n°, SLPG-B07 and SLPG-B08). The step angle is obtuse (170°) and the oblique pace presents an average value of 175 cm. The footprints are tridactyl, mesaxonic and show a rounded rear border (especially SLPG-B07 and SLPG-B08). They are preserved as concave epirelief in a recently eroded surface (Fig. 9.5).

The trackway SLPG-C (Fig. 9.5a) is constituted of three tridactyl consecutive footprints (SLPG-C09, SLPG-C10 and SLPG-C11). The step angle is obtuse (170°) and the oblique pace presents an average value of 257 cm, the longest one among the seven trackways of Ponta da Guia ichnosite. The digits are thin in their extremities and in the footprint SLPG-C09 there is a clear claw. The interdigital angles between digits II–III and III–IV are acute and the rear borders of the footprints are pointed.

Three consecutive footprints (SLPG-D13, SLPG-D14 and SLPG-D15) constitute the trackway SLPG-D (Fig. 9.5b). The step angle is 165° and the oblique pace

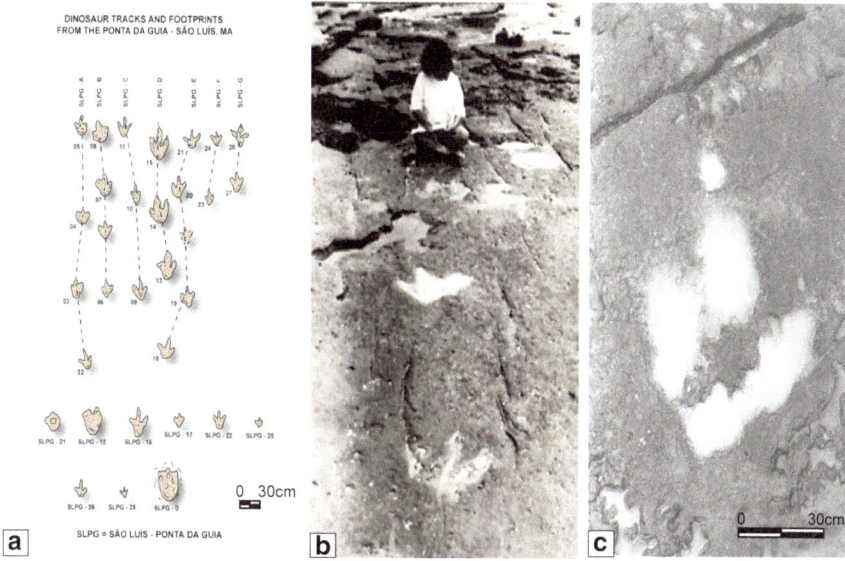

Fig. 9.5 Ponta da Guia ichnosite. **a** Trackways and isolated footprints from Alcântara Formation, Ponta da Guia; **b** The short theropod trackway (SLPG-D) with three footprints; **c** Detail of the large theropod footprint SLPG-D 15 (modified from Carvalho 1994b)

presents an average value of 238 cm. The footprints are tridactyl, mesaxonic with acute hypexes. There is a claw imprint in SLPG-D15 and the others (SLPG-D13 and SLPG-D14) present pointed digits. All they are preserved as concave epirelief (Fig. 9.5c).

The trackway SLPG-E is constituted of four non consecutive footprints (SLPG-E18, SLPG-E19, SLPG-E20 and SLPG-E21). Between the footprints SLPG-E19 and SLPG-E20 there is an erosion gap that interrupts the continuity of the bedding plane where the footprints are preserved. All of them are tridactyl, mesaxonic and with pointed digits. In the rear border of SLPG-E19 there is a prominent projection that suggests the presence of digit I. The step angle is 153° and the oblique pace presents an average value of 195 cm.

Two consecutive tridactyl and mesaxonic footprints (SLPG-F23 and SLPG-F24) are a portion of an incomplete trackway (SLPG-F). They are preserved as concave epirelief. The digits are pointed and the hypex between digits II–III and III–IV are acute. The oblique pace measures 190 cm.

The SLPG-G trackway is also constituted of two consecutive footprints (SLPG-G27 and SLPG-G28), tridactyl and mesaxonic. In SLPG-G28 there are claw imprints in all the three digits, with a very acute rear border. There are not claw imprints in the SLPG-G27, despite digit III being tapered. In both footprints the rear border is acute, indicating the probable presence of a digit I. The oblique pace measures 190 cm.

There are nine isolated footprints (SLPG-01, SLPG-12, SLPG-16, SLPG-17, SLPG-22, SLPG-25, SLPG-26, SLPG-29 and SLPG-0) in the same bedding surface of the trackways preserved as concave epireliefs (Fig. 9.5). In the SLPG-01 footprint does not preserve the digits. It has a rounded morphology and despite it being preserved as convex epirelief, the central area of the footprint is concave. SLPG-12 shows two digits with rounded extremities, without any evidence of claws. The hypex between these two digits is acute and the rear border of the footprint is rounded. SLPG-16 footprint is tridactyl, mesaxonic, with pointed digits, without claw impressions and presents a rounded rear border. Digit III is the larger and the hypexes between digits II–III and III–IV are rounded. The SLPG-17 footprint is tridactyl, mesaxonic and the extremities of the digits and hypexes are rounded. The rear border of the footprint is very acute suggesting the imprint of a digit I. The footprints SLPG-22 and SLPG-25 are tridactyl, mesaxonic and present digits of the same length with rounded anterior extremities and the rear border of the footprints are rounded. In SLPG-22 the hypexes between digits II–III and III–IV are rounded, while in SLPG-25 the hypex between digits II–III is rounded, while between digits III–IV is acute. The tridactyl and mesaxonic SLPG-29 footprint, despite showing digits with the same length, the anterior extremities are pointed, and the rear border of the footprint is acute. The hypexes between digits II–III and III–IV are acute. The footprint SLPG-26 is tridactyl, mesaxonic and the digit III is the largest. While digit III shows a pointed anterior extremity, in digits II and IV they are rounded. The hypex between digits II–III is rounded and between digits III–IV is acute. The rear border of the footprint is acute and slightly curved. The SLPG-0 is the biggest footprint in this ichnosite, with a length of 70 cm, penetrating 25 cm in the substrate. It is tridactyl, mesaxonic with short digits and rounded rear border. The hypexes between digits II–III and III–IV are wide and rounded. Digits II and IV are pointed and digit III presents a rounded anterior extremity. In the surrounding area of the footprint there is a deformation zone, probably corresponding to the displacement rim.

The succession of the strata bearing the footprints is interpreted as a tidal plain, cut by freshwater channels and bordered by aeolian dunes, under a hot and dry climate (Carvalho 1995, 2004).

A gregarious behavior was interpreted to the Ponta da Guia ichnosite based on the parallel trackways SLPG-A, SLPG-B, SLPG-C, SLPG-D and two isolated footprints (SLPG-12 and SLPG-16) that point to the same southeastern direction. Other groups of footprints are grouped in a southwestern direction (SLPG-G, SLPG-25 and SLPG-26). These preferential directions probably are a sign of herding structure as observed by other authors (Currie 1983; Leonardi 1980b, 1989; Lockley 1986, 1991; Lockley et al. 1986, 1992; Thulborn 1990).

9.3.2   Ponta do Farol Ichnosite

It was identified just one isolated, tridactyl, digitigrade and mesaxonic footprint (SLPF-01) with pointed digits suggestive of claws. It is preserved as concave epirelief. It is 35 cm in width and 40 cm in length, with acute hypexes (38°) between digits II–III and III–IV (Fig. 9.6a). Nearby there are deformation structures associated with liquefaction that are similar to enlarged tridactyl footprints. Nevertheless, short sauropod tracks recently found at this locality (Fig. 9.7) enhance Ponta do Farol ichnosite for a promising paleobiological survey.

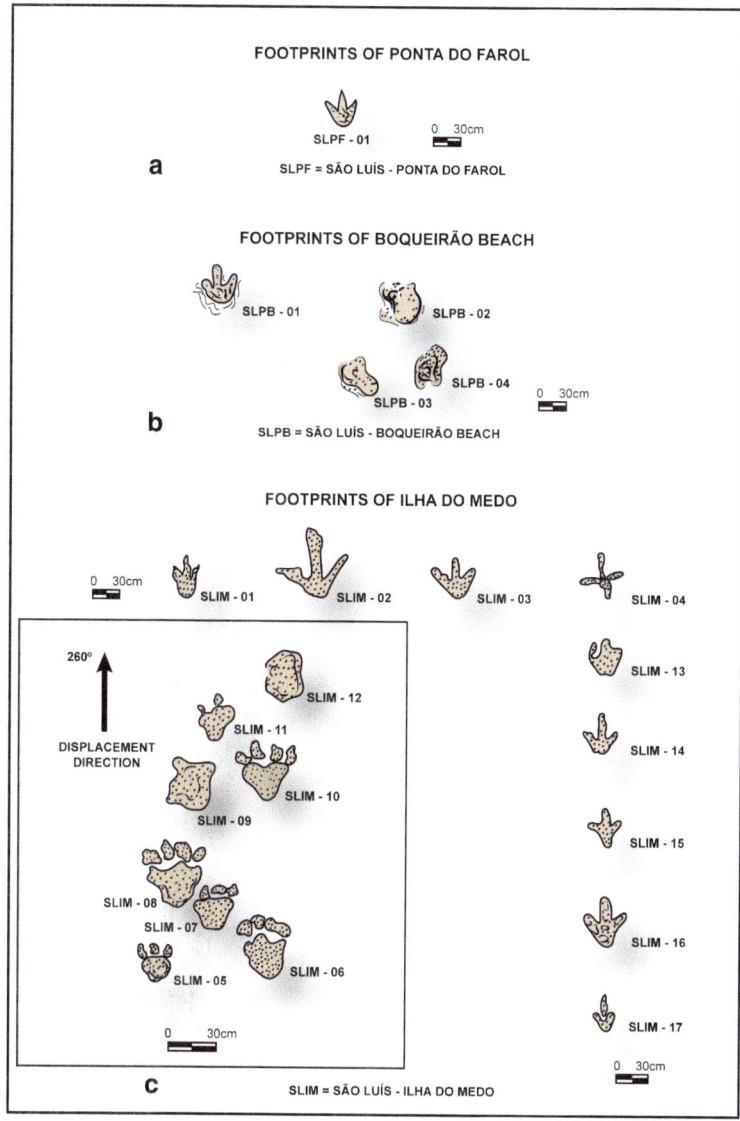

Fig. 9.6 Trackways and isolated footprints from Alcântara Formation. **a** Isolated footprint from Ponta do Farol ichnosite; **b** Theropod isolated footprint and short trackway with liquefaction footprints from Praia do Boqueirão ichnosite; **c** A short track and isolated footprints from Ilha do Medo ichnosite (modified from Carvalho 2001)

Fig. 9.7 Short sauropod tracks at Ponta do Farol ichnosite

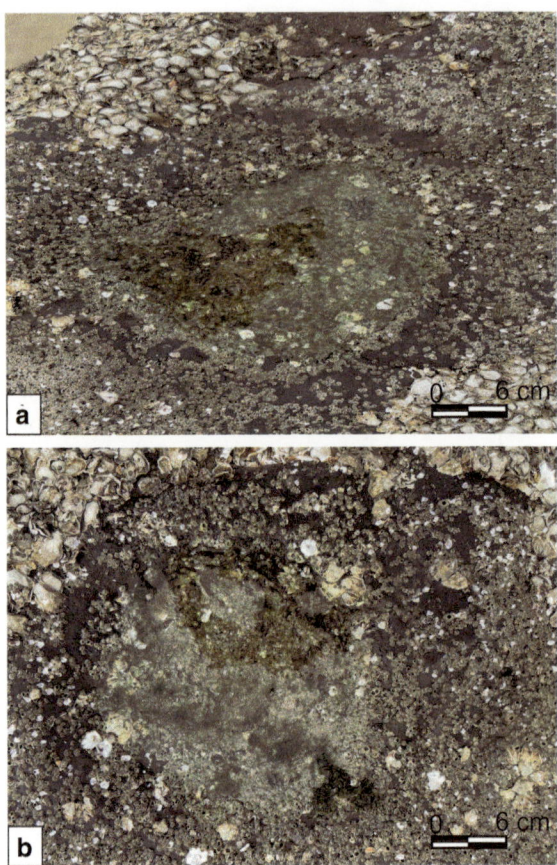

9.3.3 *Praia do Boqueirão Ichnosite*

There are four isolated footprints in this ichnosite. They are preserved as concave epirelief in a fine-grained sandstone. Liquefaction structures may occur surrounding some footprints (SLPB-01), while the matrix on the posterior borders and digits show crenulations. SLPB-01 is 45 cm in width and 48 cm in length with rounded digits. In the other footprints (SLPB-02, SLPB-03, SLPB-04) it is not possible to identify the digits, as they are circular deformations with concentric rings ranging 30–40 cm in width and 40–45 in length. These features in waterlogged substrate are interpreted as the substrate deformation induced by the load of the trackmakers (Fig. 9.6b).

9.3.4 Ilha do Medo Ichnosite

This ichnosite shows sixteen isolated footprints and one short trackway (Fig. 9.6c). The isolated footprints are generally tridactyl, mesaxonic and preserved as concave epirelief. Their sizes range from 30 to 70 cm in width and 30 to 75 cm in length. There are no evident claws despite the digits of some footprints being pointed (SLIM-01, SLIM-02 SLIM-14, SLIM-15, SLIM-17) and with acute hypexes (interdigital angles of 35–45°), suggesting theropod trackmakers. The footprints SLIM-03, SLIM-13, SLIM-16 show more rounded digits, but present V-shaped rear borders, a dubious aspect to interpret as ornithopod trackmakers. They are 30–45 cm in width and 30–48 in length. The footprint SLIM-04 (55 cm in width and 55 cm in length) shows a cruciform pattern with two right-angle hypexes (90°), side by side. The opposite smaller digit is a possible digit I.

The short trackway SLIM is composed of eight footprints (SLIM-05, SLIM-06, SLIM-07, SLIM-08, SLIM-09, SLIM-10, SLIM-11, SLIM-12), with two distinct sizes and pattern. It is possible to observe three elongated digits, isolated from the plantar portion, in some of the smaller footprints (15–20 cm in width and 20–30 cm in length). They are alternated with tetradactyl footprints (SLIM-06, SLIM-08, SLIM-10) with rounded and elongated digits, in which the rear footprint borders are more acute (Fig. 9.6c). These show width and length (30–35 cm and 33–40 cm, respectively) bigger than SLIM-05, SLIM-07, SLIM-11, SLIM-12. The oblique pace between SLIM-06 and SLIM-08 (Fig. 9.8) is 45 cm, and between SLIM-05 and SLIM-07 is 30 cm. An ornithischian could be the trackmaker of this short trackway.

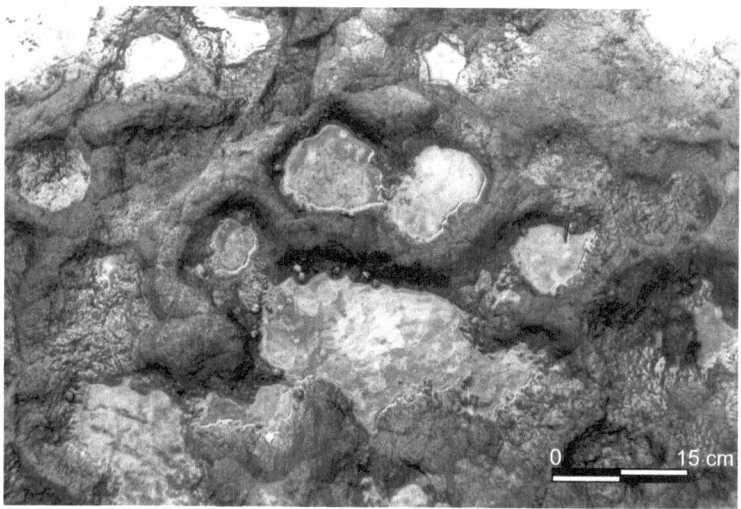

Fig. 9.8 A dubious tetradactyl footprint (SLIM-08) with rounded digits; it is interpreted as produced by an ornithischian trackmaker

9.3.5   Praia da Baronesa Ichnosite

The footprints from Praia da Baronesa ichnosite (Figs. 9.9 and 9.10) are tridactyl, mesaxonic, digitigrad, with evidence of claws and digital pads. They are smaller than the Ponta da Guia footprints, with 22–33 cm in length and the average value of 18 cm in width. There are twenty-two footprints preserved as convex epirelief in fine-grained sandstone, many of them showing a more reddish color (distinct from the greenish color of the surrounding matrix) and fluidization structures, surrounding the digits and rear borders (Carvalho and Gonçalves 1994). There are six isolated footprints (ALBA-11, ALBA-12, ALBA-13, ALBA-14, ALBA-15, ALBA-18) with pointed digits, some with claw impressions (ALBA-13, ALBA-14, ALBA-15, ALBA-18). The hypexes are acute and the interdigital angles between digits II–III and III–IV range from 30 to 45°. ALBA-15 is the smallest footprint of this set with 15 cm in width and 20 cm in length. The others (ALBA-13, ALBA-14 and ALBA-18) present a range of 30–45 cm in width and 30–48 cm in length.

There are also short tracks with two (ALBA-01 and ALBA-02, ALBA-06A and ALBA-06B, ALBA-07 and ALBA-08, ALBA-09 and ALBA-10) three (ALBA-03, ALBA-04, ALBA-05) and five (ALBA-16, ALBA-17, ALBA-18, ALBA-19, ALBA-20) consecutive footprints. These present a wide range of sizes (ALBA-09 and ALBA-10—12 cm in width, 15 cm in length; ALBA-01 and ALBA-02—50 cm in width and 60 cm in length). Some footprints of distinct tracks (ALBA-03, ALBA-04 and ALBA-05; ALBA-06A and ALBA-06B; ALBA-07 and ALBA-08; ALBA-16) show crenulations of the matrix surrounding the footprints, that is an evidence of liquefaction. The trackway with the longest paces (ALBA-16, ALBA-17, ALBA-18, ALBA-19, and ALBA-20) shows an average oblique pace of one meter, and footprints with 15 cm in width and 25 cm in length. The footprints present pointed digits and acute hypexes with interdigital angles between digits II–III and III–IV ranging from 35 to 38°. ALBA-16 presents a liquefaction feature surrounding the footprint, with the crenulations of the matrix (Figs. 9.10a, 9.11).

A common feature in the Praia da Baronesa footprints is the contrasting colors from the surrounding substrate. There is a range from reddish to blue-gray colors, contrasting with the light greenish hue of the substrate (Fig. 9.11). Kuban (1991a) also observed this feature in dinosaur tracks of the Glen Rose Formation (Lower Cretaceous, Texas—USA) and explained this taphonomic aspect as the result of secondary sediment infilling on the original track depressions and oxidation of iron on the surface of infilling material. It is likely that the pressure from the foot on the substrate had the consequence of expelling water and perhaps salts or oxides in solution or suspension, thus locally changing the characteristics—and later the color—of the compressed soil. The deformation of the substrate with the liquefaction structures is the result of a "dinostatic pressure" in water-saturated and low cohesive sediments. Such substrate aspect is corroborated by the metatarsal impressions in many footprints (Carvalho 1994c). Kuban (1991b) considered that this preservation character could be indicative of a behavior response to a soft substrate, a low posture

Fig. 9.9 **a** The outcrop of Praia da Baronesa ichnosite. **b** This is the locality where dinosaur remains and footprints are found in the same bedding plane; theropod tooth with probable dromeosaurid affinities found in the same surface of the dinosaur footprints

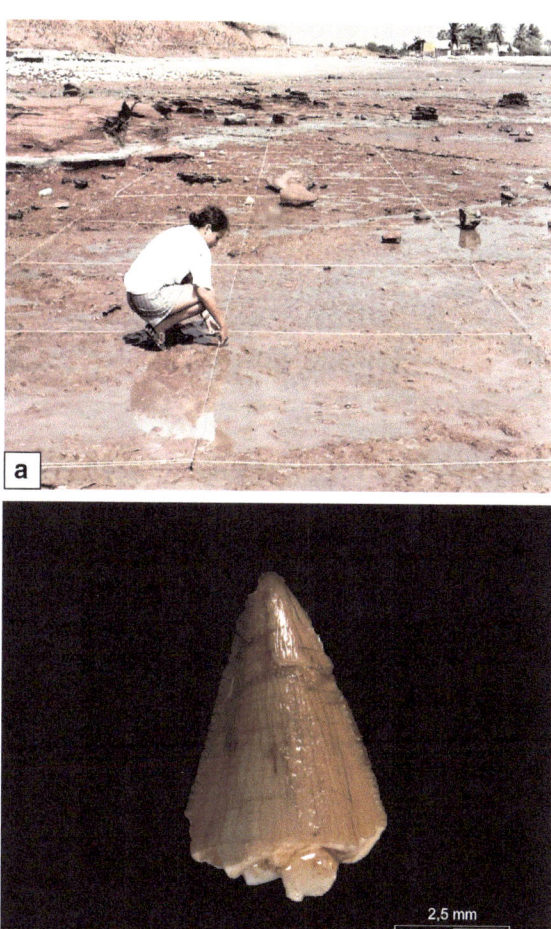

assumed whenever a dinosaur foraged in mud flats or shallow water for small food item, stalking large prey, or while approaching other dinosaurs.

9.3.6   *Praia Prefeitura de Alcântara Ichnosite*

In this ichnosite (Fig. 9.10b) there are two tridactyl footprints (ALPR-01, ALPR-02) and one with two digits (ALPR-03) preserved as concave epirelief. ALPR-01 shows more pointed digits than ALPR-02 and ALPR-03. They range from 20 to 30 cm in

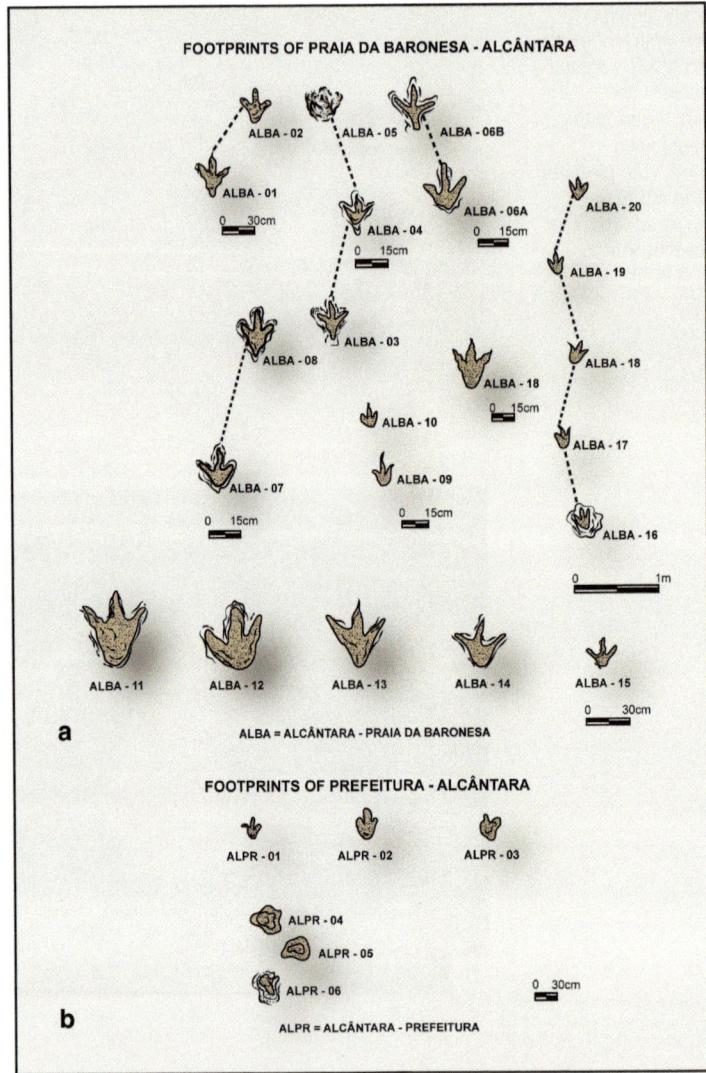

Fig. 9.10 Trackways and isolated footprints from **a** Praia da Baronesa and **b** Praia Prefeitura de Alcântara ichnosites (modified from Carvalho 2001)

width and 30–45 cm in length. Acute hypexes, with the angles between digits II–III ranging from 30 to 45° and digits III–IV 35 to 40°.

There are also four isolated rounded structures (Fig. 9.12), with concentric rings, that range from 30 to 40 cm in width and 40–45 in length. It is also possible to identify three consecutive similar structures (ALPR-04, ALPR-05, ALPR-06), with an average oblique pace of 60 cm, that are probably part of a short trackway. These

Fig. 9.11 Footprints from Praia da Baronesa ichnosite. The deformation of the substrate with the liquefaction structures is the result of a "dinostatic pressure" in water-saturated and low cohesive sediments. **a** A theropod footprint (ALBA 06B) surrounded by a liquefaction structure; **b** In some footprints, the crenulations also occur inside the imprint (ALBA 04); **c** A reddish color footprint surrounded by a greenish matrix with liquefaction structures

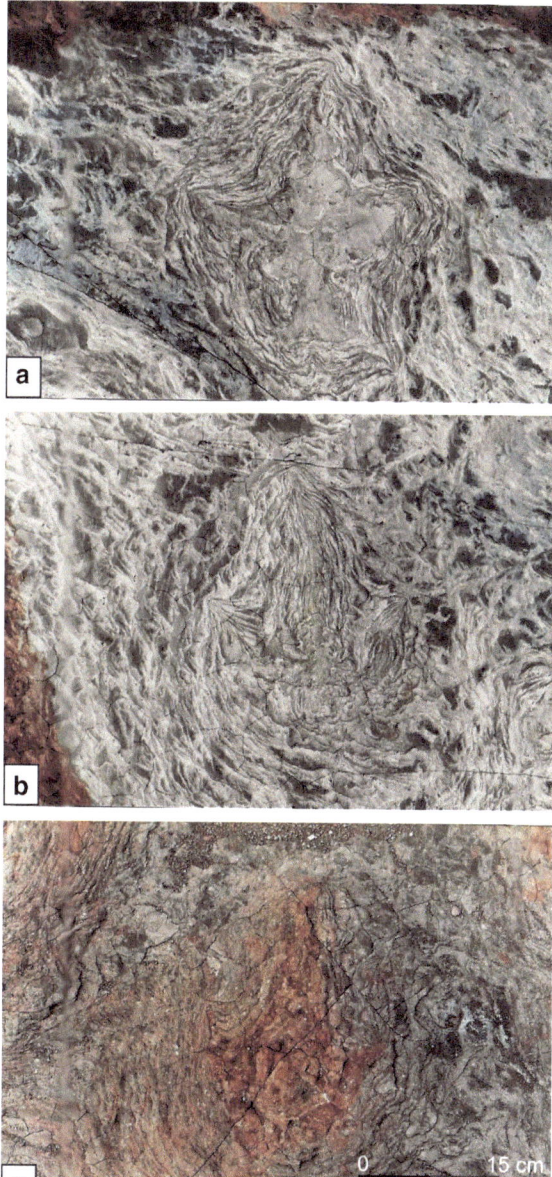

Fig. 9.12 Footprint from
Praia Prefeitura de Alcântara
ichnosite. Isolated rounded
structure, with concentric
rings, interpreted as
liquefaction structures
induced by a sauropod
trackmaker

rounded structures are interpreted as liquefaction structures induced by sauropod trackmakers.

9.3.7 Paleobiological Interpretation

The footprints found at Ponta da Guia ichnosite were produced by two different dinosaur clades: Theropoda and Ornithopoda. The ones attributed to theropods are large-sized footprints. Four trackways present parallel orientation, a possible evidence of a gregarious behavior. The ornithopod footprints were assigned to hadrosaurians (Carvalho and Pedrão 1998). It would then be rather primitive hadrosaurs. Probably , during the Cenomanian, there was the dispersion of dinosaur faunas between the two American continents.

In Ponta do Farol ichnosite, the single isolated footprint is considered to be produced by a large-sized theropod trackmaker. At Praia do Boqueirão ichnosite the footprints are load deformations in the substrate and only one footprint denotes digits and the plantar impression, produced by a probable ornithopod trackmaker. The Ilha do Medo ichnosite presents isolated footprints of large-sized theropods, ornithopods and a probable short track of a quadrupedal ornithischian. A small-sized theropod and sauropod footprints are possible to be identified in the Praia Prefeitura de Alcântara.

The trackways at Ponta da Guia ichnosite show a set of four large-sized theropod trackways (SLPG-A, SLPG-B, SLPG-C and SLPG-D), that present high morphological similarity, moving in SE direction. The angular range of movement directions is only 20° (between 120 and 140°), and there is a quite standard intertrackway spacing between three of these trackways (SLPG-A, SLPG-B, SLPG-C). The regular space between adjacent trackways suggests animals walking in some kind of regular formation (Lockley 1991), that constitute good evidence to postulate the gregariousness among the producers (Fig. 9.13). Because there are tracks in different directions, it is possible to consider the absence of physical barriers in the configuration of the landscape that could have controlled this main movement direction of individuals. Despite gregarious behavior occurs more frequently among sauropods and ornithopods (Carvalho 1989; Leonardi 1980b, 1981; Lockley et al. 1986, 1992; Nadon 1993; Myers and Fiorillo 2009; Castanera et al. 2011, 2014; Lockley et al. 2012; Piñuela et al. 2016; Paik et al. 2020), there are many examples of parallel theropod trackways as an evidence of gregariousness among theropod dinosaurs (Leonardi 1984, 1989; Moreno et al. 2012; García-Ortiz and Pérez-Lorente 2014; McCrea et al. 2014; Lockley et al. 2015; Heredia et al. 2020; Hernández et al. 2023).

The other main theropod footprint assemblage is located at Praia da Baronesa ichnosite. It is composed of randomly oriented trackways and isolated small and medium-sized footprints. This is interpreted as the record of a "foraging area" for theropods, searching for food in small lagoons and channels of a tidal flat environment. During low-tide periods, subaerial exposure of the sediments allowed the dinoturbation (Carvalho and Pedrão 1998). The footprints are always associated with fluidization structures and present superficial color stains (blue-gray, green or red) which can be interpreted as produced in low cohesive sediments and a soft substrate (Carvalho and Leonardi 2021; Kuban 1991b). Such substrate aspect is corroborated by the metatarsal impressions (Carvalho 1994c; Carvalho and Pedrão 1998) in many footprints (e.g., ALBA-03, ALBA-04, ALBA-06B, ALBA-08). The elongate plantigrade footprints would be explained by a low posture assumed whenever a dinosaur foraged in mud flats or shallow water for small food items, stalking larger prey or while approaching other dinosaurs.

The fossils from the Alcântara Formation indicate a diverse vertebrate community that lived in the coastal forested areas surrounded by a dominantly dry environment. They indicate that during the mid-Cretaceous the northeastern South American and northern African continental fauna were more similar than to the austral South American one. This aspect allows us to interpret that faunal interchanges between these two continents may have persisted until the early Cenomanian through continental bridges (Calvo and Salgado 1996; Popoff 1988; Maisey 2000; Medeiros et al. 2014).

The main dinosaur fauna from the Alcântara Formation, as regards the body-fossils, comprises *Carcharodontosaurus* sp., Spinosauridae, noasaurids with *Masiakasaurus*-like teeth, Diplodocoidea and Titanosauridae (Medeiros and Schultz 2001, 2002, 2004; Medeiros et al. 2007, 2014; Lindoso et al. 2012, 2013) that can be related to the Cenomanian record of the Kem Kem beds, Morocco and to the Bahariya

Fig. 9.13 The regular space between adjacent trackways suggests animals walking in some kind of regular formation, that is a good evidence to postulate the gregariousness among some of the large theropod trackmakers from Ponta da Guia ichnosite. Art by Guilherme Gehr

Formation, Egypt (Stromer 1915; Lapparent 1960; Benton et al. 2000; Richter et al. 2013; Medeiros et al. 2014; Ibrahim et al. 2020). The Theropoda were mainly identified through isolated teeth (Fig. 9.14) allowing the classification as *Charcarodontosaurus* sp., Spinosauridae, Dromeosauridae, Unenlagiinae and Noasauridae (Vilas Bôas 1999; Vilas Bôas and Medeiros 1997; Vilas Bôas et al. 1999; Medeiros 2006; Lindoso et al. 2012; Letizio et al. 2022). There is also a Spinosauridae named as *Oxalaia quilombensis* (Kellner et al. 2011).

The trackmakers from the Alcântara Formation can be related to the body-fossilsfound in this lithostratigraphic unit. Fossils of theropods are more diverse than sauropods, marked by the occurrence of Spinosauridae and Carcharodontosauridae (*Carcharodontosaurus* sp.) and the probable existence of more than one species of the former (Medeiros 2006; Richter et al. 2013; Medeiros et al. 2014). These two groups are good options as trackmakers to the large-sized footprints of the Ponta da Guia tracksite. Otherwise, small to medium-sized footprints from Praia da Baronesa and

Fig. 9.14 Dinosaur remains from the bone bed Laje do Coringa, Cajual Island, Alcântara Formation, São Luís Basin. **a** Sauropoda tooth; **b** *Carcharodontosaurus* sp. tooth; **c** Spinosauridae tooth

Praia Prefeitura de Alcântara ichnosites, could be related to dromeosaurids, unenlagiinids or noasaurs. In both cases they are present as body-fossils in the Alcântara Formation, sometimes in the same stratigraphic surface as in the case of Praia da Baronesa ichnosite.

There is only one specimen of Rebbachisauridae sauropod formerly described to the Alcântara Formation, *Itapeuasaurus cajapioensis* (Lindoso et al. 2019). It is the northernmost record of Diplodocoidea in South America, and it seems that rebbachisaurids outnumbered titanosaurs in the early Late Cretaceous in northern South America, right before the steep decline of Rebbachisauridae (Barrett and Upchurch 2005; Lindoso et al. 2019). Nevertheless, there are also bone fragments and osteoderm assigned as Titanosauria (Lindoso et al. 2013; Medeiros et al. 2014); then it is very probable that the footprints interpreted as sauropods in the Praia Prefeitura de Alcântara ichnosite could be related to these two groups.

9.4 Paleogeographical Distribution of the Footprints

The ichnofossiliferous localities of São Luís Basin are in a paleoenvironmental and temporal context distinct from the other occurrences of Cretaceous footprints in Brazil. They record dinosaur ichnofaunas of a megatracksite (Lockley 1991) as they occur in stratigraphic correlated surfaces, with wide geographic distribution. The São Luís Megatracksite presents footprints preserved in a coastal environment, on a low gradient tidal flat, under hot and dry climate conditions (Pedrão et al. 1993; Carvalho 1994a, b, c; Carvalho and Pedrão 1998).

The low gradient coastal plain allowed the establishment of specific dinosaur groups, with a probable ecologic "segregation" of large-sized (Spinosauridae and Carcharodontosauridae) and medium to small-sized theropods (Dromeosauridae,

Unenlagiinae and Noasauridae), throughout the exposed Cenomanian surfaces of the nearshore environments (Carvalho and Pedrão 1998).

The distribution of footprints along the outcrops of São Luís Bay is temporally related. The levels with teeth, bones and footprints of dinosaurs, besides other vertebrates, can be used to correlate the strata, as they are frequently identified in the outcrops of the basin. The outcrops with footprints record a same temporal event.

In coastal environments, footprints can be preserved in the intertidal zone when the tide begins its ebb. Thus, those originated early, in the ebb phase, will remain for a cycle of approximately 12 h, while those carried out in the tidal filling phase will not remain longer than 6 h until they are covered. A sandy beach with a steep slope is not the most appropriate environment for the preservation of tracks, as the rapid "washing" of the surface at each tidal cycle would be an agent of destruction of the footprints left by dinosaurs. On the other hand, in low gradient coastal plains, the ebb and flow of each tidal cycle occurs more slowly, favoring the preservation of trackways and footprints (Lockley 1991).

Fragmented bones and teeth of fishes, crocodyliforms, chelonians and dinosaurs at Praia da Baronesa ichnosite allow us to consider them as the accumulation of debris in a tidal flat environment. The random footprints distribution in this ichnosite could indicate a "foraging area" for the theropods, where they wandered across searching for food (Carvalho 2001). The theropods would search for fishes, turtles and other organisms, foraging food in the exposed channels of the tidal plain. During low-tide periods, sub aerial exposure of the sediments allowed the preservation of their tracks (Carvalho and Pedrão 1998).

There is also a brief description of theropod tracks in the neighboring Parnaíba Basin, in the Triassic deposits of Sambaíba Formation (Fig. 9.15). However, the age of this unit is dubious and the morphology and large diversity of theropod tracks illustrated by Assis et al. (2010) is inconsistent with a Triassic age, being more probable a Cretaceous age. It is necessary a review of these outcrops and their tracks.

9.5   Paleoenvironmental and Paleoclimatic Contexts

The bedding surfaces with footprints are found in the context of the "upper succession" of the Alcântara Formation (Rossetti 1994, 1996a, b). This succession consists of tidal-dominated deposits assigned to channel, sand plain, delta, and bay-fill in an estuarine environment. Along the coast of a shallow sea an abundant dinosaur fauna roamed, recording their footprints on the seashore environments.

The strata at Ponta da Guia are interpreted as the result of tidal flat and aeolian sedimentation. The track-bearing strata are fine-grained sandstones, interbedded with argillaceous siltstones, that show small-sized channel and tabular cross-stratification, ripple-marks, mud-cracks and clay-balls, accumulated in a sand flat depositional environment. The theropod footprints were probably produced in the supra-tidal region of a low-gradient tidal flat, where the preservation potential is greater (Carvalho 1995; Carvalho and Gonçalves 1994).

Fig. 9.15 Trackways and isolated theropod footprints from Fortaleza dos Nogueiras, south of Maranhão State, Parnaíba Basin

The fossil tracks and skeletal remains found in the fine-grained sandstones and siltstones of Praia da Baronesa ichnosite are related to a tidal channel setting (Fig. 9.16). During low-tide periods, subaerial exposure of the bedforms allowed them to be subject to dinoturbation. Ancient estuarine paleochannels also have high potential for burial and preservation of vertebrate skeletal remains (Eberth and Brinkman 1997). In the same track bearing surface of fine-grained sandstones occur many skeletal remains of a diversified fauna (Eugênio 1994) including fishes (Myliobatidae, Semionontidae, Enchodontidae and Sparidae families) and reptiles (Pelomedusidae testudines, Mosasauridae and Theropoda). The theropod remains (isolated teeth) are indicative of a clade with Dromeosauridae affinities (Vilas-Bôas and Medeiros 1997).

Through palynological analyses (Pedrão 1995) a better understanding was obtained of some environmental aspects under which the dinosaur fauna of São Luís Basin lived. The palynological content of outcrops in the east portion of the basin includes miospores, protozoans and polychaetas (Pedrão et al. 1993, 1995).

Fig. 9.16 Praia da Baronesa ichnosite shows small- and medium-sized theropod tracks living in a low-gradient coastal plain. The randomly-oriented trackways and isolated footprints are interpreted as the record of a "foraging area" for theropods. Art by Guilherme Gehr

Continental palynomorphs are pteridophyte spores (*Petrotriletes*), gymnosperm pollen grains (*Classopollis major*, *Equisetosporites ambiguus*, *Equisetosporites* spp., *Steevesipollenites* spp.), angiosperms (*Afropollis jardinus*, *Hexaporotricolpites emilianovi*, *Cretacaeiporites polygonalis*, *Cretacaeiporites mulleri*) and some taxa of doubtful botanical affinity (*Elateroplicites africaensis*, *Galeacornea causea* B, *Elaterocolpites castalaini*, *Elaterosporites* aff. *Klaszi* and *Sofrepites legouxae*). The main pollen group in this assemblage derives from gymnosperms (*Classopollis*) akin to plants of the Cheirolepideaceae family. Secondarily, there occur *Equisetosporites* and *Steevesipollenites* pollen grains. All of them are indicative of a hot and dry climate. The angiosperm pollen grains are rare, although the presence of *A. jardinus* indicates an equatorial climate. The occurrence of perisporate trilete spores in this assemblage also points to the same climate, and in addition, it is indicative of a fluvial influence in the depositional area (Carvalho and Pedrão 1998).

In this palynological assemblage is found marine plankton, including unicellular algae (Division Pyrrophyta, dinoflagellates) such as gonyaulacoids (*Spiniferites*), peridinioids (*Subtilisphaera cheit*) and condensates (*Florentinia* spp., *Oligosphaeridium* aff. *O. breviconispinum*). There are also chitinous remains of palynoforaminifera and scolecodonts (Fig. 9.17).

9.6 Conclusions

The fossil footprints from the São Luís Basin are a single record of the Cenomanian dinosaur fauna which inhabited the Brazilian equatorial margin during the first stages of the Equatorial Atlantic opening. The stratigraphic succession where the footprints are found consists of tidal-dominated deposits assigned to channel, sand plain, delta, and bay-fill in an estuarine environment. Along the coast of a shallow sea an abundant dinosaur fauna roamed, recording their footprints on the seashore environments. There are six ichnosites on the outcrops surrounding the São Luís bay distributed in the São Luís and Alcântara counties: Ponta da Guia, Ponta do Farol, Praia do Boqueirão, Ilha do Medo, Praia da Baronesa and Praia Prefeitura de Alcântara. These ichnosites record a same temporal event, in the context of a surface with wide geographic distribution, the São Luís Megatracksite.

Trackways and isolated footprints were interpreted as related to large and small theropods, sauropods and ornithischians (e.g., ornithopods) trackmakers. The dominance of large-sized theropod footprints is detected in the southern area of the basin, which includes the Ponta da Guia region. To the north, the ichnosites such as that one of Praia da Baronesa and Praia do Boqueirão show small- and medium-sized theropod tracks. The environmental context of a low-gradient coastal plain favored the establishment of specific dinosaur communities, with a probable ecologic "segregation" of large and other smaller dinosaurs. A gregarious behavior for some large-sized theropods is deduced from the analysis of the same oriented direction trackways from Ponta da Guia ichnosite. The randomly oriented trackways and isolated footprints from Praia da Baronesa ichnosite were interpreted as the record of a "foraging area" for theropods which, in this particular tidal plain environment, probably lived on a rather varied diet, composed of small tetrapods and invertebrates, fish, mollusks and so on. The palynological data of Cenomanian age is consistent with the inference of fluvial, lagoonal and shallow marine depositional environments under dry and hot climate. A tidal flat of a low-gradient coastal plain is the most likely environment inhabited by the trackmakers.

Fig. 9.17 The Cenomanian palynomorphs from the Alcântara Formation are consistent with the inference of a tidal flat of a low-gradient coastal plain under dry and hot climate inhabited by the dinosaur trackmakers. **a** *Classopollis major*, **b** *Equisetosporites ambiguous*, **c** *Galeacornea subtilis*, **d** *Elateroplicites africaensis*, **e** *Galeacornea causea*, **f** *Sofrepites legouxae*, **g** *Elaterosporites* aff. *klaszi*, **h** *Afropollis jardinus*, **i** palynoforaminifera, **j** dinoflagellate, **k** *Subtilisphaera cheit*, **l** scolecodonts, **m** *Oligosphaeridium* sp., **n** *Florentinia* sp. (modified from Carvalho and Pedrão 1998)

References

Arai M (2001) Palinologia de depósitos cretáceos no Norte e Meio-Norte do Brasil: histórico e estado-de-arte. In: Rossetti DF, Góes AM, Truckenbrodt W (eds) O Cretáceo na Bacia de São Luís e Grajaú. Coleção Friedrich Katzer, Museu Paraense Emílio Goeldi, Belém, pp 175–189

Aranha LGF, Lima HP, Souza JMP, Makmo RK, Figueiras JM (1990) Evolução tectônica e sedimentar das bacias de Bragança-Vizeu, São Luís e Ilha Nova. In: Raja Gabaglia GPR, Milani EJ (eds) Origem e evolução das bacias sedimentares. Petrobras, Rio de Janeiro, pp 221–234

Assis FP, Macambira JB, Leonardi G (2010) Dinossauros terópodes do Ribeirão das Lajes, primeiro registro fóssil da Formação Sambaíba (Neotriássico-Eojurássico), Bacia do Parnaíba: Fortaleza dos Nogueiras, Maranhão-Brasil. 2010. In: 45° Congresso Brasileiro de Geologia. Belém—PA. Sociedade Brasileira de Geologia, Boletim de resumos, p 1720

Barrett PM, Upchurch P (2005) Sauropodomorph diversity through time: paleoecological and macroevolutionary implications. In: Wilson JA, Curry Rogers KA (eds). Evolution and Paleobiology. University of California Press, The Sauropods, pp 125–156

Benton MJ, Bouaziz S, Buffetaut E, Martill D, Ouaja M, Soussi M, Trueman C (2000) Dinosaurs and other fossil vertebrates from fluvial deposits in the Lower Cretaceous of southern Tunisia. Palaeogeogr Palaeoclimatol Palaeoecol 157:227–246

Calvo JO, Salgado L (1996) A land bridge connection between South America and Africa during Albian-Cenomanian times based on sauropod dinosaur evidences. 39° Congresso Brasileiro de Geologia, Salvador. Sociedade Brasileira de Geologia, Anais, pp 392–393

Carvalho IS (1989) Icnocenoses continentais: bacias de Sousa, Uiraúna-Brejo da Freiras e Mangabeira. Rio de Janeiro, Dissertação de Mestrado, Universidade Federal do Rio de Janeiro 167 p

Carvalho IS (1994a) As ocorrências de icnofósseis de vertebrados na Bacia de São Luís, Cretáceo Superior, Estado do Maranhão. In: 3° Simpósio sobre o Cretáceo do Brasil, UNESP—Rio Claro, Boletim, pp 119–122

Carvalho IS (1994b) As pistas de dinossauros da Ponta da Guia (Bacia de São Luís, Cretáceo Superior—Maranhão, Brasil). An Acad Bras Ciênc 67(4):413–431

Carvalho IS (1994c) Contexto tafonômico das pegadas de terópodes da Praia da Baronesa (Cenomaniano, Bacia de São Luís). In: 38° Congresso Brasileiro de Geologia, Camboriú, 1994, Boletim de Resumos Expandidos 3: 211–212

Carvalho IS (1995) As pistas de dinossauros da Ponta da Guia (Bacia de São Luís, Cretácio Superior—Maranhão, Brasil). An Acad Bras Ciênc 67(4):413–431

Carvalho IS (2000) Geological environments of dinosaur footprints in the intracratonic basins from Northeast Brazil during South Atlantic opening (Early Cretaceous). Cretac Res 21:255–267

Carvalho IS (2001) Pegadas de dinossauros em depósitos estuarinos (Cenomaniano) da Bacia de São Luís (MA), Brasil. In: Rossetti DF, Góes AM, Truckenbrodt W (eds) O Cretáceo na Bacia de São Luís-Grajaú. Museu Paraense Emílio Goeldi. Coleção Friedrich Katzer. MPEG Editoração, pp 245–264

Carvalho IS (2004) Dinosaur footprints from Northeastern Brazil: taphonomy and environmental setting. Ichnos 11:311–324

Carvalho IS, Araújo SAF (1995) A distribuição geográfica dos fósseis e icnofósseis de Dinosauria na Bacia de São Luís. In: 47ª Reunião Anual da Sociedade Brasileira para o Progresso da Ciência. São Luís, Resumos 2: 439

Carvalho IS, Gonçalves RA (1994) Pegadas de dinossauros neocretáceas da Formação Itapecuru, Bacia de São Luís (Maranhão, Brasil). An Acad Bras Ciênc 66(3):279–292

Carvalho IS, Leonardi G (2021) Fossil footprints as biosedimentary structures for paleoenvironmental interpretation: examples from Gondwana. J South Am Earth Sci 106: 102936. https://doi.org/10.1016/j.jsames.2020.102936

Carvalho IS, Pedrão E (1998) Brazilian theropods from the equatorial Atlantic margin: behavior and environmental setting. Gaia 15:369–378

Carvalho IS, Borghi L, Leonardi G (2013) Preservation of dinosaur tracks induced by microbial mats in the Sousa Basin (Lower Cretaceous), Brazil. Cretac Res 44:112–121

Carvalho IS, Leonardi G, Rios-Netto AM, Borghi L, Paula Freitas A, Andrade JA, Freitas FI (2021) Dinosaur trampling from the Aptian of Araripe Basin, NE Brazil, as tools for paleoenvironmental interpretation. Cretaceous Res 117: 104626

Castanera D, Barco JL, Díaz-Martínez I, Gascón JH, Pérez-Lorente F, Canudo JI (2011) New evidence of a herd of titanosauriform sauropods from the lower Berriasian of the Iberian range (Spain). Palaeogeogr Palaeoclimatol Palaeoecol 310(3–4):227–237. https://doi.org/10.1016/j. palaeo.2011.07.015

Castanera D, Vila B, Razzolini N, Santos V, Pascual C, Canudo J (2014) Sauropod trackways of the Iberian Peninsula: palaeoetological and palaeoenvironmental implications. J Iber Geol 40(1):49–59. https://doi.org/10.5209/rev_JIGE.2014.v40.n1.44087

Castro DF, Toledo CEV, Sousa EP, Medeiros MA (2004) Nova ocorrência de *Asiatoceratodus* (Osteichthyes, Dipnoiformes) na Formação Alcântara, Eocenomaniano da bacia de São Luís, MA, Brasil. Revista Brasileira de Paleontologia Revista Brasileira de Paleontologia 7(2):245–248

Currie PJ (1983) Hadrosaur trackways from the Lower Cretaceous of Canada. Acta Paleontologica Polonica 28(1–2):63–73

Eberth DA, Brinkman DB (1997) Paleoecology of an estuarine, incised-valley fill in the dinosaur park formation (Judith River Group, Upper Cretaceous) of Southern Alberta, Canada. Palaios 12:43–58

Elias FA, Bertini RJ, Medeiros MA (2007) Pterosaur teeth from the Laje do Coringa, middle Cretaceous, São Luís, Grajaú Basin, Maranhão State, Northern-Northeastern Brazil. Revista Brasileira De Geociências 37(4):1–9

Eugênio WS (1994) Aspectos Paleontológicos do Cretáceo da Baía de São Marcos, Maranhão, Brasil. Dissertação de Mestrado (Universidade Federal do Rio de Janeiro/Instituto de Geociências), 77 p (unpublished)

García-Ortiz E, Pérez-Lorente F (2014) Palaeoecological inferences about dinosaur gregarious behaviour based on the study of tracksites from La Rioja area in the Cameros Basin (Lower Cretaceous, Spain). J Iber Geol 40(1):113–127. https://doi.org/10.5209/rev_JIGE.2014.v40.n1. 44091

Heredia AM, Días-Martínez I, Pazos PJ, Comerio M (2020) Fernández DE (2020) Gregarious behaviour among non-avian theropods inferred from trackways: a case study from the Cretaceous (Cenomanian) Candeleros Formation of Patagonia, Argentina. Palaeogeogr Palaeoclimatol Palaeoecol 538:109480. https://doi.org/10.1016/j.palaeo.2019.109480

Hernández JPF, Hernández AT, Gascó-Lluna F, Valverde MC, Melón JMV, Miranda AR (2023) Analysis of gregarious behavior in theropod dinosaurs at the Peñaportillo tracksite (Munilla, La Rioja, Spain). In: Estraviz-López D et al. (eds) Palaeontological publications, Abstracts book of XXI EJIP/6th IMERP, 3: 40

Ibrahim N, Sereno PC, Varricchio DJ, Martill DM et al (2020) Geology and paleontology of the Upper Cretaceous Kem Kem Group of eastern Morocco. ZooKeys 928:1–216. https://doi.org/ 10.3897/zookeys.928.47517

Kellner AWA, Azevedo SAK, Machado EB, Carvalho LB, Henriques DDR (2011) A new dinosaur (Theropoda, Spinosauridae) from the Cretaceous (Cenomanian) Alcântara Formation, Cajual Island, Brazil. An Acad Bras Ciênc 83:99–108. https://doi.org/10.1590/S0001-376520110001 00006

Klein VC, Ferreira CS (1979) Paleontologia e estratigrafia de uma fácies estuarina da Formação Itapecuru, Estado do Maranhão. An Acad Bras Ciênc 51(3):523–533

Kuban GJ (1991a) Color distinctions and other curious features of dinosaur tracks near Glen Rose, Texas. In: Gillette DD, Lockley MG (eds) Dinosaur Tracks and Traces. Cambridge University Press, Cambridge, pp 427–440

Kuban GJ (1991b) Elongate dinosaur tracks. In: Gillette DD, Lockley MG (eds) Dinosaur Tracks and Traces. Cambridge University Press, Cambridge, pp 57–72

Lapparent AF (1960) Les dinosauriens du "continental intercalaire" du Sahara central. Mémoires la Société Géologique France 88(A): 3–56

Leonardi G (1980a) Dez novas pistas de dinossauros (Theropoda Marsh, 1881) na bacia do Rio do Peixe, Paraíba, Brasil. In: 2° Congreso Argentino de Paleontología y Bioestratigrafía, 1° Congreso Latino Americano de Paleontologia, Buenos Aires, 1978. Actas, Buenos Aires, 1980, Asociación Paleontologica Argentina, 1: 243–248

Leonardi G (1980b) Ornithischian trackways of the Corda Formation (Jurassic), Goiás, Brasil. In: 2° Congreso Argentino de Paleontología y Bioestratigrafía, 1° Congreso Latino Americano de Paleontología, Buenos Aires, 1978. Buenos Aires, 1980, Asociación Paleontologica Argentina, Actas 1: 215–222

Leonardi G (1981) Ichnological data on the rarity of young in North East Brazil dinosaurian populations. An Acad Bras Ciênc 53(2):345–346

Leonardi G (1984) Le impronte fossili di dinosauri. In: Ligabue G (ed). Sulle orme dei dinosauri. Venezia: Erizzo Editrice, pp 165–186

Leonardi G (1989) Inventory and statistics of the south American dinosaurian ichnofauna and its paleobiological interpretation. In: Gillette DD, Lockley MG (eds) Dinosaur Tracks and Traces. Cambridge University Press, New York, pp 165–178

Leonardi G (1994) Annotated Atlas of South America tetrapod footprints (Devonian to Holocene) with an appendix on Mexico and Central America. CPRM (Serviço Geológico do Brasil) Brasília

Leonardi G, Spezzamonte M (1994) New tracksites (Dinosauria: Theropoda and Ornithopoda) from the Lower Cretaceous of the Ceará, Brasil. Studi Trentini di Scienze Naturali. Acta Geologica 69: 61–70

Leonardi G, Carvalho IS (2021) Dinosaur Tracks of Brazil: A Lost World of Gondwana. Indiana University Press, Bloomington, U.S.A.. p 462

Leonardi G, Santos MFCF, Barbosa FHS (2021) First dinosaur tracks from the Açu Formation, Potiguar Basin (mid-Cretaceous of Brazil). An Acad Bras Ciênc 93(Suppl. 2):e20210635. https://doi.org/10.1590/0001-3765202120210635

Lindoso RM, Elias, FA, Medeiros MA, Santos RAB, Pereira AA (2011) Pterosaur teeth from the Alcântara Formation, Cretaceous of Brazil. In: Calvo J, Porfiri J, Riga BG, Santos D (eds) Editorial de La Universidad de Cuyo, pp 171–177

Lindoso RM, Medeiros MA, Carvalho IS, Marinho TS (2012) *Masiakasaurus*-like theropod teeth from the Alcântara Formation, São Luís Basin (Cenomanian), northeastern Brazil. Cretac Res 36:119–124

Lindoso RM, Marinho TS, Santucci RM, Medeiros MA, Carvalho IS (2013) A titanosaur (Dinosauria: Sauropoda) osteoderm from the Alcântara Formation (Cenomanian), São Luís Basin, Northeastern Brazil. Cretac Res 45:43–48

Lindoso RM, Medeiros MAA, Carvalho IS, Pereira AA, Mendes ID, Iori FV, Sousa EP, Arcanjo SH, Silva TCM (2019) A new rebbachisaurid (Sauropoda: Diplodocoidea) from the middle Cretaceous of northern Brazil. Cretaceous Res 104:10419. https://doi.org/10.1016/j.cretres.2019.104191

Lockley MG (1986) The paleobiological and paleoenvironmental importance of dinosaur footprints. Palaios 1:37–47

Lockley MG (1991) Tracking dinosaurs: a new look at an ancient world. Cambridge University Press, Cambridge, MA, 238pp

Lockley MG, Houck KJ, Prince NK (1986) North America's largest dinosaur trackway site: implications for Morrison Formation paleoecology. Geological Soc Am Bull 97(10): 1163–1176

Lockley MG, Hunt A, Holbrook J, Matsukawa M, Meyer C (1992) The dinosaur freeway: a preliminary report on the Cretaceous megatracksite, Dakota Group, Rocky Mountain front range, and high plains, Colorado, Oklahoma and New Mexico. In: Flores RM (ed) Field Guidebook—Mesozoic of the Western Interior. SEPM 1992 Theme Meeting. Rocky Mountain Section, Colorado State University, Publications and Creative Services. Fort Collins, Colorado, pp 39–54

Lockley MG, Huh M, Gwak SG, Hwang KG, Paik IS (2012) Multiple tracksites with parallel trackways from the Cretaceous of the Yeosu City Area Korea: implications for gregarious behavior

in ornithopod and sauropod dinosaurs. Ichnos 19(1–2):105–114. https://doi.org/10.1080/104
20940.2011.625793

Lockley MG, McCrea R, Alcala L, Cart K, Martin J, Hadden G (2015) A preliminary report on
an assemblage of large theropod tracks from the Cretaceous Dakota Group, Western Colorado:
evidence for gregarious behavior. In: Sullivan RM, Lucas SG (eds) New Mexico museum of
natural history and science. Bulletin 67, Fossil Record 4: 179–183

Letizio LA, Bertini RJ, Medeiros MA (2022) New evidence of putative Unenlagiinae (Deinony-
chosauria, Theropoda) in the São Luís-Grajaú Basin, Albian– Cenomanian, State of Maranhão,
Brazil. Revista Brasileira de Paleontologia 25(2):157–164

Maisey JG (2000) Continental break up and the distribution of fishes of Western Gondwana during
the Early Cretaceous. Cretac Res 21:281–314

McCrea RT, Buckley LG, Farlow JO, Lockley MG, Currie PJ, Matthews NA, Pemberton SG (2014)
A 'terror of Tyrannosaurs': the first trackways of tyrannosaurids and evidence of gregariousness
and pathology in Tyrannosauridae. PLoS ONE 9(7):e103613. https://doi.org/10.1371/journal.
pone.0103613

Medeiros MA (2006) Large theropod teeth from the Eocenomanian of northeastern Brazil and the
occurrence of Spinosauridae. Revista Brasileira de Paleontologia 9(3):333–338

Medeiros MA, Schultz CL (2001) Uma paleocomunidade de vertebrados do Cretáceo médio, Bacia
de São Luís. In: Rossetti DF, Góes AM, Truckenbrodt W (eds) O Cretáceo na Bacia de São Luís
e Grajaú. Coleção Friedrich Katzer, Museu Paraense Emílio Goeldi, Belém, pp 209–221

Medeiros MA, Schultz CL (2002) A fauna dinossauriana da Laje do Coringa, Cretáceo médio do
Nordeste do Brasil. Arquivos do Museu Nacional 60(3):155–162

Medeiros MA, Schultz CL (2004) *Rayososaurus* (Sauropoda, Diplocoidea) no meso-Cretáceo do
Norte-Nordeste Brasileiro. Revista Brasileira de Paleontologia 7(2):255–279

Medeiros MA, Freire PC, Pereira AA, Santos RAB, Lindoso RM, Coêlho AFA, Passos EB, Júnior
ES (2007) Another African dinosaur recorded in the Eocenomanian of Brazil and a revision of
the Laje do Coringa site. In: Carvalho IS, Cassab RCT, Schwanke C, Carvalho MA, Fernandes
ACS, Rodrigues MAC, Carvalho MSS, Arai M, Oliveira MEQ (eds) Paleontologia: Cenários de
Vida, Rio de Janeiro: Interciência 1: 413–423

Medeiros MA, Lindoso RM, Mendes ID, Carvalho IS (2014) The Cretaceous (Cenomanian) continental record of the Laje do Coringa flagstone (Alcântara Formation), northeastern South
America. J S Am Earth Sci 53(2014):50–58. https://doi.org/10.1016/j.jsames.2014.04.002

Moraes-Santos HM, Melo CCS, Toledo PM, Rossetti DF (2001) Ocorrência de Pleurodira na
Formação Alcântara (Albiano-Cenomaniano), Bacia de São Luís, MA. In: Rossetti DF, Góes
AM, Truckenbrodt W (eds) O Cretáceo na Bacia de São Luís e Grajaú. Coleção Friedrich Katzer,
Museu Paraense Emílio Goeldi, Belém, pp 235–244

Moreno K, Valais SD, Blanco N, Tomlinson AJ, Jacay J, Calvo JO (2012) Large theropod dinosaur
footprint associations in western Gondwana: behavioural and palaeogeographic implications.
Acta Palaeontologica Polonica 57(1): 73–83. https://doi.org/10.4202/app.2010.0119

Myers TS, Fiorillo AR (2009) Evidence for gregarious behavior and age segregation in sauropod
dinosaurs. Palaeogeogr Palaeoclimatol Palaeoecol 274(1–2):96–104. https://doi.org/10.1016/j.
palaeo.2009.01.002

Nadon GC (1993) The association of anastomosed fluvial deposits and dinosaur tracks, eggs, and
nests: implications for the interpretation of floodplain environments and a possible survival
strategy for ornithopods. Palaios 8:31–44

Oliveira PE (1958) Notas preliminares e estudos sobre a idade do calcário de Ponta Grossa, Estado
do Maranhão. Departamento Nacional da Produção Mineral/Divisão de Geologia e Mineralogia,
1–7 (Boletim 107)

Paik IS, Kim HJ, Baek SG, Seo YK (2020) New evidence for truly gregarious behavior of
ornithopods and solitary hunting by a theropod. Episodes 43(4): 1045–1052. https://doi.org/
10.18814/epiiugs/2020/020069

Pedrão E (1995) Palinoestratigrafia e evolução paleoambiental de rochas sedimentares aptianas-cenomanianas das bacias de Bragança-Viseu e São Luís (Margem Equatorial Brasileira). Dissertação de Mestrado (Universidade Federal do Rio de Janeiro/Instituto de Geociências), 215 pp (unpublished)

Pedrão E, Arai M, Carvalho IS, Santos MHB (1993) Palinomorfos cenomanianos da Formação Itapecuru - análise palinológica do afloramento da ponta do Farol, município de São Luís (MA). In: 13° Congresso Brasileiro de Paleontologia, 1° Simpósio de Paleontologia do Cone Sul, São Leopoldo, 1993, Companhia de Pesquisa de Recursos Minerais, Boletim de Resumos, p 61

Pedrão E, Carvalho IS, Martins FJC, Santos MHB (1995) Palinoestratigrafia e análise quantitativa de amostras de superfície da Formação Itapecuru, Bacia de São Luís. PETROBRAS-CENPES, Rio de Janeiro, 5 p., 2 figs., 2 plates

Pedreira da Silva AJ, Lopes RC, Vasconcelos AM, Bahia RBC (2003) Bacias Sedimentares Paleozóicas e Meso-Cenozóicas Interiores. In: Bizzi LA, Schobbenhaus C, Vidotti RM, Gonçalves JH (eds) Geologia, Tectônica e Recursos Minerais do Brasil. Brasília: CPRM, pp 55–85

Piñuela L, García-Ramos JC, Romano M, Ruiz-Omeñaca JI (2016) First record of gregarious behavior in robust medium-sized Jurassic Ornithopods: evidence from the Kimmeridgian trackways of Asturias (N. Spain) and some general considerations on other medium-large ornithopod tracks in the Mesozoic record. Ichnos 23(3–4): 298–311. https://doi.org/10.1080/10420940.2016.1178640

Popoff M (1988) Du Gondwana à l'Atlantique Sud: les connexions du fossé de la Bénoué avec les bassins du Nord-Est brésilien jusqu'à l'ouverture du golfe de Guinée au Crétacé Inférieur. Journal of African Earth Sciences (and the Middle East) 7(2):409–431. https://doi.org/10.1016/0899-5362(88)90086-3

Richter U, Mudroch A, Buckley LG (2013) Isolated theropod teeth from the Kem Kem beds (early Cenomanian) near Taouz. Morocco. Paläontologische z. 87(2):291–434

Rossetti DF (1994) Morphology and internal structures of mixed tide and storm—generated bedforms: Upper Cretaceous Itapecuru Formation, Northern Brazil. In: 14th International Sedimentological Congress, Recife, 1994, Abstracts pp A21–A22

Rossetti DF (1996a) Genesis of large-scale cross beddings from the Cenomanian Upper Itapecuru Formation. Northern Brazil. In: 4° Simpósio sobre o Cretáceo do Brasil, Águas de São Pedro, 1996, Sociedade Brasileira de Geologia, Boletim pp 147–149

Rossetti DF (1996b) Depositional evolution of two estuarine successions: the Upper Itapecuru Formation, São Luís Basin, Northern Brazil. In: 39° Congresso Brasileiro de Geologia, Salvador, 1996, Sociedade Brasileira de Geologia, Anais 1: 187–188

Rossetti DF (1996c) Sequence stratigraphy and depositional evolution of the Itapecuru Formation (Late Cretaceous) in the São Luís Basin, Northern Brazil. Acta Geologica Leopoldensia 19:111–126

Rossetti DF (1997) Facies analysis of the lower succession of the Upper Itapecuru Formation, São Luís Basin, northern Brazil. In: Costa ML, Angelica R (eds) Contribuições à Geologia da Amazônia, Belém, Falângola, pp 241–284

Rossetti DF (1998) Facies architecture and sequential evolution of incised valley estuarine fills: the Cujupe Formation (Upper Cretaceous to lower Tertiary), São Luís Basin, northern Brazil. J Sediment Res 68(2):299–310

Rossetti DF, Truckenbrodt W (1997) Revisão estratigráfica para os depósitos do Albiano-Terciário Inferior(?) na Bacia de São Luís, Maranhão. Boletim do Museu Paraense Emílio Goeldi, Série Ciências da Terra 9:29–41

Stromer E (1915) Ergebnisse der Forschungsreisen Prof. E. Stromers in den Wüsten -Ägyptens. II- Wirbeltier-Reste der Baharîje-Stufe (unterstes Cenoman). 3-Das original des Theropoden *Spinosaurus aegyptiacus* nov. gen., nov. spec. Abhandlungen der Königlich Bayerischen Akademie der Wissenschaften 28(3):1–32

Thulborn T (1990) Dinosaur tracks. Chapman and Hall, St Edmundsbury Press Ltd., Great Britain, 410pp

Vilas Bôas I (1999) Dentes de terópodes e associação fossilífera da Praia da Baronesa, Alcântara, MA (Formação Itapecuru, Bacia de São Luís). In: 16° Congresso Brasileiro de Paleontologia, URCA, Crato, Boletim de Resumos, p 125

Vilas Bôas I, Carvalho IS (1999) Mosasaurios de la Formación Itapecuru (Cretácico Superior), playa de la Baronesa, Alcântara (Estado de Maranhão), Brasil. In: XV Jornadas Argentinas de Paleontología de Vertebrados, 1999. Ameghiana, La Plata, Argentina, 36(4):24 R

Vilas Bôas I, Carvalho IS (2001) Répteis marinhos (Mosasauria e Plesiosauria) do Cretáceo Superior da Bacia de São Luís (Maranhão, Brasil). In: Rossetti DF, Góes AM, Truckenbrodt W (eds) O Cretáceo na Bacia de São Luís-Grajaú. Coleção Friedrich Katzer, Ed. Museu Paraense Emílio Goeldi, pp 223–233

Vilas Bôas I, Medeiros MA (1997) Theropod teeth (dromeosaurids?) from the Itapecuru Formation, Maranhão—Brasil. In: Jornadas Argentinas de Paleontologia de Vertebrados, 1997, La Rioja. Programa y Resumenes, Ameghiniana 13: 35

Vilas Bôas I, Carvalho IS, Medeiros MA, Pontes H (1999) Dentes de *Carcharodontosaurus* (Dinosauria, Tyrannosauridae [sic]) do Cenomaniano, Bacia de São Luís (Norte do Brasil). An Acad Bras Ciênc 71(4):846–847

Chapter 10
New Steps and New Challenges to the Brazilian Dinosaur Track Researches

Ismar de Souza Carvalho and Giuseppe Leonardi

10.1 Introduction

The history of Brazilian dinosaur ichnology begins in 1924 with the description of the first dinosaur footprints from Rio do Peixe Basins. The discovery of the tracks is due the research of Luciano Jacques de Moraes (1896–1968), a Brazilian mining engineer, working for the DNOCS (Departamento de Obras contra as Seccas, the Department of Works against the Drought), surveying the Sousa County, in the Paraíba State, northeast Brazil. Moraes (1924) discovered two dinosaur trackways in the mudstones of the riverbed of Rio do Peixe. Based on these trackways it was possible the first approach of the age (Comanchean, Lower Cretaceous) of the mudstone succession, an important datum to the understanding of the geological framework of the basins. These tracks were not further studied or published, except for a brief mentions (Price 1961; Cavalcanti 1947; Haubold 1971). Giuseppe Leonardi, from December 1975, began a series of thirty-three expeditions to the Rio do Peixe basins (1975–2016) and Ismar de Souza Carvalho, over the last twenty-seven years (from April 1986) with informal groups of researchers, undergraduate and graduate training programs periodically visited the ichnosites of these basins, twice a year, to study the dinosaur tracks. The result has been a great number of studies concerning the research on dinosaur ichnology and the creation of a natural park of the "Dinosaur Valley"

I. S. Carvalho (✉)
CCMN/IGEO, Departamento de Geologia, Universidade Federal Do Rio de Janeiro, 21.910-200 Cidade Universitária, Ilha Do Fundão, Rio de Janeiro, Estado Do Rio de Janeiro, Brazil
e-mail: ismar@geologia.ufrj.br

Centro de Geociências, Universidade de Coimbra, Rua Sílvio Lima, 3030-790 Coimbra, Portugal

G. Leonardi
Istituto Cavanis, Dorsoduro 898, 30123 Venezia, Italy

CCMN/IGEO, Departamento de Geologia, Universidade Federal Do Rio de Janeiro, Cidade Universitária, Ilha Do Fundão, Rio de Janeiro, Estado Do Rio de Janeiro 21949-900, Brazil

(Leonardi and Carvalho 2021). Marcelo A. Fernandes and Luciana B. R. Fernandes collected the Botucatu flagstones with ichnofossils establishing one of the largest collections of fossil footprints in Latin America. In 2006, these were incorporated into the institutional collection of the Federal University of São Carlos, enabling the formation of two local museums.

Many other ichnosites were searched for and discovered since then, especially in the Paraná Basin. In the Triassic successions of Rio Grande do Sul State were revealed many of the oldest dinosaur tracks of the world (Silva et al. 2007a, b, 2008, 2012). In the Jurassic and Cretaceous fluvial and aeolian deposits of the Guará and Botucatu formations a great diversity of new specimens was discovered (Francischini et al. 2015, 2020; Fernandes and Carvalho 2007; Leonardi 1994).

10.2   A Footprint in the Past and a Step into the Future

The discovery of a new tracksite is mainly based on the detail observation of natural and artificial outcrops on river banks and dry beds, cliffs, mountain ranges, road pavements, rail-road outcrops and quarries, mines and tunnels. The description, illustration and interpretation of the footprints and trackways changed through time due the new technologies and the comprehension of the importance of their sedimentological and stratigraphic context.

The historical methods of field and laboratory research can sometimes be considered vintage. Nonetheless the mapping of the footprints using compass, metric tape, and strings with the gridded quadrant system and graph paper (Leonardi 1977) or the direct reproduction on PVC sheets are very efficient and allow an excellent approach of the distribution and spatial position of the footprints in the outcrops. Other methodologies, especially concerning the photographic reproduction using individual printed photographs to be later associated manually in a photomosaic is completely outdated due the facilities of the digital cameras and programs for merging photographs (Falkingham et al. 2014b).

After mapping all the footprints they are manually drawn at a small scale, with numberless measurements, for statistical analyses. Although these data can be questioned due the subjectivity of landmarks in footprints, the traditional measurements can yield useful results (Leonardi 1991; Farlow 2018). Despite the correct interpretation of a footprint or trackway is always a complex task, the understanding of these facts is a step forward in their correct understanding (Falkingham et al. 2016a, b).

10.3   New Technologies and New Challenges

Techniques using laser scanning permit imaging with higher detail and precision the shape and position of the footprints, producing digital rock outcrops that can be used to compare with other outcrops and preserved for future studies (Medeiros et al.

2007; Romilio et al. 2017; Leonardi and Carvalho 2021). Also the use of drones (Romilio et al. 2017; Xing et al. 2018; Petti et al. 2018) permit to cover in detail large areas and steep cliffs.

Other important point is the use of the 3D photogrammetry techniques to produce the topographic images allowing a more precise delimitation of the contours and perception of the footprints morphology. The 3D photogrammetry uses softwares to combine multiple photos of an object taken from different directions and angles into one 3D digital model of that object (Vitkus et al. 2023). This technique should be used in the description of the new localities with dinosaur footprints and also as a tool of geoheritage preservation to the known ichnosites.

All new digital technologies of photography and surveying, by means of manned or unmanned aircrafts (today especially by drones), although extremely valuable, do not exempt us from the field work. The preliminary study of geological maps and previous studies to locate the most favorable areas to specific research, the accurate examination of outcrops are essential to the success of new discoveries. This is also the case when the fossil footprints are, as is often the case, on inclined or even vertical rocky walls, on the roofs of mines or tunnels (Staines and Woods 1964; Meyer et al. 1999; Moreau et al. 2020; Belvedere et al. 2008), generally on places of difficult access. Experience has shown that publications of ichnologic material made solely on the basis of photographs taken by others or from far away, almost always lead to make gross mistakes.

10.4   Looking Forward

The ichnotaxonomy presents many problems concerning the trackmakers behavior and the substrate physical properties. Therefore, a prudent and limited use of ichnotaxonomy is recommended, that is, it is advisable to establish new taxa names for new kinds of fossil footprints, only after statistical analysis (Belvedere et al. 2018; Belvedere and Farlow 2016) and only in the case that such tracks have a good or excellent state of impression and morphological preservation (Falkingham et al. 2014a, b; Marchetti et al. 2019; Razzolini et al. 2014, 2016). It should also be avoided to establish new taxa for isolated footprints, even if of excellent quality of preservation. It is therefore suggested to reserve the establishment of new taxa for trackways of at least three footprints (or pairs of footprints) in sequence, including the trackway parameters in the diagnosis. This allows a better understanding of the trackmaker. The methods of geometric morphometric analysis for the dinosaurian ichnodiversity (Castanera et al. 2015, 2018) and trackmakers' behavioral patterns (Citton et al. 2017) should be tested in the Brazilian ichnosites. It is of special importance to follow the standard protocol on these issues put forward by Falkingham et al. (2018).

Other aspect to be evaluated is the land-vertebrate ichnofacies (Lockley 2007; Lockley et al. 1994, 2007; Meyer et al. 1999) and the stratigraphic correlation of distinct outcrops in a same basin or neighboring basins, to apply the concept of megatracksites established by Lockley and Meyer (2023) or as a tool of stratigraphic correlation (Carvalho et al. 2019).

The perspectives in studying the Brazilian fossil tracks include new methods for the documentation, such as laser scanning, aerial and close-range photogrammetry, three-dimensional (3-D) models and biplanar X-ray 3-D motion analysis (Belvedere et al. 2012; Breithaupt and Matthews 2012; Petti et al. 2018; Costa-Pérez et al. 2019; Gatesy and Ellis 2012, 2016; Matthews et al. 2016; Romilio et al. 2017; Wings et al. 2016; Leonardi and Carvalho 2021). It is also important petrographic studies and clay mineral analysis of the matrix where the footprints occur. The petrography will allow recognizing the role of microbial mats in the preservation (Carvalho et al. 2013), and the clay mineral analyses (Dai et al. 2022; Rodrigues et al. 2023), the influence of the clay mineralogy in the morphological pattern of the footprints and paleoenvironmental interpretation.

The experimental analysis (Falkingham and Gatesy 2014; Falkingham et al. 2010; Gatesy and Ellis 2016; Marty 2005, 2012; Marty et al. 2009) are important methods to evaluate the relationships between the physical properties of the substrate, the behavior of the trackmakers and the induced morphological patterns. Previous studies on experimental analysis should be considered in any new analysis of the Brazilian dinosaur footprints. Besides the control of the geological context of the occurrences, the observation of the recent tracks can be very useful to comparisons to the paleoenvironmental interpretations (Carvalho and Leonardi 2021).

The use of AI (Artificial Intelligence) opens new perspectives to ichnology concerning a faster and detailed classification of the footprints. An example of this potential was presented by Ha and Kim (2023) to validate ichnotaxa, employing convolutional neural network-based Xception transfer Learning. This technique allowed to automatically classify ornithopod dinosaur tracks. These machine-learning techniques open new perspectives to verify the ichnotaxonomic assignments and to compare a great number of samples, with the establishment of a global relationship of the ichnofaunas.

A challenge to the future is the in situ and ex situ preservation of trackways and fossil footprints. The in situ preservation is always subject to vandalism, erosion, weathering and limitations to urban expansion. In tropical environments the vegetal covering and weathering induces a great impact in the aspects of the bedding planes with footprints just after few years. The wearing of the surface, chemical weathering and cracking are recurrent aspects that lead to the destruction of the footprints. The employ of airborne and handheld high-resolution LIDAR (light + radar, an acronym for light detection and ranging) for characterization and conservation of fossil tracks (Platt et al. 2018) could be very useful as a tool to the control of natural and anthropogenic destruction of outcrops with footprints. On the other hand, the ex situ preservation in museums, universities, research centers and open air are also difficult. The dimensions, weight and friability of the samples with trackways and footprints are a problem to handle and to house adequately in the paleontological collections (Carvalho 2004). The open-air exhibition in an urban environment, as in the sidewalks of the Araraquara or in São Carlos cities (São Paulo State, Brazil) are also a challenge to preservation as a geoheritage, with the addition of abrasion by the shoes of the passers-by (Francischini et al. 2020). Although they show a great potential for education, these flagstones are exposed to the same damages of the in situ trackways and fossil footprints found in the outcrops.

Another dare is to improve the interest of the "pure" geologists (stratigraphers, tectonists, geophysicists), who normally do not deal with paleontological data, especially in the results of vertebrate ichnology. It is important to present the information and interpretation obtained from the ichnological studies as a scientific data to the increase of the geological knowledge of a region, instead of objects of curiosity.

In the perspective of ichnology popularization, the use of smartphones and tablets to apply augmented reality is a powerful tool for instruction. The possibility to connect the real surfaces with fossil tracks and the reconstruction of life sceneries of the trackmakers has a great educational potential. Other strategies to the dissemination of the dinosaur tracks information from Brazil should include the artificial intelligence, an important new approach to the science vulgarization.

10.5   New Perspectives for Field Exploration in Brazil

Brazil is a sub-continental country, with its 8,510,000 km^2 of area, greater than the area of Europe with the exclusion of European Russia (6,031,000 km^2), little less than that of the United States of America (9,834,000 km^2) and Canada (9,985,000 km^2); and greater than the area of Australia (7,688,000 km^2). Field research often still takes place, in Brazil, in physical environments that are difficult to reach and explore. Several sedimentary basins remain to be partially explored and others totally.

As a result, there are many sedimentary basins whose exploration needs to be increased and/or improved. However, it should also be remembered that, especially in the most densely populated areas, most fossils in general and especially fossil tracks have appeared as secondary products of excavations: for tunnels, mines, highway and rail-road cuts, quarries, building and bridge foundations, water wells, and canals. In Brazil, the complex of Amazonian basins (Acre, Solimões, Alto Tapajós, Amazonas and Marajó, and minor ones) has only been touched, under the aspect of the vertebrate ichnology, with some results in the Tacutu Basin (Barros et al. 2023, 2024a, b, c) and at the margins of the Marajó Basin (Ferreira et al. 1979). There is an immense area to be explored, although the wilderness environments make searching difficult.

The large Parnaíba Basin has a similar situation. Only three ichnosites have been reported: São Domingos of Itaguatins, in the Tocantins State (Leonardi 1980, 1994; De Valais et al. 2015), Fortaleza dos Nogueira locality (Assis et al. 2010) and a site along the Itapecuru river, in the Maranhão State (Menezes et al. 2019). Other large areas have been visited by ichnologists, but the results for now are limited to this, and there is much opportunities for further research. The Paraná Basin has been much more systematically visited, with remarkable results, but there are still good chances of new discoveries.

The Northeastern interior basins, intensely explored by ichnologists, should be periodically revisited, since erosion can highlight new outcrops of layers with fossil footprints. In the basins of the Rio do Peixe, remains to visit the westernmost part of the basin of Sousa, and the northern border of the Triunfo Basin.

All these perspectives are important tasks to the development of the vertebrate ichnology in Brazil. However, there is an unsolved question in the Brazilian pale-ontological studies that is the recognition and value of the "invisibles of science" as defined by Carvalho and Leonardi (2022). The Brazilian paleontologists cannot continue to be blind about the imperative importance to be aware of the native, enslaved, riverside populations, workers operating in mines and quarries, the popu-lation of the villages where fossils are found, field or laboratory assistants who have volunteered or contractually contributed to the assistance of scientists. Their impor-tance should be recognized as relevant to the advancement of knowledge of the science of fossils.

10.6   Conclusion

The fossil footprints should be analyzed in their stratigraphic context and new technical procedures are needed to represent them in the outcrops, including 3D photogrammetry and laser scanning techniques, to their representation in the outcrops. They should be studied neither just like isolated biogenic structures nor just as a record of the passage of individual dinosaurs but rather as populations and associations in their whole paleobiological, sedimentological, and stratigraphic settings. This will allow a better use of them in paleoenvironmental, paleoclimato-logical, and paleogeographical interpretations. Other important point is to develop new strategies for the geoheritage conservation, including ex situ and in situ condi-tions. A good strategy is the diffusion of the scientific knowledge through formal and non-formal education.

The new challenges also include the searching for, the discovery and the preser-vation of outcrops before they disappear due the urban expansion and changes in the mining or quarrying techniques, that nowadays are no more a manual rock extraction. The in situ or ex situ preservation of the dinosaur footprints is certainly the great challenge to the future of this Brazilian geoheritage.

References

Assis FP, Macambira JB, Leonardi G (2010) Dinossauros terópodes do Ribeirão das Lajes, primeiro Registro fóssil da Formação Sambaíba (Neotriássico-Eojurássico), Bacia do Parnaíba: Fort-aleza dos Nogueiras, Maranhão-Brasil. In: 45° Congresso Brasileiro de Geologia, Belém, PA, Brasil, Anais p 1720. http://sbg.sitepessoal.com/anais_digitalizados/2010-BEL%C3%89M/2010-BEL%C3%89M.zip'
Barros LS, Vieira CEL, Souza V, Pinheiro FL (2023) Resultados preliminares da primeira ocorrência de icnofósseis de Dinosauria na porção Nordeste da Amazônia, Cretáceo Inferior da Bacia

do Tacutu, Roraima. In: 12° Simpósio Brasileiro de Paleontologia de Vertebrados, Sociedade Brasileira de Paleontologia, Santa Maria: Boletim de Resumos, pp 28–29

Barros LS, Souza V, Vieira CEL, Zaranza GS, Pinheiros FL (2024a) Challenges to the paleoich-nology research on the Tacutu Basin, Central/Northeast Amazonia, Roraima (Brazil). In: Ichnia 2024. The 5th International Congress on Ichnology, Florianópolis, Brazil, Abstract Book, pp 32–33

Barros LS, Souza V, Vieira CEL, Zaranza GS, Pinheiros FL (2024b) Evidence of sauropod tracks and other traces in the Barremian–Albian from the Tacutu Basin, Roraima, Brazil. In: Ichnia 2024. The 5th International Congress on Ichnology, Florianópolis, Brazil, Abstract Book, pp 37–38

Barros LS, Souza V, Vieira CEL, Zaranza GS, Pinheiros FL (2024c) Possible in situ dinosaur excrements from the Serra do Tucano Formation (Barremian–Albian), Tacutu Basin, Roraima, Brazil. In: Ichnia 2024. The 5th International Congress on Ichnology, Florianópolis, Brazil, Abstract Book, pp 34–36

Belvedere M, Farlow JO (2016) A numerical scale for quantifying the quality of preservation of vertebrate tracks. In: Falkingham PL, Marty D, Richter A (eds) Dinosaur tracks: the next steps. Bloomington, Indiana University Press, chap.6, pp 92–98

Belvedere M, Avanzini M, Mietto P, Rigo M (2008) Norian dinosaur footprints from the "Strada delle Gallerie" (Monte Pasubio, NE Italy). Studi Trent. Sci Nat, Acta Geol, 83: 267–275

Belvedere M, Baucon A, Furin, S, Mietto P, Felletti F, Muttoni G (2012) The impact of the digital trend on ichnology: ichnobase. In: Richter A, Reich M (eds) Dinosaur Tracks 2011. An Interna-tional Symposium. Obernkirchen, April 14–17, 2011. Göttingen, Göttingen Universitätsdrucke, 2012, Germany

Belvedere M, Bennett MR, Marty D, Burdka M, Reynolds SC, Bakirov R (2018) Stat-tracks and mediotypes: powerful tools for modern ichnology based on 3D models. PeerJ 6ce4247. https://doi.org/10.7717/peerj.4247

Breithaupt BH, Matthews NA (2012) Neoichnology and photogrammetric ichnology to interpret theropod community Dynamics. In: Richter A, Reich M (eds) Dinosaur tracks 2011. An interna-tional symposium. Obernkirchen, April 14–17, 2011. Göttingen, Göttingen Universitätsdrucke, 2012, Germany

Carvalho IS (2004) Curadoria Paleontológica. In: Carvalho IS (Ed) Paleontologia, 2nd edn. Editora Interciência, Rio de Janeiro, 2: 3–15

Carvalho IS, Leonardi G (2021) Fossil footprints as biosedimentary structures for paleoenviron-mental interpretation: examples from Gondwana. J S Am Earth Sci 106:102936. https://doi.org/10.1016/j.jsames.2020.102936

Carvalho IS, Leonardi G (2022) The invisibles of science and the paleontological heritage: the Brazilian study case. Geoheritage 14:107. https://doi.org/10.1007/s12371-022-00737-1

Carvalho IS, Borghi L, Leonardi G (2013) Preservation of dinosaur tracks induced by microbial mats in the Sousa Basin (Lower Cretaceous), Brazil. Cretaceous Res 44:112–121

Carvalho IS, Rios-Netto AM, Borghi L, Leonardi G (2019) Dinoturbation structures from the Aptian of Araripe Basin, Brazil, as tools for stratigraphic correlation. STRATI 2019. In: 3rd international stratigraphic congress. Milano, Abstract book, p 262

Castanera D, Belvedere M, Marty D, Paratte G, Lapaire-Cattin M, Lovis C, Meyer CA (2018) A walk in the maze: variation in Late Jurassic tridactyl dinosaur tracks—a case study from the Late Jurassic of the Swiss Jura mountains (NW Switzerland). PeerJ Preprints. https://doi.org/10.7287/peerj.preprints.3506v1

Castanera D, Colmenar J, Sauqué V, Canudo JI (2015) Geometric morphometric analysis applied to theropod tracks from the Lower Cretaceous (Berriasian) of Spain. Palaeontology 58(1):183–200

Cavalcanti DF (1947) Pegadas de Dinosáurios no Rio do Peixe, Paraíba. Ceará, Secretaria de Agricultura e Obras Públicas, Boletim 1(1): 45–49

Citton P, Romano M, Carluccio R, Caracciolo FA, Nicolosi I, Nicosia U, Sacchi E, Speranza C, Speranza F (2017) The first dinosaur tracksite from Abruzzi (Monte Cagno, Central Apennines, Italy). Cretac Res 73(2017):47–59

Costa-Pérez M, Moratalla JJ, Marugán-Lobón J (2019) Studying bipedal dinosaur trackways using geometric morphometrics. Palaeontologia Electronica 22.3. pvc_3: 1–13. https://doi.org/10.268 79/980

Dai X, Du Y, Ziegler M, Wang C, Ma Q, Chai R, Guo H (2022) Middle Triassic to Late Jurassic climate change on the northern margin of the South China plate: insights from chemical weathering indices and clay mineralogy. Palaeogeogr Palaeoclimatol Palaeoecol 585:110744. https://doi.org/10.1016/j.palaeo.2021.110744

De Valais S, Candeiro CR, Tavares LF, Alves YM, Cruvinel C (2015) Current situation of the ichnological locality of São Domingos from the Corda Formation (Lower Cretaceous), northern Tocantins State, Brazil. J S Am Earth Sci 61:142–146. https://doi.org/10.1016/j.jsames.2014.09.023

Falkingham PL, Marty D, Richter A (2016a) Introduction. In: Falkingham PL, Marty D, Richter A (eds) Dinosaur tracks: the next steps. Indiana University Press, Bloomington, pp 2–27

Falkingham PL, Marty D, Richter A (eds) (2016b) Dinosaur tracks: the next steps. Indiana University Press 428 p ISBN-10: 9780253021021

Falkingham PL, Bates KT, Avanzini M, Bennett M, Bordy EM, Breithaupt BH, Castanera D, Citton P, Diaz-Martinez I, Farlow JO et al (2018) A standard protocol for documenting modern and fossil ichnological data. Palaeontol, Front Palaeontol 2018:1–12

Falkingham PL, Bates KT, Farlow JO (2014a) Historical photogrammetry: bird's Paluxy river dinosaur chase sequence digitally reconstructed as it was prior to excavation 70 years ago. Public Library of Sci ONE 9(4):e93247. https://doi.org/10.1371/journal.pone.0093247

Falkingham PL, Bates KT, Margetts L, Manning PL (2014b) The "Goldilocks" effect: preservation bias in vertebrate track assemblages. J R Soc Interface 8(61):1142–1154

Falkingham PL, Gatesy SM (2014) The birth of a dinosaur footprint: subsurface 3d motion reconstruction and discrete element simulation reveal track ontogeny. Proc Natl Acad Sci USA Early Edition 111(51):18279–18284

Falkingham PL, Margetts L, Manning PL (2010) Fossil vertebrate tracks as paleopenetrometers: confounding effects of foot morphology. Palaios 25:356–360

Farlow JO (2018) Noah's Ravens: interpreting the makers of tridactyl dinosaur footprints. Indiana University Press, Bloomington

Fernandes MA, Carvalho IS (2007) Pegadas fósseis da Formação Botucatu (Jurássico Superior–Cretáceo Inferior): o registro de um grande dinossauro Ornithopoda na Bacia do Paraná In: Carvalho IS, Cassab RCT, Schwanke C, Carvalho MA, Fernandes ACS, Rodrigues MAC, Carvalho MSS, Arai M, Oliveira MEQ (eds) Paleontologia: Cenários de Vida. Interciência, Rio de Janeiro 1: 425–432

Ferreira CS, Cunha FLS, Leonardi G (1979) Evidências do Mesozoico na região dos Rios Guamá e Capim, PA. O registro das pegadas fósseis de répteis (Prov. Triássico-Jurássico) no arenito aflorante do Rio Guamá (Projeto ABC/FINEP). Anais da Academia Brasileira de Ciências 51(2), 360

Francischini H, Dentzien-Dias PC, Fernandes MA, Schultz CL (2015) Dinosaur ichnofauna of the Upper Jurassic/Lower Cretaceous of the Paraná Basin (Brazil and Uruguay). J S Am Earth Sci 63:180–190

Francischini H, Fernandes MA, Kunzler J, Rodrigues R, Leonardi G, Carvalho IS (2020) The ichnological record of Araraquara sidewalks: history, conservation, and perspectives from this urban paleontological heritage of Southeastern Brazil. Geoheritage 12:50.

Gatesy SM, Ellis RG (2012) Tracks as 3-D particle trajectories. In: Richter A, Reich M (eds) Dinosaur tracks 2011. An international symposium. Obernkirchen, April 14–17, 2011. Göttingen, Göttingen Universitätsdrucke, 2012, Germany

Gatesy SM, Ellis RG (2016) Beyond surfaces: a particle-based perspective on track formation. In: Falkingham PL, Marty D, Richter A (eds) Dinosaur tracks: the next steps. Indiana University Press, Bloomington, chap. 5, pp 82–91

Ha Y, Kim S-S (2023) Classification of large ornithopod dinosaur footprints using Xception transfer learning. PLoS ONE 18(11):e0293020. https://doi.org/10.1371/journal.pone.0293020

Haubold H (1971) Ichnia Amphibiorum et Reptiliorum Fossilium. Handbuch der Paläeoher-petologie, pt. 18. Stuttgart, Germany and Portland, USA. Gustav Fischer, vii + 124 pp

Leonardi G (1977) Two simple instruments for ichnological research, principally in the field of vertebrates. Dusenia 10(3):185–188

Leonardi G (1980) Ornithischian trackways of the Corda Formation (Jurassic) Goiás, Brazil. In: 1st Congreso Latinoamericano de Paleontología, Buenos Aires, Argentina, Abstract 1: 215–222

Leonardi G (1991) Inventory and statistics of the South American dinosaurian ichnofauna and its paleobiological interpretation. In: Gillette DD, Lockley MG (eds) Dinosaur Tracks and Traces. Cambridge University Press, Cambridge, pp 165–178

Leonardi G (1994) Annotated Atlas of South America tetrapod footprints (Devonian to Holocene). CPRM, Brasília

Leonardi G, Carvalho IS (2021) Dinosaur Tracks from Brazil: a Lost World of Gondwana. Indiana University Press, Bloomington, 445p

Lockley MG (2007) A tale of two ichnologies: the different goals and potentials of invertebrates and vertebrate (tetrapod) ichno-taxonomy and how they relate to ichnofacies analysis. Ichnos 14(1–2):39–57

Lockley MG, Meyer CA (2023) The megatracksite phenomenon: implications for tetrapod palaeo-biology across terrestrial-shallow-marine transitional zones. Geol Soc, London, Special Publ 522:285–324. https://doi.org/10.1144/SP522-2021-164

Lockley MG, Hunt AP, Meyer C (1994) Vertebrate tracks and the ichnofacies concept: implications for paleoecology and palichnostratigraphy. In: Donovan S (ed) The paleobiology of trace fossils. Wiley, Chichester, UK, pp 241–268

Lockley MG, Cart K, Foster J, Lucas SG (2017) Early Jurassic Batrachopus-rich track assem-blages from interdune deposits in the Wingate sandstone, Dolores valley, Colorado, USA. Palaeo 491(2018):185–195. https://doi.org/10.1016/j.palaeo.2017.12.008

Menezes MN, Araújo-Júnior, HI, Dal'Bó PF, Medeiros MA (2019) Integrating ichnology and pale-opedology in the analysis of Albian alluvial plains of the Parnaíba Basin, Brazil. Cretaceous Res 96: 210–226.https://doi.org/10.1016/j.cretres.2018.12.013

Moraes LJ (1924) Serras e montanhas do Nordeste; pp. 43–58 in Inspectoria de Obras Contra As Seccas. Geologia. Rio de Janeiro. Ministério da Viação e Obras Publicas. (Série I. D. Publ. 58). 2nd ed. Coleção Mossoroense, 35(1). Fundação Guimarães Duque, Rio Grande do Norte, Brazil

Marchetti L, Belvedere M, Voigt S, Klein H, Castanera D, Díaz-Martínez I, Marty D et al (2019) Defining the morphological quality of fossil footprints. Problems and principles of preservation in tetrapod ichnology with examples from the Palaeozoic to the present. Earth Sci Rev 193:109–145

Marty D (2005) Sedimentology and taphonomy of dinosaur track-bearing Plattenkalke (Kimmerid-gian, Canton Jura, Switzerland). Zitteliana B 26:20

Marty D (2012) Formation, taphonomy, and preservation of vertebrate tracks; p. 37. In: Richter A, Reich M (eds) Dinosaur tracks 2011. An international symposium. Obernkirchen, April 14–17, 2011. Göttingen, Göttingen Universitätsdrucke, 2012, Germany

Marty D, Strasser A, Meyer CA (2009) Formation and taphonomy of human footprints in microbial mats of present-day tidal-flat environments: implications for the study of fossil footprints. Ichnos 16:127–142

Matthews N, Noble T, Breithaupt B (2016) Close-range photogrammetry for 3-D ichnology: the basics of photogrammetric ichnology. In: Falkingham PL, Marty D, Richter A (eds) Dinosaur tracks: the next steps. Indiana University Press, Bloomington, pp 28–55

Medeiros MAM, Della Favera JC, Reis MAF, Vieira e Silva, TL, Oliveira EB, Biassusi RA, Silveira RG (2007) O laser scanner e a paleontologia em 3D. Anuário do Instituto de Geociências 30(1): 94–100

Meyer CA, Lockley MG, Leonardi G, Anaya F (1999) Late Cretaceous vertebrate ichnofacies of Bolivia—facts and implications. J Vertebrate Paleontol 19 (suppl. to #3): 63A

Moreau J-D, Trincal V, Fara E, Baret L, Jacquet A, Barbini C, Flament R, Wienin M, Bourel B, Jean A (2020) Middle Jurassic tracks of sauropod dinosaurs in a deep karst cave in France. J Vertebrate Paleontol e1728286. https://doi.org/10.1080/02724634.2019.1728286

Petti FM, Petruzzelli M, Conti J, Spalluto L, Wagensommer A, Lamendola M, Francioso R, Montrone G, Sabato L, Tropeano M (2018) The use of aerial and close-range photogrammetry in the study of dinosaur tracksites: Lower Cretaceous (upper Aptian/lower Albian) Molfetta ichnosite (Apulia, Southern Italy). Palaeontologia Electronica 21.3.3T 1–18. https://doi.org/10.26879/845

Platt BF, Suarez CA, Boss SK, Williamson M, Cothren J, Kvamme JAV (2018) LIDAR-based characterization and conservation of the first theropod dinosaur trackways from Arkansas, USA. PLoS ONE 13(1): e0190527. https://doi.org/10.1371/journal.pone.0190527

Price LI (1961) Sobre os dinossáurios do Brasil. An Acad Bras Ciênc 33(3–4):28–29

Razzolini NL, Vila B, Castanera D, Falkingham PL, Barco JL, Canudo JI, Manning PL, Galobart A (2014) Intra-trackway morphological variations due to substrate consistency: the El Frontal dinosaur tracksite (Lower Cretaceous, Spain). PLoS ONE 9(4):e93708. https://doi.org/10.1371/journal.pone.0093708

Razzolini NL, Vila B, Díaz-Martínez I, Manning PhL, Galobart A (2016) Pes shape variation in an ornithopod dinosaur trackway (Lower Cretaceous, NW Spain): new evidence of an antalgic gait in the fossil track record. Cretac Res 58(2016):125–134

Rodrigues IC, Mizusaki AMP, Queiroga GN, Michelin CRL (2023) Rios FR (2023) Mica in a sedimentary feature as evidence of humid conditions for a classic desert in Paraná Basin, southern Brazil. J S Am Earth Sci 128:104443. https://doi.org/10.1016/j.jsames.2023.104443

Romilio A, Hacker JM, Zlot R, Poropat, G, Bosse M, Salisbury SW (2017) A multidisciplinary approach to digital mapping of dinosaurian tracksites in the Lower Cretaceous (Valanginian–Barremian) Broome Sandstone of the Dampier Peninsula, Western Australia. PeerJ (March 21, 2017): 26. https://doi.org/10.7717/peerj.3013

Silva RC, Carvalho IS, Fernandes ACS, Ferigolo J (2007a) Preservação e contexto paleoambiental das pegadas de tetrápodes da Formação Santa Maria (Triássico Superior) do Sul do Brasil. In: Carvalho IS et al (eds) Paleontologia: Cenários da Vida 1. Interciência, Rio de Janeiro, pp 525–532

Silva RC, Carvalho IS, Schwanke C (2007b) Vertebrate dinoturbation from the Caturrita Formation (Late Triassic, Paraná Basin), Rio Grande do Sul State, Brazil. Gondwana Res 11:303–310

Silva RC, Carvalho IS, Fernandes ACS (2008) Pegadas de dinossauros do Triássico (Formação Santa Maria) do Brasil. Ameghiniana 45:783–790

Silva RC, Barboni R, Dutra T, Godoy MM, Binotto RN (2012) Footprints of large theropod dinosaurs and implications on the age of Triassic biotas from Southern Brazil. J S Am Earth Sci 39:16–23. https://doi.org/10.1016/j.jsames.2012.06.017

Staines HRE, Woods JT (1964) Recent discovery of Triassic dinosaur footprints in Queensland. Aust J Sci 27:55

Vitkus A, Chin K, Lockley M (2023) Fossil footprints through geologic time III. How do we study fossil tracks? University of Colorado. https://ucmp.berkeley.edu/science/trackways/trackways4.php

Wings O, Lallensack GN, Mallison H (2016) The Early Cretaceous dinosaur trackways in Münchehagen (Lower Saxony, Germany): 3-D photogrammetry as basis for geometric morphometric analysis of shape variation and evaluation of material loss during excavation. In: Falkingham PL, Marty D, Richter A (eds) Dinosaur tracks: the next steps. Indiana University Press, Bloomington, chap. 3, pp 56–70

Xing L, Lockley MG, Klein H, Zeng R, Cai S, Luo X, Li C (2018) Theropod assemblages and a new ichnotaxon *Gigandipus chiappei* ichnosp. nov. from the Jiaguan Formation, Lower Cretaceous of Guizhou Province, China. Geosci Front. https://doi.org/10.1016/j.gsf.2017.12.012

Index

A

Abaeté formation, 7, 123
Abaeté sub-basin, 124
Abelisaurid, 66, 134, 189, 223, 228
Abelisauridae, 99, 130, 131, 137
Abelisauroidea, 129–131, 133, 137, 229
Aborted rifts, 16, 181, 217, 219
Acre basin, 12
Açu formation, 16, 223, 226
Aeolian deposits, 11, 19, 95, 97, 114, 123, 125, 264
Aeolian dunes, 5, 9, 21, 50, 114, 240
Aeolian sandstones, 8, 20, 21, 111, 116, 135, 137
Aeolian system, 7, 8, 97, 133, 138
Africa, 6, 14, 20, 22, 41, 46, 161, 202, 203, 205, 217, 219, 227–229
Afro-Brazilian depression, 22, 161
Agathoxylon africanum, 65
Alagoas basin, 225
Alagoas local stage, 150, 151, 161
Albian, 15, 22, 151, 162, 215, 219, 226
Alcântara formation, 16, 19, 234, 235, 237, 239, 241, 249–252, 256
Alemoa member, 39, 42, 53
Aliança formation, 224, 225
Alkaline lakes, 15, 148, 166
Allosaurus, 49, 73
Alluvial deposits, 124
Alluvial fan, 7, 15, 17, 22, 114, 123, 151, 163, 170, 182, 184, 185, 205
Alter do Chão formation, 11
Alto Tapajós basin, 11, 12, 267
Amazonas basin, 3, 11–13
Amazonas province, 11
Amazonian basins, 267

Andes, 12
Angaturama limai, 152
Ankylosaur, 75, 77, 81, 84, 85, 181, 187, 190, 201
Ankylosaur tracks, 75, 76, 82
Anomoepus scambus, 132
Antenor Navarro formation, 15, 17, 184–189, 192, 193, 199–202, 204, 207, 219
Aptian, 7, 11, 15, 16, 18, 75, 124, 131, 133, 137, 150–152, 161–163, 166, 167, 169, 170, 225–227, 235
Aquidauana formation, 94
Aracoaraichnium leonardii, 94
Araraquara, 20, 93–95, 97–99, 109, 115, 266
Aratu, stage, 18
Araucarioxylon, 41, 44
Archosauromorpha, 46
Areado group, 7, 8, 112, 123, 125, 126, 128, 129
Arenicolites, 42, 67, 83
Arenito Mata, 41
Arid climate, 4, 23, 55, 163, 164, 168, 170, 205
Arid environment, 134, 136, 227
Aridity, 19–21, 95, 114, 134–136, 169, 205, 207
Arroio do Beco, 68, 83
Artificial Intelligence (AI), 266, 267
Atlantic margin, 3, 13, 16, 227
Atlantic ocean, 16, 21–23, 95, 147, 161, 162, 169, 181, 182, 206, 215, 219, 220, 227, 233
Australestheria, 65